# TRANS-DIMENSIONAL UNIFIED FIELD THEORY

GEORGE JAMES DUCAS

Library of Congress Control Number: 2011902545

ISBN:
Softcover 978-1-4568-7110-9
Hardcover 978-1-4568-7111-6



This book was printed in the United States of America.

To order additional copies of this book, contact:
Xlibris Corporation
1-888-795-4274
www.Xlibris.com
Orders@Xlibris.com

# TABLE OF CONTENTS

# List of Figures

# Chapter 1

## A New Universal Code, the Keys to the Universe

$D^5 = MC^2$

Energy and Mass are related through Velocity and Time which makes up all things. All space is localized by inertial distance D of a Trans Dimensional Inertial System.

Most of the things we reason come from observing emergent principles from within the universe. When we have a transcendent intuitive thought, the likelihood of its justification is hard to prove due to the abundance of emergent rational that excludes transcendent maxims where proof becomes the science of limit. At this point we have to discuss metaphysics and philosophy, but then again it is also possible through rigorous discourse to make a relationship between the two and see all reality as connected by systems of mathematics. The question about God exists whether or not there is a big bang, but any beginning brings to mind the notion of a first cause. Hierarchy means there is a first cause. The principle of symmetry and cause-effect suggest that in the nature of the effect is also connected to the nature of the cause. And hence all that is by connection is life, consciousness, represented by mathematics. I cannot feel love without being loved; I am directly connected universally to the source of everything I experience. What is real is shared. The bang of a thought in my head is no different than the emergence of the universe in the mind of God. We illustrate a pattern throughout all reality which is an extension of God. The idea refers to the source, which is the true light.

## Introduction

Our universe is a multidimensional universe where processes and procedures involving natural physics relate and exist simultaneously in multiple dimensions. Natural occurrences are multidimensional. Historically, we identify our existence within three dimensions or vectors of space. However, the matrix of space needs to be redefined as a periodic table of "components" or "vectors," which build up space-time and relate all physics within the relationship of space "component vectors" and "component matrixes." For simplification, these components will be called components or vectors, not to be understood as our modern use of the term *energy*. The basic building blocks of space "vectors" are velocity ($V$), time ($T$), and dimension ($D$), which will also be defined later as superstrings. Each represents a separate vector or component, the basic of these components is $D$ or dimension, and from this all the relationships of physics are established by its combination with velocity and time and multiplications of said components. This theory proposes to unite all of physics within a unified field of mathematical relationships, building on the components of velocity, time, and dimension. When these components are combined in various dimensional states, they form the more complex systems of natural occurring phenomena of physics. Hence, the universe is defined by *Trans Dimensional Unified Field Theory*, where components can be explained in terms of mathematical relationships to the same or other components, and its action and existence acts simultaneously in different dimensional planes.

The composition of the natural order of this universe is founded upon on basic vectors $VT$ or components that are repeatedly added or better "multiplied" in each dimension $D^N$ and properties altered by such combinations. A single alteration or multiplication of $VT$ to a higher dimension is a minute step that increases the degree of complexity of the properties of that combination or dimension. This will be explained and expressed below by a periodic table of composition. Evolution in complexity of components and dimensions is the nature of the dynamic universe and also in the basic structure of space. The philosophy of composition from basic elements to complexity is a universal principle.

**The rules of a transdimensional unified field universe are as follows:**

1. Basic components (superstrings) or vectors are combined and properties altered in steps to form simple to more complex systems by multiplying a $VT$ to each successive and higher dimension. This is one of the periodicities of this theory. $VT$ or $D$ is the "thing" that everything is made from. If there is a God particle, it is superstring $D$.
2. All naturally occurring phenomenon and components is representative of basic mathematical relationships that are connected via unified field of physics-mathematics.
3. All natural phenomena and components in the construct of space-time are related and can be explained mathematically.
4. The universe is a multidimensional world where processes can be studied and viewed simultaneously in space-time.
5. All dimensions exist in stability, and hence their effects are experienced as naturally occurring events.
6. Space and matter are symbiotic, additive, and stable, where one coexists with the other. Space is proportionately related to matter by the cube root. So to divide the matter nearly doubles the space by proportion. Matter generates a force. Space generates a force. Matter creates space. Space creates matter.
7. $C$ is a limiting factor for all dimensional relationships and represents balance in the universe. $C$ represents the elastic collision of $D^3$. Space is an elastic superconductor.
8. An expanding universe suggests the "big bang" or creation is still in process. An expanding universe is explained by space as components created or related to matter.
9. The universe is a mechanical process analogous to cellular growth where divisions in matter accelerate the expansion of the universe.
10. A dimension is so by reference to another and including something of the other. In the case of the theory I propose, a dimension is tied to another by relationship or mathematics and, in this case, by a defined order or set of rules. Basic superstrings $V$, $T$, and $D$ are multiplied dimension per dimension and create varied properties from simplicity (low energy and dimension) to complexity (high energy and dimension).

## The Periodic Table of the Universe Energies and Dimensional Relationships

$\boldsymbol{D^1 = VT}$: Dimension 1 represents a vector of velocity and a vector of time working together in complementary balance or polarity. As demonstrated in relative physics, velocity is inversely proportional to time, with the constant *C* or the speed of light as the limiting factor. So as time increases, velocity diminishes. Time is a polar velocity. The basic building elements of the universe are the constant *C* for velocity of light and *D*. Space ($D^3$ in three dimensions) constitutes the building block of all elements, components, and matrixes. Hence $\boldsymbol{D^1 = CT}$.

By space I mean space as several applications; $D^1$ is space in dimension 1, $D^2$ are space as area, and $D^3$ are volumetric space. Space $D^3$ is a limited extent in one, two, or three dimensions. To understand these dimensions 1 through 3, you have to know that in each, a component of *VT* is multiplied. The addition of these components vastly changes the nature of the resultant, yet it is possible to understand the most complex components in terms if the simplest working back through the dimensions. This unified theory takes you from simplicity to complexity in twenty-three steps, and in doing so; you never really get away from the simple while explaining the complex. That makes it all the more challenge in understanding all the steps and what is suggested. Just as revolutionary as $\boldsymbol{E = MC^2}$ is $\boldsymbol{D^5 = MC^2}$ or $\boldsymbol{M = D^3T^2}$, so mass is explained as volume, time, velocity; going up in dimensions, we see mass related to electricity and a whole host of subatomic nuclear events. These components and their relationships, although invisible, create the particles that we see.

$D^{10}$ is gravity as volume, which induces by rotational motion $D^7$ gravity as density, which also, I might add, is related to the force of an object in $D^4$. $D^{10}$ induces $D^7$, so then I interpreted circular (area) motion of an electron in a helical orbit to generate gravity. The task is in the interpretation. The qualities of gravity extends to a black hole in $D^{20}$ where gravity is squared ($G^2$). It may be that the highest dimension is $D^{21}$ where the force of $E^4$ is singularity.

Also remember that as we add VT to each dimension, new units need to be created. Take for example $\boldsymbol{D / T = V}$. We know that **distance / time = velocity**. Now $\boldsymbol{M = D^3 \times T^2}$. If I only knew *D* and *T*, what is to tell me what the combinations of those are, namely *M* is mass? So you see, as we progress up the scale, it will be a discovery to create new units because by each dimension up, the nature of the dimension changes drastically the properties of that dimension, even to the point at first glance you might say they are not related, but they are. And by the discovery that they are, the true physics emerges, and understanding of how the most remote concepts and phenomena are connected. The analogy of the atom (neutron, proton, electron, three parts in one) is to show *D* as neutral. *D* is space in one dimension, and in three dimensions, *D* is volumetric space. I would go on to describe space as any combination of powers of *D* such as vector (*D*), area ($D^2$), and volume ($D^3$) and continuing up. Just as I describe gravity as vector, area, and volume, which is the same as a gravitational vector (force), area, and volume, I can do so for other concepts. As we know, space, time, and velocity are all related, and at the most basic level, they are connected. As a concept I imagine $\boldsymbol{D = CT}$, where *T* can be converted to *D* multiplied by constant *C*. So *D* and *T* are equivalent just like matter and energy is traditionally viewed. *V* and *T* are inversely proportional; when one increases, the other decreases. I like to call that polar, yet acting together. *D* and *T* are different manifestations of the same thing. Very small amounts of time can be converted to very large amounts of *D* and vice versa. This is particularly true in hyperspace to be discussed later.

The relation of *V* and *T* means if you have motion, you have time. You can understand the relationship of these more clearly by the concept of induction by rotation in this theory. The rotation of $T^4$ induces $T^1$. $T^4$ includes space, so the rotation of $D^3$ (space) induces *T*. $V^4$ induces $V^1$, $\boldsymbol{V^4 = D^3 \times A / T^2}$, and interpreted, I understand that the rotation and acceleration of space induces *V*. So the difference in *V* and *T* is solely due to rate of change. In the first instance its D affected by V to induce time, and in the second it's D affected by T to induce velocity.

How to see this visually: Well, the *D* particle is neutral, and does it incorporate circular motion in this picture as it appears that everything we see or imagine has this characteristic? Yes! Perhaps all components are related by rotational motion. Or perhaps this *D* is a line or a string spinning or rotating on itself like a sphere while inducing *T* or another line or string like a sphere that is also rotating and produces a third string that is related to the other two, which we call $V$? Yes! We may have a geometry of three intersecting spheres, which induces three-dimensional space in $D^3$ and is the construction of space, which is similar to the tetrahedral structures used by Buckminster Fuller. These geometries may subdivide into higher complexity in the higher dimensions. $D^3$ dimension may allow both Cartesian and radial geometries by the relationship of geometries of three combinations of 3 ($D^3$), which is three combinations of $VT$ (***D* = *VT***) or nine intersecting and rotating strings. $D^3$ and consecutive dimensions upward may be fractal geometries and the geometries of space. *D*, *T*, and *V* are the three basic superstrings, and when combined in three dimensions and also in positive, negative, and neutral, results in twenty-seven strings for $D^3$ in a tetrahedral arrangement. Strings form the construct of space. Space in interference and packed rotating spherical geometries produce particles.

I would better prefer imagining two vectors together *VT* and the combination of those we call *D* to simplify the balance of polar components or polarity and for the sake of simple discussion. Then I can imagine space more in a Cartesian way rather than spherical, but the geometric combination of three appears to be tetrahedral and multiple. Their nature may allow various geometries to superimpose over another.

### Meaning of a Dimension

By dimension I mean the *VT* (velocity × time) or powers thereof, not items selected as different, like mass. So $\boldsymbol{D^5 = MC^2}$ means that there are five dimensions of *VT*. Each dimension increases by a power of *VT*. The periodicity of the dimensions here is the increase or multiplication of dimension per dimension and, also as they do, create the complexity of different phenomena we know but are all related. There are other periodicities that will be discussed later. When Einstein said mass and energy are different manifestations of the same "thing," he did not say what the "thing" was, but he used the word *thing*, which is a half solution and not very scientific. Now I propose to you that the "thing" is *VT* (velocity × time). Mass and energy are different manifestations of *V* (velocity) and *T* (time). *VT* is the "thing" from which everything is made. If Einstein realized this, he would have had his unified field theory. The construct of all that exists is a manifestation of motion. All dimensions are distance or spatial dimensions. The point of talking about *VT* is equivalency. Now you can talk about mass (*M*) in terms of *D*, *V*, and *T*, where before you talked about mass as *M* with its separate and unique units, such as kg. In this theory, you can define everything by superstrings *V*, *T*, and *D*, from mass to gravity and beyond. The point of *VT* is that it is the multiplier dimension per dimension, and creating vastly different characteristics as we move up, so much so that upon inspection, they may seem unrelated.

### Other equations for $D^1$ are as follows:

**$D^1$ = *A* (acceleration) × density**

**$D^2 = VTVT = V^2$ (squared) $T^2$ (squared)**. This second dimension represents two components of *VT* acting in two directions as in an area.

**$D^3 = VTVTVT = V^3$ (cubed) $T^3$ (cubed) $= A^3$ (acceleration) × density$^3$**. This represents the third dimension where we experience movement in three dimensions, with components of velocity and time also in three dimensions. Movement in all directions is allowed by a radial and spherical tetrahedral space matrix.

**$D^4 = VTVTVTVT = F = MA = A^3$ (acceleration) × $D$ × density$^3$**. The fourth dimension is represented by *VT* acting on the third dimension. This *VT* on the third dimension is essentially a resultant force. So the fourth dimension is force or forces. This is not to be mistaken with gravity as naturally occurring phenomena, which Newtonian physics has done. $\boldsymbol{D^4 = MA}$, which means that force is a dimensional property or a space field property. I believe that when two more strings of space are added in three dimensions, you get asymmetry in the field of space, which causes the force to be.

**$D^5$ = *VTVTVTVTVT* = $MV^2$ (squared) = *E*. This is $MV^2$ (squared)** typically identified as kinetic energy; however, I will explain this "$MV^2$ squared" as a specific scientific relationship, and the word *components* are now used liberally to replace my word *energies*. All elements within this table are to be considered components. Here we can associate the equation with $MC^2$ (squared), but it is a five-dimensional phenomena. I am exchanging *E* for $D^5$ to fifth power, correcting Einstein if you will. You cannot have a unified field theory by presenting parts. Even so, a theory is an idea that needs to be tested, and from it you establish laws. A law would be **$D^5$ to fifth power = $MC^2$ squared**. Let's use $D^5$ instead of *E*. The units are in kg meters and seconds; when you calculate *D*, it is in meters. *C* is in meters and seconds. Take the equation **$D^5 = MC^2$**

Sunflower Spiral-Radial & Spherical Geometry in Nature

and calculate $D$ for the mass of the Sun to measure the extent of $D$ or space vector. So run the numbers, mass of the Sun, speed of light, and you have a distance $D$ near to Neptune's orbit. That's the whole point of the theory: it combines everything in the equations and notices that in each dimension, you add a $V$ and a $T$, a periodic table, you add dimensional vectors. The correct equation is **$D^5$ fifth power = $MV^2$ squared**, but I use the limit $C$ to determine a maximum. The fact that one might say the units don't go together is what's puzzling; they do, and the relationships have been redefined. Suffice it to say since everything is a subset of one string $D$, the units can all be thrown out and expressed in magnitude until more accurate labels for units are derived.

A joule is the product of a force of one newton moving one meter. In this theory, $D^4 = MA$; so **$D^5 = D^4$ (force) × distance (meter)**. So a question about units should now move down to **$D^4 = MA$**. There is no mystery then in $D^5$, but the task is to reevaluate $MA$ or mass in terms of distance, time, and velocity, and when you get there, you have the strings that make up all matter and correct units expressed.

There are three strings that are distance, time, and velocity; these three combine to form all space and all matter, and they are transparent until they superimpose, and the interference patterns of special superposition are the particles that we see. The relationship of string dimensions is also mathematical in inter-dimensional relationships. Particles are the minor portion of reality.

A basic idea of this theory was to reinterpret mass mathematically into a more basic component of space, if it was possible. As you jump from dimension to dimension, the "nature" of the units of each dimension varies drastically. Then you have the questions surrounding units, such as the joule for $D^5$, which is a typical and fundamental question of a unified field theory; why does the "nature" of units vary so much from simplicity to complexity? This is a mystery in physics, but also the same happens similarly in the periodic table of elements, though perhaps, as in the periodic table, instead of addition, perhaps multiplication was the ladder to complexity, and the component was velocity and time (VT [D]). So to get to $D^4$, conceptually I thought of a vector VT (D) in space $D^3$ and got **$D^4$ = ma. $D^5 = mv^2$** was a naturals step, then came **$D^5 = mc^2$**. Distance is in meters and mass in kilograms.

**$D^6 = F \times D^2$ (squared) = *VTVTVTVTVTVT*.** Force acting over an area is magnetism. As an electron is moving, the subsequent magnetic field acts over an area.

**$D^7 = F \times D^3$ (cubed) = *VTVTVTVTVTVTVT*.** The seventh dimension is force acting on the third dimension, which is gravity density. This is a naturally occurring force.

So for the seventh dimension, an explaining equation will be **$G$ (gravity density) = $F$ (force) $D^3$ (cubed) or $FV$ (volume) = gravity density**.

From **$FV = G$ (density)**, we can determine the $F$ or force of gravity ($G$ density) and volume ($V$) are inversely proportional. Hence, $F$ is inversely proportional to 1/volume. To understand these equations correctly, remember that components on the same line of an equation are inversely proportional and directly proportional when divided.

So if on Earth there is a hundred-pound person and we know that Saturn has a gravity/mass 95 times of Earth and a volume 769 times of Earth, from this equation we can calculate the weight of a man one hundred pounds on Earth to be 108 on Saturn.

95 inversely proportional to **1/769 = 0.1235**
95 inversely proportional to **1/.1235 = 8.0**

So 100 pounds > 108 pounds by the ratio. So the equation has demonstrated the relationship of volume to gravity that gravity diminishes in proportion to the volume of a planet when the mass is constant. By these equations, gravity is related to the diameter of the planet and came up with the weight of a man on Saturn. Saturn is near the gravity of Earth and is related by density and diameter and expressed here as a vector.

Now my gravity equation was arrived by progression, $\boldsymbol{D^7 = F \times}$ **volume** ($\boldsymbol{D^3}$) **or gravitational density**. Now is this not a more complete understanding of gravity, a naturally occurring force in space? With that assumption, I get a relationship that tells me that gravity is influenced by volume, mass, and acceleration. The current gravity equations need to be rewritten in terms of observable components like real entities of physics, not abstract things like radius. Using radius is like describing atoms in terms of half electrons. And going further, what about the connection to electricity, magnetism, and so on? If you don't have a pure or true equation, you'll never have a unified field theory. You have to eliminate all the abstract shortcuts like radius but talk in real terms such as volumes. Now if you run the equations, you find that "gravity density" for Saturn is approximately 8,000 times greater than Earth, attributed to two things: the first is the volume of the planet, and the second is the mass. These two variables are density. Small $g$ or, a as I call it, $A$ (the component of acceleration) is nearly the same. At $D^{10}$, we find the true value for gravity, volume, or energy, then $D^7$ becomes the density and $D^4$ the force. And by ratio, we can establish the relative gravity vectors between planets. But the true nature of gravity is something much larger and multidimensional and has a force component in this dimension. What is true in a larger dimension has a three-dimensional shadow, and for gravity in this dimension, it is force.

So you ask how to apply. Well, this theory will explain the true relationships, from which you can derive the Newtonian rules of thumb or secondary equations to run comparative analysis. The equations are connected, and can be, but the understanding of them depends on use. The equations will reveal avenues of research. I think the gravity equations presented here fully explain gravity and tell me volume is a component and mass and acceleration and how it is reinterpreted in various dimensions.

***The dimensions keep rising in relationship to further complexities:***

$\boldsymbol{D^8}$ = ***VTVTVTVTVTVTVTVT***. The eighth dimension is $F^2$ (squared), which I associate with mechanical-chemical reactions or $MV^2$ (squared) acting on $D^3$ (cubed) or third dimension. This could be associated with nuclear forces that hold matter together. Kinetic energy in three dimensions is also an issue here since it combines $D^5$ or fifth dimension ($MV^2$ squared) with $D^3$ or third dimension ($D$ cubed).

$\boldsymbol{D^9}$ = ***VTVTVTVTVTVTVTVTVT***. The ninth dimension is magnetism $D^6$ acting on $D^3$ (cubed), which is related to electricity. This is also $E \times F$ or force of energy.

$\boldsymbol{D^{10}}$ = ***VTVTVTVTVTVTVTVTVTVT***. Tenth dimension is $MV^2$ (squared) to an additional (squared) or $E^2$ (squared), which could be a nuclear reaction or kinetic nuclear orbital motion. Associated with this is electromagnetic force or $D^4$ or fourth dimension acting with $D^6$ or sixth dimension (magnetism). This could be electromagnetic force due to orbiting electrons. What we measure as gravity is what we find useful. I think gravity goes beyond the way it is understood. I believe that the nature of gravity is different in its various dimensional states. Gravity has components of mass, acceleration, and volume according to this theory, so it is more than force here. Acting in three dimensions, gravity is $E^2$. The density component of gravity ($D^7$) is 8,000 times on Saturn than Earth. The force component is what we know about gravity, which is very limited. The value of acceleration, or what you know as $g$, is nearly the same due to density. Why does Saturn have so many moons? Could it be that all the moons are because of the larger component of volumetric gravity in $D^{10}$? And such a component would be very large and far reaching, as gravity is. Perhaps also contributing to its additive strength as mass accumulates. And also, something so large would be far reaching even with an inverse square law. I did say volumetric gravity is energy squared. $\boldsymbol{E^2 = D^{10}}$ = **gravity** ($\boldsymbol{D^7}$) × $\boldsymbol{D^3}$ **(space)**, $E$ would be a kinetic linear line or orbit; perhaps $E^2$ is circular orbital motion along with a pattern of small circles as in a helical orbit in 3-D. So the energy is acting across an area(s) as it moves in 3-D. What I am trying to explain is that there may be multiple levels in understanding gravity. I do believe the understanding of gravity and relationship to electricity has to do with understanding the orbits of the electrons around a nucleus. When we normally think of a particle, we think it has components that interact with other particles. So the relationship of particles causes the actions, but what if the relationship of actions results in the particle? Perhaps the interaction of components or the interface of these is the particle, so opposite to our current thinking, the actions create the particle rather than the particle creates the actions. So the forces between the mass creates the particle rather than vice versa. So what is *D*? Well, it is a component that, by relationship, creates mass or the thing we see. At the most basic level, *D* is *T* or time and balanced with *V*. Components balanced or inversely proportional exist together, perhaps as polarity or polar-balanced vector. Opposite sides of a coin is a model. I think of time as a velocity in polarity with velocity as a balancing countermotion. And distance nearly the same as time, but where the two meet or at the border. Time is compressed distance ($\boldsymbol{D^1 = C \times T}$), just like mass is compressed energy. Now to jump to a wilder thought, maybe each *D* dimension is a particle, or at least particles may be linked to these equations somehow as interfaces. I don't claim to be absolutely correct, but I believe I am asking the right questions.

$D^7$ = **space ($D^3$) × force ($D^4$) or mass × acceleration × $D^3$**. So you have a factor of density in gravity. Gravity at this level is a vector, interpreting force as a vector. Here force and volume are inversely proportional as mentioned before.

$D^{10}$ = **gravity × $D^3$** is volumetric and could be in consideration of gravity beyond its source as a volume through space and its effect far away. When the components are on the same line in an equation, they are inversely proportional. So $\boldsymbol{D^{10}}$ = **$D^4$ (force) × volume$^2$ ($D^6$)**. Hence gravitational force is inversely proportional to **volume$^2$** = ($\boldsymbol{D^3}$)$^2$.

***Other equations for $D^{10}$ are as follows:***

1. $\boldsymbol{D^{10} = M^3 \times A}$ **/ density$^2$**
2. $\boldsymbol{D^{10} = A^4}$ **× density$^2$ ×** $\boldsymbol{M^2}$
3. $\boldsymbol{D^{10} = F}$ **× volume$^2$**
4. $\boldsymbol{D^{10} = A^4}$ **× density$^4$ × volume$^2$**

$\boldsymbol{D^{11}}$ = ***VTVTVTVTVTVTVTVTVTVTVT***. The eleventh dimension is gravitational force ($D^7$ or seventh dimension) in four dimensions ($D^4$) or force. $\boldsymbol{D^{11} = D^8 \times D^3}$ could be chemical reaction in space that induces $D^8$.

$\boldsymbol{D^{12}}$ = ***VTVTVTVTVTVTVTVTVTVTVTVT***. The twelfth dimension is dimension 8 or nuclear force acting in $D^4$ or fourth dimension or force *F*. This *F* is also

**Gravity ($D^7$) and kinetic energy ($D^5$) = magnetism ($D^9$) and space ($D^3$).**

**Magnetism ($D^9$) = gravity ($D^7$) × $D^2$ = gravity ($D^7$) × $V^2T^2$; $T^2$ squared = density or magnetism ($D^9$) = gravity ($D^7$) × $C^2$ squared × density**, which implies that electricity can be harvested from a gravitational field in vast quantities. There could be free energy in unlimited amounts for all.

$D^{12}$ induces $D^9$ is electricity inducing magnetism respectively.

$\boldsymbol{D^{13}}$ = ***VTVTVTVTVTVTVTVTVTVTVTVTVT*** **= magnetism ($D^9$) × force ($D^4$)**, which is equivalent to electromagnetic force.

The dimensions keep rising and, I suppose, represent more complex and fundamental forces acting at the subatomic level. The higher the dimension, the greater the kinetic energy required in the system. Kinetic energy in the universe is rotational and, when transferred into linear motion, is great relative to our experience. But when contained in stable rotational paths, particles exist in harmonious relationship at the subatomic level and also in relation to planetary. Hence, to take matter and break it up in a supercollider is to go from higher dimensions (subatomic) to lower dimensions (exoatomic); the lowest particle is space fabric or *D*. So it may seem a waste of energy to use high amounts of energy to bring us from high-dimensional subatomic to the smallest particle, which is space or $D$ of the first dimension. But I believe that is where the analysis of supercollider will lead.

***Dimension$^{23}$ would be***

$D^{23}$ = $E^4$ **(fourth power)** × $D^3$ or what I call the force of energy $E^4$ in the third dimension, which perhaps might be the end of the periodic chart. This table leaves out the physics of polarity, which includes antimatter and subsequent relationships opposite to the balance or natural construct of nature that we experience in the universe of twenty-three dimensions. To discover these polarities would be the door to time travel and antigravity. A black hole would be the most complex arrangement of space at $D^{23} = E^4 \times D^3$, which means that $D^{23}$ = **gravity**$^2$ × $D^3$. This means to me that the amount of energy is increased by a power of 4 and gravity by the power of 2. Gravity is accelerating in space, and the resultant energy is enormous. The geometry of space is accelerating into the black hole, and thus pulling the fabric of space (the medium through which light travels) away from you. Hence, it is black since the medium through which light travels or space ($D^3$) is falling away faster than the light that moves through it to get out of the black hole. There is order even in a black hole. We might assume that higher dimensions are higher spatial densities. And so the dimensions progress from $D^{23}$ to $D^1$ in an ordered relationship of physics; hence, we can conclude that where there are generators of space, there are black holes that generate space, maybe even in each atom, center of galaxy, and planet. The universe is from singularity to singularity, from the singularity of $D^1$ to the infinite singularity of $D^{23}$.

This universe is a particular qualification of infinity. And perhaps here the space matrix is in its highest superposition and fractal pattern approaching infinity. This is either a point of fractal subdivision toward infinity. Here $E^4$ = **gravity**$^2$ **or** $E^2$ = **gravity**. The very nature of a black hole suggests that the space generated by its great mass at $E^4$ could be generating space and gravitational field at a value square compared to a natural star.

***Black Hole***

A black hole is the superposition of space in twenty-three dimensions. At this level and as has been discussed, the amount of energy is increased by a power of 4 and gravity by a power of 2. Hence, the normal manner of estimating gravity, generation of space, and energy has changed. Space has compressed to limits where the strings that make up the matrix of space are in geometry of the highest compression and complexity. In three-dimensional spaces, the nodes of the geometry of space are photons. The elastic collision of three-dimensional spaces can be described as photonic current similar to electricity. When this elasticity merges with space of a higher dimension, there is some reflectivity at the boundary of two adjacent and different space dimensions. In the space of a black hole, there is no reflectivity, and hence light is not apparently exiting or is trapped. Hence the elasticity upon entering the space of a black hole is either super elastic or inelastic, where the motion of current is totally absorbed. It is very possible as another option that the conversion or superposition of space into twenty-three dimensions does not allow waves to exit with linear reflectivity but rather rotational motion, and hence the black hole appears black. It may be that in a black hole, the space matrix is approaching singularity and does not allow elastic collisions of the space matrix to occur as in other lower dimensions.

***Review***

By dimension 4, I mean $D^4$ or $D$ to the fourth power. Let's say dimension 1 is a distance vector or $D^1$. As a measurement, let's use meters.

$D^5 = MC^2$. What we know as $E$, calculate $D$ for the Sun, and you have the distance to nearly Neptune. I prefer to understand $D$ as the relationship of space or vector $d$ to mass. So the Sun projects an umbrella of space to Neptune; after that, deep space is under the umbrella of space that extends from the center of mass of our galaxy. So density of space varies.

$\boldsymbol{D^{10}} = \boldsymbol{E^2}$. I believe when you get to this level, you are considering subatomic realities that begin at $D^5$.

$\boldsymbol{D^{23}} = \boldsymbol{E^4} \times \boldsymbol{D^3}$, so there is a lot of energy from 23 to 1. I'm not sure how many dimensions there are, but this one seems tidy.

**Periodicity**

The periodic table of dimensions has recurring themes just as the periodic table of elements as follows:

***Dimension per Dimension***

Here a *VT* is multiplied as you increase one *D* (*VT*) power in dimension.

***Squared***

There are qualities of the square of a component that repeat through the sequence of dimensions.

***Groups of 5 and 3***

$D^1$ power is line
$D^2$ power is area
$D^3$ power is volume
$D^4$ power is force
$D^5$ power is energy

This relationship of 1 to 5 involves interpretation of combinations of power according to corresponding meaning. So gravity would have expression in these various magnitudes or powers. The basis for the concept is line, area, volume, force, and energy and relationships between these that include a concept such as density. So the concept of density can be expressed again through the dimensions as a qualification of that dimension. Gravitational density and electrical density are different and express different dimensions, but the similarity is an expression of information in space. Thus the groups could show similarity in one of the categories of 5 and every three from point or power of consideration. I left out 6 since it might be covered better by the squared relationship. Dark energy is due to *D*, the primal superstring.

***Group of 3 or Induction***

Here we have higher dimensions inducing components of three dimensions lower. In the relation of threes, we also have similar characteristics grouped as chemical/space, gravitational /space and electromagnetic/space. There may be others.

**Induction**

There may be a relationship of velocity components, vectors, and rotational motion components, which I call induction by a factor of three dimensions from higher to lower dimension, although there is also a relationship of lower to higher. So the higher dimensions induce lower dimensions by a factor of three, related by rotational motion. There is a relationship between dimensions by induction due to rotational motion. The dimensions of space energy matrix form relationships due to rotational motion.

The relationship between $D^9$ (magnetism) and $D^{12}$ (electricity) is related by rotational motion three dimensions away and in an angular inductive relationship. From this there are projected relationships between various dimensions such as the following:

1. **$D^1$ and $D^4$** or one-dimensional vector $VT$ ($D^1$) and the fourth dimension ($D^4$) or force; rotating space indices $V$ and $T$. I believe acceleration is required for inducing $V$.
2. **$D^2$ and $D^5$** or second dimension and kinetic energy or $E$ ($D^5$).
3. **$D^3$ and $D^6$** or third dimension and magnetism, perhaps rotation motion in 3-D relates to magnetic fields. **$D^6$ = space squared $(D^3)^2$**; space has a connection with electricity and magnetism.
4. **$D^4$ and $D^7$** or force and gravity density.
5. **$D^5$ and $D^8$** or kinetic energy and nuclear relationship.
6. **$D^6$ and $D^9$** or magnetism and electricity and space.
7. **$D^7$ and $D^{10}$** or gravity density and electromagnetic force or gravity energy ($D^{10}$): According to this theory, gravity or $D^7$ is induced by $D^{10}$ or $E^2$. So motion of electrons around a proton creates gravity. The idea of induction in my theory is that rotation induces a vector, and a vector induces rotation. Gravity, for me conceptually, must be the accumulation of many vectors to the center of a proton. The motion then of an electron must be many circles orbiting a nucleus where the circles area or planar face center is tangent to the nucleus. Would that be a helical orbit? The electron makes many small circles as it orbits the nucleus, and each circle creates a vector of gravity (one dimension) to the center of the nucleus. So perhaps the electron makes $x$ amount of circles while orbiting the nucleus. This motion induces accumulative vector of gravity to the center. So the perceived effect of gravity is many vectors to the center, as in a spherical action. This action must be additive with relation of an atom to another. Perhaps by reversing the arrangement or reversing polarity where the nucleus is an electron and the orbiter is a proton, you might have an anti-vector(s) of gravity. What I would do is study mapping of a helical orbit of small circles around a sphere to create a helical pattern around a sphere. A computer would be good for that. After mapping geometry, I would figure out a way to run protons around a central negative (electron) core in a helical pattern and artificially reverse the construct of an atom. The generated field should produce a crude antigravity vector outward. Once this is accomplished, it will not only verify electron path and model of the atom, but could be refined in more subtle ways in the future to produce floating spacecraft.
8. **$D^8$ and $D^{11}$** or nuclear relationship and chemical reaction.
9. **$D^9$ and $D^{12}$** or electromagnetism and electrical density.
10. **$D^{10}$ and $D^{13}$** or gravitational energy and nuclear force.

The relationships continue to $D^{23}$. Within those are recurrent dimensional patterns. Circular or rotational motion is a step in the direction of exploring inter-dimensional induction, which is a departure from Newtonian mechanics where static systems are explored.

## Photon

The smallest particle-vector is space or $D^1$ and combined in various states to form all physical elements, properties, and relationships. A photon could be the result of an elastic collision wave of the $D$ particle matrix that transfers kinetic energy at the speed of light. There is no vacuum. Hence it is hard to travel in a universe where you try to exceed the elastic collisions of the matrix. It would be like trying to travel fast through water without tearing a hole in the medium you are traveling in; the same is with space. As the density of space varies, the natural motions of bodies are affected by the interaction of their space fields. Bodies generate space. Our planet is in the umbrella of space created by our Sun and the center of the galaxy. You will notice all galaxies of space universe are evenly distributed by proportion to their size and the space they generate. You have trouble exceeding the speed of the fastest event of the space matrix we know, the elastic collision of the matrix in the phenomenon known as light.

Since light or a photon is the result of an elastic collision of the space matrix $D$ and because crystal can bend light, we can assume that crystals can bend space. At first thought I would say that a laser is compressed space and that crystals can be used to compress and bend space. I would be interested in the space compression experiment to prove space can be manipulated either by compression or rotation of objects. And that space can be altered by crystals suggest that the universe might be a complex and fantastic hologram.

An experiment to measure the compression of space: In a vacuum, do the following. Use crystals to direct light into focused bands, which should be compressed space. Then send a particle through it to see if its course is altered, whether by direction or velocity.

If a photon is a phenomenon that is a result of the elastic collision of space and if a crystal can bend light, then a crystal can bend space. Perhaps a laser is essentially elastic collisions in compressed space. If Ron Mallet turns a laser in a spiral, the path is essentially bent or curved space. That a crystal can bend space, time is induced by rotating space. This seems to agree with Ronald Mallet's idea of creating a laser beam circular path to create a time machine. In the equations there is support for the idea since $T^4$ (rotation) induces $T^1$. Well, $T^4$ is essentially what's in the third dimension or components thereof (space, momentum, density, time, gravity). Ronald Mallet may have something in his experiment. Although for the experiment to be correct, space needs to bend the light. Bouncing light in a circle is not enough.

According to this theory, the speed of light is the speed of the elastic collision of the matrix of space. A photon is the effect of the elastic collision of space. Space has a matrix or geometry, and that matrix forms all things as it becomes denser in higher dimensions and more space components are multiplied together. The particles we see, such as electrons and other subatomic particles, are interference nodes in moving space, something like the interference waves on the surface of the water. Now think of a motorboat on the water. At some point, the water will not allow the boat to pass without destroying the very medium of water it is passing through. So we cannot travel faster than the matrix of space will allow an elastic collision in the same space or the speed of light. Likewise in the water, the boat is limited by the medium of water to the point that if it tries to go too fast, it rips up the medium. Now the boat can go faster at some point by jumping above the surface of the water. I believe this same jumping is possible in space and is the jump to hyperspace or dimension 4 ($D^4$). There is a balance in space between the components of time and velocity, and this does not allow too much of one without sacrificing the other, so there is entropy toward equilibrium in third dimension between time and velocity. But there is also entropy between dimensions. The universe is a perpetual motion machine that allows entropic action from equilibrium to disequilibrium and vice versa. And traveling fast to the limits of this machine is in violation of the limits allowed in this perpetual motion. You cannot break the clock, but you can transform through the clock. That is what traveling to hyperspace is and transformation above the matrix to where energy is, allowing relatively to motion at a higher rate.

Dimension 1 is singularity, the point, which is a projection from the infinite. Dimension 23 is the infinite or circle or a black hole. So from the infinite, you get the creation or projection of this world to the point or singularity. The things of this world approach the infinite but never quite make it (circle) because the energy to reach it is infinite. But look at it in reverse; the infinite projects the finite to the point in the circle. The two are connected, and thus the universe is infinite. And there is order even in a black hole. Nothing is chaos, everything is order. You can travel any speed you want; all you need to know are the limits and dimensions of the mediums through which you pass.

If you want to travel faster than the speed of light, you need a spaceship that can transform its surface into four dimensions ($D^4$) while retaining three dimensions ($D^3$) internally. Hence, in dimension 4 ($D^4$), you are traveling relatively at the speed of light squared ($C^2$) compared to this dimension. In just a few seconds in hyperspace, you can travel to the center of the galaxy, and maybe in sixteen days for sixteen billion light-years. We are very limited to this solar system in our present technology, but soon this will change. The galaxy is much larger than we know and cannot be assessed with telescopes.

### *Space Matrix*

Space is a geometric matrix. A photon is the result of an elastic collision of the matrix nodes of space. The photon does not travel just as an electron does not travel at the speed of light. The elastic collision of the matrix travels at the speed of light and communicates the impulse to the destination photon as is similar in electricity. The nodes of the three-dimensional space matrix or the overlap of intersecting string spin is the photon. This matrix is composed of three dimensions of *VT*, and the nodes of these three *VT* could be neutrinos electron, muon, and tau. These three neutrinos combine to form a photon in three dimensional spaces. Hence, space is the smallest particle-wave construct. But the construct is transparent.

In the same manner that electricity provides energy via current and via elastic collisions, there is a similar but neutral current in operation in the solar system. The Sun generates elastic photon collision through the space matrix to the planet Earth. It might be feasible to create a completed circuit that harnesses the photon current similar to electron current. This vast supply could be a source of free energy for not only space travel, but to power all energy on the planet Earth.

The interference pattern between waves "is" the construction of particle. However, it's a space wave or string node as opposed to an elastic collision wave. Imagine a jump rope, now ten in a line with a person spinning the rope on each side. Now imagine a cube for simplicity and people with six hands. The stings are the rope that has spun and appear as a sphere. The people are the photons or where the spheres overlap. When a wave or elastic collision hits the matrix, the people or photons vibrate, and that vibration through ten ropes goes at the speed of light, something like electricity, and the photon on the receiving end or person is what you see. This is similar to a wave of current in a wire. When this space matrix spins on itself and superimposes and its dimensional properties change, then you get an electron and other particles or nodes in the space matrix, and the strings slow down; and the electron is a heavier node than the previous photon. But the lattices of space components that make up all particles or mass are from the same primary space components. In reality, I believe these spheres or strings overlap more than I have identified in my visual explanation. And also, at the most fundamental overlapping of these spinning strings, the interference patterns that create space particles such as neutrinos and photons are transparent to other space densities. This allows further superposition that creates the denser particle matter that we measure.

The speed of light represents a perfect elastic collision in the matrix of space. This matrix is composed of strings, which are **distance, time, and velocity** in negative, positive, and neutral. Hence in three dimensions, you have **3 × 3 × 3 = 27** superstrings, and the monopole or node of the matrix is the photon. Strings are mathematical in relationship and form the construct of the universe. What we see are the interference patterns of the spinning strings and their active rotation in spherical and three-dimensional matrixes. The form of the geometry is spiral as a toroid. The overlapping superstrings form the photon in a three-dimensional tetrahedral, symmetrical, and spherical geometry. What we see is the vibration of that matrix on the receiving end of vision. But at the most minute level of the matrix, all is light. (Coldea: 177-180).

In a perfect elastic collision, energy is conserved; however, even after a long journey of billions of miles through space, light may seem to dim. Current theory accepts the dilemma of the double-slit experiment without clarification and assumes that the contradictions in current theory are satisfactory to explain truth. A model that space has structure resolves the issue. Next time you go to the beach, imagine that it is space, where the water is not moving to you are the speed of water. Molecules collide to transmit kinetic energy. And so why would electrical or photonic current (elastic collision) be the exception? The speed of light means elastic collision; unlike the ocean, space is a perfect elastic medium of which the photon is a constituent part. And if the photon has no time and is essentially everywhere, then the concept of structure makes sense. The elastic wave of the light speed can decay due to the shift in space density or creation/subtraction of space or compression/expansion of space, but ideally the matrix is elastic. The photon does not decay.

### *Now the Double-Slit Experiment*

If you see space as fractal geometry or matrix where a photon is the result of an elastic collision along the matrix or the transfer of kinetic energy, then you can resolve the conflict in the experiment that light is particle and wave. Light is kinetic energy transferred through the matrix of space and transfers motion. Yes, so the photon is an impulse with no mass, except for the transfer of the kinetic energy, which we detect. So this matrix is affected by optics and barriers, and through the slit(s), kinetic energy is transferred depending on how the space matrix is either blocked or distorted. Kinetic energy of an elastic collision can be diverted by as it passes through slits or barriers. Any distortions in the matrix will affect the path of kinetic motion. If you see the photon as "the effect" (light) of an elastic collision and not the "product" or light, you can resolve the duality of the experiment. Particles are the results of circular or spin wave interference patterns, so what we see are the effects of the real component waves. That is not to say that the universe is waves. It is composed of strings that are related in groups of multidimensions of space by motion, and the interference patterns are the particles we see. Hence, space creates matter, components; those things unseen are the substance that creates the particles we see, and hence, that is why there is so much more space than particles because the relationships are established by dimension of space components. It may be that the interference nodes or monopoles of the space matrix are "photons," and the elastic collisions of these nodes transfers the "light" in the same manner the elastic collision or electrons transfers the "electricity in a wave at the speed of light." However, the photon does not traverse space, but the motion of collision is at the speed of light. Hence, the arrival of a photon is the arrival of a node of the matrix on the receiving end.

The dual nature of light is the dual nature of the experiments. But the true nature of light is very different. Light is a monopole of three-dimensional space matrixes, which is a matrix and superconductor. The dual nature is only the experiments that claim a dual nature, not the true nature of light. All is infinite in the matrix of space and the monopole photon, and you are but a reference point. The matrix is also transparent to denser matrix forms, dimensions of space, and heavier and slower interference patterns. As the interference patterns of the space matrix build, particles result, and motion slows to sub–light speed. As particles acquire mass, their transparency decreases. The discrepancy between quantum equations and relativity is that quantum described the relationship of particles, and relativity refers to the behavior of strings. Strings form the matrix of space and the interference patterns due to rotation, superposition, and compression of the matrix creates particles. So there are no particles, only an effect of the behavior of strings, a shadow of the main event of string processes, or better expressed as a minute portion. In the double slit, imagine space with nodes as in a structure that when you strike it, the vibration of photons goes though the slits. So this explains the duality of wave and particle. The elastic collision is the wave, and the transmission at the end of the vibration is the particle. The solution is in the geometry; a lattice or matrix permeates and travels through the slits. In the model here as in the double slit for the photon, what we measure is the effect, not the source. The photon is not travelling through. The effect is travelling through communicated to the photons on the receiving end. Light is neither a wave nor particle as we understand particles or waves; it is an elastic collision in the matrix of space. It is a monopole (particle of interference or overlap pattern) of the matrix of space, and in space the matrix is an elastic superconductor where the maximum transmission of information is the speed of light. Photons have volume, have mass from **mass =** $D^3T^2$. Photons do not travel through space but elastically collide as monopoles of the space matrix and what you see are on the receiving end.

## Entropy

Entropy and the expansion of the universe are similar actions. The fabric of space limits motion since it is difficult to exceed the limits of space or the speed of light. The retraction of the universe is disentropy. Equilibrium or motion between entropy and disentropy that is established between dimensions and the limits of motion that are permitted in each dimension by natural occurrences results in perpetual motion. Entropy and the expansion of the universe are similar actions. Disentropy and the retraction of the universe are also similar actions. The motion between entropy and disentropy is perpetual motion. The separation of matter due to entropy is the expansion of space. The union of matter and retraction of space is disentropy. There is a limit to the generation of space due to its relationship through dimensions (to 23 or black hole) just as there is a limit to motion relative to the speed of light. There is also a limit to the retraction of space, which reverses and causes or explodes into an expansion. Balance and order exist, even in a black hole. Closed systems cannot predict the function of the universe. Telescopes cannot predict the size. The reality of entropy can only be expressed within a unified field theory.

## Relationship of Mass and Space

Let us explore $M = D^3$ **(cubed)** $T^2$ **(squared)**, so *D* is proportional to the cube root of the mass. By this relationship, we know that space is in relationship to mass by the cube root. So for a mass 125, cube root is 5; now half the mass to 62.5 each and take cube root of each, and you will have 4 and 4, which are 8. So in the explosion of a galaxy, when you half the mass of a single-bodied system, you nearly double the space and probably by a factor of $C^2$ squared. The relationship of mass and space and gravity is illustrated.

The other point I would like to make about space is that there is a ratio of matter to space. The two have a symbiotic relationship. Matter does not exist in space, but the two create one another. So at the creation of the universe or big bang, as the pieces separated, or matter separated, you get a doubling of space every time the amount of matter breaks into two halves. And you perhaps get spaces adding to one another, as gravity does. So with many pieces, you get quite an acceleration of space. Recently, science has discovered that the universe is accelerating outward, and that it is caused by an "invisible energy," what I presume in my theory proposed to be space itself and its relation to matter. But as matter comes together and forms a black hole with increasing gravity, you have the reduction of space.

An equation explaining the relationship of space to mass is $D^3$ **(cubed)** $T^2$ **(squared) = mass**, so mass is related to space by the cube root. This means that if you half the mass, you more than double the space, perhaps multiplied by some constant. That may explain the ratio of mass to space in the universe. So an exploding star would create vast amounts of space while a black hole would diminish space.

*D* is the smallest particle or component. Let us use string terminology to describe *D*. *D* is a string from which everything is made. I would rather say that each dimension is made of vector that combines *v* and *t*. Just like we add electrons in the periodic table, which is also similar to addition and subtraction, the universe is multiplication. Each dimension adds two vectors of *V* and *T*, which results in *D*. Now in three dimensions, you have three vectors. Now you ask, what does it mean? That's what is so incredible about the whole thing. These vectors combine to form matter, gravity; things upon observation seem totally unrelated. How can you jump four dimensions from $D^3$ and have $D^7$ or gravity? Well, think of molecules and how different they are just to add an electron (addition), but now instead of addition, multiplication. So now in our third dimension, you have a matrix of *D* in three directions and the elastic collision of that matrix is a photon. If a crystal can bend it, then it must be like a fantastic hologram. What supercolliders will find if they could is *D*, *V*, and *T*, the basic components of all construction.

What I understood by the $M = D^3 \times T^2$ **= space × density**. At first I thought that *D* or the vector component of *d* or space related to mass was influenced by the cube root. So my interpretation was that if you had a star and it exploded into two pieces, you would more than double the space. So if there was a big bang, the acceleration of space or expansion of it was a mighty unfolding. Mass is a relationship to space and density ($T^2$). Half the mass or a large object into two separate pieces more than doubles the space.

**Another equation more accurately explains the distance of space surrounding a mass:**

***D = root to fifth (MC2 squared)1***

If you calculate the mass of the Sun and solve *D*, the distance for space *D* is just inside or short of Neptune. Note that planets beyond Neptune have a more eccentric orbit and are not orbiting in the same plane as other planets. Take for example Pluto. As Pluto escapes the umbrella of space generated by the Sun, it travels through less dense space, and thus the path of Pluto is more eccentric, deviated from the ecliptic plane, and asymmetric about the Sun, caused by moving through a medium of space of varying density. Our solar system must be in the umbrella of space caused by the center of our galaxy or Milky Way. So a planet such as Neptune is affected by space matrixes from the center of the galaxy, from the

Sun, and that space generated by it and interacting with other spaces. So for any particular piece of matter, space is infinite since you can never travel to the limit of space that is generated since you are at the center. Space itself must be the dark matter that keeps galaxies apart and in equilibrium or evenly distributed in space. Of course, smaller matter would have more freedom to move between the larger generators of space. Also, it may be more difficult to travel between galactic space generators as in the thin space areas between them, where space may be more distorted rather than homogeneous as in our solar system. Due to rotational motion of bodies, space has many twists in addition to varying density.

### *Sun and Orbits*

In the theory of space generation, if mass generates space, then the Sun with a mass of 2 × 10(thirtieth power) and *C* plugged into the equation ***D* (fifth) = fifth power cube root of (*mc* [squared]) = a distance of 3 × 10 (ninth power)**, which is the space generation of the Sun. Neptune is just on the edge of this at 4.5 (ninth power). After Neptune, outer solar orbits such as Pluto show more deviation from the ecliptic and asymmetry in orbit, where the layers of space thin out and vary density. You are under an umbrella of the center of the galaxy at Pluto. So Pluto travels through light and dense space matrix/matter, in and out of our Sun's umbrella. From this you might assume well, should not the space of the Sun extend further out to include Neptune in a model of planets in the space of the Sun aligned within the plane of the ecliptic? Well, maybe the Sun's mass is a bit more, or maybe it was at one time but has expended this via nuclear fusion, and Neptune is in a stable space density zone now for some time since its original orbit. When I refer to eccentricity of planets beyond Neptune being more eccentric, beginning with Neptune, I refer to the "deviation from the ecliptic plane" and asymmetry in orbit. What I am trying to convey is that a less dense space does affect orbits differently. In the case of Mercury, it is very possible that the compression of space close to the Sun is also affecting its orbit at the opposite extreme; space there is more distorted. Compression and expansion of space due to density, a concept not even considered, affected by its originator and the spin of the source. (Page:1-36)

It could be that the eccentricities of Neptune and Mercury are also due to variations in space density, the density being greater near the Sun. Deviation from the ecliptic and asymmetry in orbit are results of lighter-space density bordering greater-space density. Neptune orbit begins to be more asymmetric to a center. Neptune is within the limit; however, believe its orbit was established when the Sun had more mass and is now receding and effecting its orbit slowly. Beyond Neptune, orbits get wild. Every component of the universe can be explained by other components, so you can equate any component of physics to others components. Hence you have a Trans Dimensional Unified Field Theory to explain all these relationships.

### *Motion and Density*

Space by these equations is proportionately related to motion and density. Note the following:

1. *D* = (cube root) of $M / T^2$ (squared)
2. Acceleration × density = *D*
3. Force / $D^3$ (cubed) = *D*

It could be that spaces generated by various bodies are additive and overlap as gravity is additive. Hence there are different densities of space surrounding and in the path of different planetary bodies. This could explain the eccentric, asymmetric, and a-ecliptic orbits of planets where more eccentric, asymmetric, and a-ecliptic orbits result from greater fluctuations in space density in the orbital path. Perhaps Pluto's orbital path is the least homogeneous.

### *Space Density*

In the theory discussed, mass generates space. The space created by the Sun is just up to Neptune. The space beyond is covered by the center of the galaxy, so it is less dense. It may be that planetary bodies prefer to exist in areas of greater space density. So in a sense, the counterbalance to entropy is found where the dispersal of energy is reversed. So we see the result as a spiral galaxy where bodies tend to concentrate in spiraling arms from the center of the galaxy in arms of greater space density. It may be easier to travel in areas of greater space density, which is opposite to our normal thinking of traveling through higher densities. As the galaxy spins about its center and pulls its space, the resultant effect is a spiral from its center. Entropic inertia keeps the forms in spiral rather than circles with condensed bands of space.

According to this theory, planets or bodies create space. The density of space varies. So the Sun creates an umbrella of space just short of Neptune; from the equation $\boldsymbol{D^5 = MC^2}$ **squared**, Einstein left out $D^5$ to the fifth. So the center of our galaxy creates space that passes through our solar system. The density of space varies. Maybe that explains why the orbit of bodies is more asymmetric from Neptune to beyond. Think of an object traveling from oil into air. If all things are related, why not space? I'm thinking that the smallest particle is space. So the things affecting planetary orbits are multiple to include the fabric of space itself. This also could explain why objects exiting our solar system into deep space diverge in their paths because the gravity of other celestial objects could affect them differently in a different space density. Gravity equations will have to be altered to include space density.

It is possible that early in the evolution of the solar system, the space generated by our Sun was a bit farther away than Neptune and perhaps just beyond Pluto but has slowly been receding. It may be that the shift is space density from a higher density of our solar system to a lower density of interstellar space is also responsible for the creation of the heliosphere. This area of space-density transition would create a transitional density zone between solar space and interstellar space.

Each dimension of space is a different field or matrix type of space. The monopoles of that space matrix are different. It could be that the space from different celestial generators overlap, affecting the dimension of space such as the space of $D^3$. Also, the space of a gravitational field of a higher dimensional matrix may blend with the space of $D^3$. Hence the density of space may vary according to blend of different space matrixes and types. And the extension of these spaces from the celestial generators may also vary. Hence the reality of space may be a complex blend of densities and matrixes of the same and different types. Since these matrixes are transparent to the human eyes, we perceive space as a homogeneous energy field, which it is not.

**Definition of the Atom, Gravity, and Mass**

$\boldsymbol{F}$ **= magnetism /** $\boldsymbol{D^2}$ **(squared) =** $\boldsymbol{MA = D^4}$ **(fourth power)**

**Gravity force vector = (MAGNETISM) ×** $\boldsymbol{D = F \times D^3}$ **(cubed)**. The idea of gravity intrigued me. It is true that an atom does not exert a magnetic field, which is a mystery, but I claim that it does, but the nature of it being one dimensional, and this is expressed as gravity in my opinion. Hence the connection to electromagnetism is discovered in the atom. And perhaps to recreate the mechanism of the atom would actually create a gravity generating device, and once this was understood, the counter principle could be engineered to create a gravitational field of opposite polarity. Maybe this is what flying saucers use. From the equation, we have determined the relationship of magnetism and gravity. Gravity is the result of magnetism acting as a one-dimensional vector. Since all atoms each exert gravity, we have to look to the atom and see what component of the atom generates a magnetic field restricted to a single vector or point. Since moving electrons create a magnetic field, we can assume that magnetism is caused in each atom but restricted to a one-dimensional component by the moving electron-proton interaction. This magnetic field, restricted by its interaction with the positive proton and in one dimension, is the gravitational component. Hence, gravity is magnetism, but of dimension 1. By this we can now know how to create a machine that can create gravity. We need to map the path of an electron around a nucleus and create a machine with electric current that does the same, but of greater magnitude. And also, the reverse of such a process in polarity would be to create an anti-gravitational field. A reverse in polarity means a reverse in the direction of gravitational vector or field. A reverse in polarity would then be a reverse in the relationship of orbiting charge.

According to this theory, the smallest particle is space, and a photon is the result of an elastic collision along the matrix of space. Photon nodes collide to transfer the collision wave, and the effect is light, but light is not a traveling substance much like electricity. Hence, you have a particle and a wave, the elastic collision being the particle, and the matrix of space the wave. But since the matrix of space is transparent (penetrating all matter), it can be bent and turned; such is the phenomena of light. The fastest speed that the collision can occur is the speed of light. So it is hard to travel faster than the matrix of space will allow without disturbing it. Space represents the smallest dimension (1) and particle. As relationships build to twenty-three dimensions, you get subatomic forces and particles of great energy. No wonder a supercollider has to work so hard to transform twenty-three dimensions into one; is it possible, isn't it comical to look for what is in front of you already? The relationship of complex geometric/dimensional structure allows far less flexibility at high dimensions, such as twenty-three dimensions (at the subatomic level) than at low level exoatomic level of space. Hence you get the behavior of atoms attracting and repelling to conform within the limits of the subatomic parameter dimension of high order.

## Time and Time Travel

I am exploring time. To me it appears a velocity. Five dimensions of time are resulting in mass. In the space-time, we have three vectors of time. We measure our clocks with one dimension. **$M = D^3$ (cubed) $T^2$ (squared) = $V^3$ (cubed) $T^5$ (fifth power). $T^2$ squared = density.**

**$T^1$ = time = $(MV)$ $D^3$ cubed / gravity ($D^7$).** Density means time is defined by momentum in three dimensions divided by gravity. Hence, momentum, space, and gravity define time. So to increase $T$ directly or by induction would increase momentum, speed up time, and reduce gravity. So a spinning ball (high rpm) would rise faster and fall faster than a non-spinning ball.

**$T^2$ = time$^2$ squared = density**

**$T^3$ = $T$ in third dimension**, but we measure $T$. But in $D^3$, there are three vectors or superstrings of $T$.

**Gravity (density) = $V^7$ seventh × $T^7$ seventh = $D^7$ seventh.** Now take a **limit of $C$ and**
**Time = seventh root of $G$ / $C$ = seventh root of $G$ / $V$**, where $G$ approaches 1, and hence **$T = V$.**
So time is nearly a velocity, inversely proportional, but in some instances equivalent.

**$T = D / V$; $T = D / C$**, and so at the speed of light, time approaches 0.

### *Future Time Travel or Forward Time Travel*

Where the gravity is greater than 1, time decreases with velocity, but at the subatomic level (higher dimension such as fourth or fifth dimension) where gravity is less than 1, time speeds up, or the reverse happens relatively. So the implications for time travel exist, but in such a scenario, time is moving forward, as in a river, with slow boats and fast boats. Time travel in the reverse would work in a physics that is working in reverse. This would exist particularly in the physics of polarity yet to be discovered in the area of antigravity and antimatter. However, the actual existence in such a dimension may be impossible. There is a crossing point where time and velocity are equivalent between exo and subatomic worlds. In the exo world or our world, when you speed up, time slows down relatively; this is known. At the subatomic level, when you speed up, time speeds up relatively or faster into the future according to the equations. This means a leap into a higher dimension can propel you into a faster time rate, such as faster-than-light travel or hyperspace. This is discussed in further detail at $T^4$ (time to fourth power) below.

### *Past Time Travel*

Time travel into the past would require traveling in dimension of polarity. In the case of polarity in the exo world or polar exo world, you are traveling back in time. As you speed up in the polar exo world, time goes back slower, and at the subatomic exo world, time is faster as it moves backward; both scenarios are moving backward in time and opposite to our normal world (exo and subatomic). So a spaceship, to time travel to the past if possible, has to enter polarity at a lower dimension (polar exoatomic or third dimension like our own) for normal time or at a higher subatomic dimension for faster hyper time (opposite to here). So the ship must have the capability to enter exo polarity at a lower dimension and move to a higher dimension to go backward faster but return via higher "normal" subatomic dimension to speed forward. The question is how effective is the time spread involving inter-dimensional travel, meaning how fast can you move back or forward or backward in hyperspace; and of course, traveling via other dimensions, the time displacement will be different. This is complicated physics and requiring much advancement in the relative motion of different dimension, or more exactly relative time ($T$), distance ($D$), and velocity ($V$). You may not be able to travel in a polar dimension. I do not believe there are any paradoxes in physics or time. Time is a velocity, like a river with motion or a motion picture. The river has a destination, and you can ride a boat to the destination to the point that you depart; going forward seems to be unknown. If there is a loop back, it is just that, still a part of the river.

Time travel can be induced because time can be induced.

### *By principle of induction, $T^4$ (fourth power) induces $T^1$ (first power)*

**$T^4 = T \times$ density $\times$ ($MV$) $D^3$ cubed / gravity force ($D^4$)**; a relationship construct of these in velocity and rotational motion will induce time ($T$). Space is constructed by the smallest particle $D$. And a photon is an elastic collision along the matrix of space, at the speed of light, which is the limit for the matrix of space. Can the galaxies expand faster than the speed of light? Or a better question is can space expand faster than the elastic collisions it allows? No, since we would not detect the phenomenon of light under these conditions. Since space permeates or passes through everything, it is pliable. Since it is possible to redirect light through a prism, we can assume that space elastic collisions are directionally altered in a prism. The control of light is the bending, compression, or turning of space and redirection of the elastic collisions that are taking place. Thus, by principle of induction, we can rotate space by rotating light in the third dimension and induce a vector or component of time. Hence, this could be a possible method of inducing time travel. Note the equation also implies other events. The rotation of space not only involves the induction of time but 1/gravity, which could be a polarity of antigravity. The bending of space could induce not only time but antigravity. This has nothing to do with photons, but the photons can show the direction of space since photons demonstrate or are the elastic collision of space. Reference the $T^1$ equation above. There are multiple dimensions in each superstring $D$, $T$, and $V$.

### *Other Possibilities*

**$T^4$ = density squared**, a fluctuation in the density of space will induce $T$.

**$T^4 = D^4 / V^4 = D^4 / C^4$**; at the speed of light, **$T^4$ = force or fourth dimension ($D^4$)**, which might be the portal for time travel. Since the distances are so great in the galaxy, time travel may be necessary to get around. Perhaps time as we see it is different from what it really is.

**$T^4$ = space ($D^3$) $\times$ $M$ $\times$ density / magnetism ($D^6$)**; manipulation of magnetic construction of space will vary the density of space and time.

### *Relative Time, Inter-dimensional Travel*

In $T^4$, time would be flowing at $C^2$ squared compared to the third dimension. Since by adding a $VT$ to the next higher dimension, you square or add a power to the result of time, velocity, and distance. So $3 \times 10$ to the eighth power kilometers for one second of time of light speed would be 9 × 10 to sixteenth power kilometers for one second of time in fourth dimension (hyperspace) relative to the third dimension, and this would even magnify to another power or $C^3$ for the fifth dimension of hyperspace. So flying a spaceship into the fourth dimension, in an instance you would travel across the galaxy. So an adjustment of relative time in hyperspace fourth dimension would result in a vast jump in space on the third dimension. Adjusting relative motion in hyperspace or fourth dimension would also be an issue of time travel since I believe adjustment of motion faster would push you forward in time and slowing down perhaps backward, but I doubt the backward component since I believe all things move forward. The key to moving forward would be to maintain relative velocity to freeze relative time, and to slightly adjust relative time to affect a change is distance travel. Or by adjusting relative distance travel without adjusting relative motion, that would affect relative time travel. Hence, time would adjust distance, velocity would adjust time, relatively speaking. And of course, the ship would have to be transformed magnetically on the exterior to conform to a multidimensional matrix.

### *Hyperspace*

According to this theory, one second in hyperspace (fourth dimension) is equivalent to 10,000 light-years away in travel. The Milky Way galaxy is 100,000 light-years in diameter and 1,000 light-years thick. So travel across our galaxy would be ten seconds. Now, the horizon of the visible universe is fourteen billion light-years away, so in the fourth dimension, travel would be 1,400,000 seconds or 388.88 hours or sixteen days in hyperspace. Multidimensional travel is definitely the way to traverse space. It wouldn't surprise me if the universe is much larger. I've heard recently that Hubble saw a horizon of seventy-six billion light-years; it would not surprise me if it was one billion times larger.

- Time adjusts distance in hyperspace.
- Velocity adjusts time in hyperspace.

I prefer the concept of time travel to the past being possible by concept of polarity where the motion of time is naturally reversed, if possible at all.

When you think of breaking the speed of sound, you have broken one, sound and air, but still are traveling in another medium: space. What happens when you pass the speed of a medium, and then there is none left behind it? Well, then you cannot. You cannot travel faster than the speed of light of the medium or space. With relationship to the speed of light, to exceed you need to disappear to the fourth dimension or hyperspace where time varies distance and velocity varies time relative to the third dimension. So before you ever break the barrier, you are going to bounce to hyperspace. In this hyperspace, your relativistic connection to your home dimension is drastically altered and will involve huge displacements of position, time, and velocity. So much so that the first explorers may, in fact, is lost in time.

### *Time Travel*

About time travel, if it is possible to travel into the past, I believe the following scenarios:

1. Time events cannot be changed. If there was a traveler who altered an event in the past, it is already known by the traveler who traveled to the past or known by someone, hence, no surprises. In other words, it's already a part of history; if I changed time such as met myself, I already knew it to happen before I left to change time.
2. Into the future is like a river with two boats; the one behind has to experience what the one ahead has seen. A better explanation is, I can age slower in the fast boat and see the future, while in the slow boat I age faster.
3. There are two ways to time travel: either you move forward, or you allow history to move forward.

### Antigravity

Gravity is the result of the helical path of an electron around a proton. The multiple helixes created by rotating and circular-helical orbit create multiple vectors of gravity to the center of the nucleus. Hence, you have gravitational forces to the center of mass from all sides or points of spherical surface. This is only possible by an orbiting electron at very high-speed energy at very small distances, and so all points of the surface of the sphere are covered almost instantaneously or simultaneously. At near the speed of light, multiple gravitational vectors toward the center of the nucleus (positive charge) is possible as almost to appear instantaneous or, more appropriately, simultaneous. It is possible to produce a gravity machine by rotating electrons in a circle or sphere at high velocity. If in a sphere as in an atom, you will get a gravitational field pulling equally from all directions to a point or inward. If in a circle, you will have a bipolar gravitational field. Understanding this mechanism of a gravitational field is the beginning to deriving a solution for the creation of an anti-gravitational field. A reverse in polarity is required where the orbiting charge at high speed is positive and may be assisted by a central negative charge. Rotational motion of interrelated positive and negative charge has a relationship to gravity. The two forces related are gravitational. And this can be applied to the construction of gravitational or anti-gravitational machines.

If you want to create antigravity, what is required is to reverse polarity of the atom "model" that creates the gravity as we know it to be. This can be accomplished by a proton current around a negative charge. How easy is that? The resulting gravitational vectors will be in a radial pattern outward rather than oriented to the center of the nucleus caused by small orbital electron circles of planar area tangent to the nucleus. A model that reverses the relationship of charge will create an antigravity model. The repulsion of the planet's gravity may be accomplished in several ways to include the following:

1. Circulating or rotating positive current surrounding a negative charge.
2. Induction-producing positive charge on a sphere surrounding a negative charge
3. Circulating and rotating positive charge ring.
4. There is a possibility that rotating space has an effect on gravity. Look at equations for $T^{\psi}$. In this theory, space is represented as a matrix, and a photon is the elastic collision of that space. Hence, rotation of space would show a rotation of photons in space.

It is my contention that to duplicate the mechanism of an atom, rotating motion is required; however, for weak fields of antigravity, it may be possible to find repulsion in various mechanisms that produce positive charge or polarity to the natural gravitation construct. Superconductive mechanisms may be a manner for creating electron orbits. Other mechanisms are also possible due to the variety of relationships in a transdimensional or multidimensional model.

***Electron Orbit***

Orbit of an electron usually describes the size of an atom. The orbit is usually $10^5$ larger than the radius of the nuclei, or something like a grapefruit seed in a soccer stadium. So by size comparison, there is much more space than matter by relationship of distance. This is something like our Sun to the farthest planet in orbit. From our perspective of distances, the distances of an atom are very small; and an electron, traveling near the speed of light, is virtually everywhere in orbit. So you have the concept of an electron cloud. And you have the possibility of simultaneity, that multiple gravitational vectors that radiate toward the nucleus center appear simultaneous due to high electron speed and small distance of orbit. Electricity and magnetism caused by the orbiting electron and relationship to proton hold the atom together and define the space of the atom. The atom's center nucleus is positively charged, and the electrons that move around this nucleus at very high speed are negatively charged, so they attract each other. The reason the force is strong is because the atom is so small. The distance between the nucleus and the electrons is about 1 angstrom; this is 0.00000001 cm ($10^{-8}$ cm) or about four billionths of an inch. Since the electric force varies with the distance between the positive and negative charges like 1 over the distance squared, the force gets large for small distances. The electron is small compared to the nucleus. One can consider that **1 neutron = 1 proton + 1 electron**. This accounts for a large part of the difference between the neutron and proton mass. The actual explanation is much more complicated when described in terms of smaller particles, but the above explanation is approximately true.

A photon is an elastic collision along the matrix of space. The components that create space and its matrices rotate at the speed of light. When the wave patterns intersect, they create interference patterns or the particles that we see that also slows down the rotation of the waves that caused the interference particles. Since the interference or superposition of space components creates the particles we see, there is more space than particles, and that is why there is more space in the universe—both exoatomic and subatomic—since particles are created by rotating components that compose space. When the patterns or space wave components rotating at the speed of light intersect or interfere to create particles with mass, their speed of rotation slows down. Hence, we have such things as electrons and other small particles moving slower than the speed of light and acquiring mass.

I like the double-slit experiment to describe the distinction between the nature of a particle and the nature of a wave. Actually, a photon is an elastic collision along the matrix of space. So by such an observation, you then reconcile the dilemma between the wave and particle perceived in a photon. A photon is not the "thing" or particle but the effect or elastic collision of the fabric of space, which will not exceed the vibratory motion of the space matrix, which is the speed of light.

With regard to the wave description of an electron, there are two waves to consider as follows:

- **The first wave is the space waves that intersect or interfere to create an electron.**
- **The second is the elastic collisions of electrons that create a transverse wave.**

The first waves (strings or spheres) and their interference patterns result in the construct of an electron. An electron is made by the interference patterns of the "strings" (or spheres) that make up space. And then the wave (as we know) or elastic collision of electron is the electricity or electrical wave. Electrons are interference points in space and the result of space interference patterns. So think of an atom as a collection of rotating strings or spheres. These interference nodes are particles, and the elastic collisions of those nodes are waves. So in a model for the atom, think of each particle comprised of the intersection of motion of strings or spheres, and this superposition of a new geometry will allow you to see what is not seen and understand what a wave is and reconcile the dilemma between first wave or space, particle and wave. Particles are made from the intersections of space waves, and the elastic collisions of those particles are the other (transverse) waves we know. So now, the particles do have their separateness and combined effect as waves. Particles like electrons are a high dimension of space but are not the waves or elastic collision we imagine, but a particle generated by a circular wave.

The centricity of gravity convinces me that the electron path shifts as it orbits a nucleus. Take it for a given that in nature, motion is orbital and causes induction of other properties (space relationships). And the rules are different for each different dimension of space. Thus, one could surmise as natural laws

- **rotation motion causes induction, and**
- **rotational motion is the relationship of all dimensions.**

The waves of space or strings or spheres, however you want to identify them, are the construct of space. These strings are distance (D), time (T), and velocity (V), and their relationship to higher dimensions increases the complexity of physical phenomena that we see. The most fundamental wave or first wave is space, so let's call that a string to separate the concept. The wave we normally think of is an elastic collision of the constructs of first wave or space. The first wave is the smallest component that builds space and particles, and their vibration of collision generates other waves.

The shift in the orbiting electron must have to do with some asymmetry in the arrangement in the nucleus and/or its relationship to the electron, so with each pass, you get a shift. Perhaps what is asymmetrical in the nucleus is that it is spinning, pulling the electron and shifts its position constantly. Because both the electron and nucleus are rotating and the charge relationship is slightly shifting, the electron shifts its orbit and, at high speed, covers the entire area of an orbital sphere, and each orbit creates a gravity vector to the center. This multiple action, in the millions, is what is perceived to be a spherical action or electron cloud and achieves a resultant effect of a concentric gravity from all points of the sphere around a nucleus.

The evidence in current gravitational experimentation so far has to do with experiments that induce gravity. Many experiments create bipolar fields because the positive superconductor in some experiments that use them is not shifting position as it spins. The goal in imitating gravity is to create a concentric three-dimensional effect, and that is only possible with a shift. Now, antigravity does not need to be three dimensional but only to repel gravity; hence, a unidirectional or bipolar device would suffice.

### *Space and Geometric Superposition*

A photon of light is the result of elastic collisions in the fabric of space. The light is the effect of the collision and not a "substance" in transit. If light can be bent by a prism, then a prism can bend space. Coherent light or the light of a laser can be accomplished by compressed space and/or geometric superposition. The geometry of space has the flexibility of compression and superposition. The geometry of space is radial, spiral, and allows for a Cartesian interpretation. A radial geometry allows the shortest line between two points. Spiral geometry is sympathetic to circular motion or spin. This flexibility in the geometry has its foundation in spherical geometries, somewhat like the double spirals of a sunflower, except in three dimensions rather than two. What we see in nature reflects the geometries of space. The nodes of this matrix of space are transparent or unseen and would be the smallest particle (photon), and the strings supporting them are traveling at the speed of light. At higher dimension or subatomic level the nodes become denser; for an example, an electron and the space of the higher dimension is superimposed on the transparency of three-dimensional spaces.

As the dimensions increase, the degree of compression and superposition is increased from the exoatomic level to the various subatomic levels. The interference patterns of various dimensions of space or the interference of spatial circular motion create the particles that we see. And these particles also decrease the motion of the component strings or waves that comprise that dimension. Hence, an electron exists in a different spatial dimension than the surrounding space. A beam of electrons exist because of the compression and superposition of space in its corresponding dimension. And by the rules of a transdimensional framework, each dimension has a relationship to others by a set of rules. Hence, magnetism and charge have relationship to all dimensions of space, including electrons and their motion or gravity or whatever you wish to discuss. Or more precisely put, a dimension of space has a relationship to all other dimensions of space. For example, all dimensions have an impact and relationship to time or all other relationships conceivable. So the most remote relationships can be discovered. An electron exists as an interference pattern of a more-fundamental circular wave pattern of space. An electron is an effect of space.

The waves or strings that make up space and all matter are an extension of the center of each galaxy or black hole, and these are connected to infinite space. To limit the explanation, space is the building block of all that we see, and the strings that comprise space are in motion at the speed of light and so have existed since creation.

If you have an electron rotating around a rotating nucleus, the path motion of the electron will definitely be altered by a positive charge in the nucleus that is constantly changing position, even if the points of positive charge or protons are dispersed symmetrically along with asymmetry in the matrix, and since you will have both strong points and weak points on the surface of the nucleus. This variation of charge and the interaction of a spinning nucleus coupled with a spinning electron will cause a shift in the position of the electron. Gravitational experiments have produced bipolar results in the

## Spherical Geometry Space Matrix

induction of gravitational fields because in the atom, the relationship is constantly shifting and has not been replicated in a lab. In a symmetrical model, where you have the positive and negative charge placed symmetrically, the relationship to the electron cannot be symmetrical if the nucleus is spinning. Therefore, the charge relationship of photon to electron is constantly changing. That means the path will change. There is plenty of current experimentation in gravitics or gravitational generating devices that says a concentric and uniform field has not been developed. I attribute this to the absence of the kind of orbital field I am explaining. If the nucleus spins, the electron path would have to change. The concentric nature of gravity suggests a shifting electron path around a nucleus. Although not related, but even planets in their orbits shift position, and near the speed of light, it's a movie fast-forward. Gravity is concentric.

Particles are the result of space relationships. Hence, it is conceivable to covert the matrix of space into a group of electrons or the matrix of space into a hydrogen atom. We normally think of fusion as combining atoms like helium to more complex atoms, but by this theory, it is possible to create any element from space, even a black hole, and to convert back to space again.

The idea that magnetism can influence the path of electrons can be restated that one dimension of space and geometric dimension have a direct influence on another space and geometric dimension since their commonality is the very same thing they have in common: the geometry of space as a building block. This is no more far-fetched than saying hydrogen is related to sulfur by a fusion reaction of common elements or particles. We can manipulate elements by knowing what the commonalities are between them. And for a general theory, this commonality is the geometry of space, whose components are distance, time, and velocity.

Electrons are not invisible since they are the interference patterns, but the elements or space components that make up an electron are not seen by current physical measurement techniques.

### *Antigravity, Reverse Polarity, and Superconductivity*

Antigravity is generated by the reverse in polarity of positive and negative charge in rotational motion. The classic view of gravity as discussed is the electron orbiting a positive nucleus. Gravity vectors are generated toward the center by an electron orbiting and shifting orbit constantly in rotation around proton(s) at 75 percent of the speed of light. Hence, the high speed of the electron in all orbital positions gives an almost instantaneous effect or electron cloud and vectors of gravity toward the center of the nucleus that cover an entire sphere. So to reverse polarity, positive charge needs to orbit and generate a field of anti-polarity or anti-gravitation—reverse model.

Substances where positively charged particles or ions can be rotated could be in superconductor rings, which would also include super fluids such as helium or other elements that are also superconductors. A ring would allow the arrangement of charge in a circle to be rotated or spun. A superfluid would allow the ration by spinning of the fluids within a superconducting ring. Rotational motion induces the gravitational fields that the experiment would be looking for. Superconductors expel negative charge from their interior, which consequently becomes positively charged; this displacement in rotation can be used to generate an antigravity field. We now have understood that the analogy is in fact much closer than originally thought: not only is the diamagnetic behavior of superconductors similar to that of atoms, but also the charge distribution of superconductors is qualitatively similar to that of atoms—more positive charge near the center, more negative charge near the surface. Perhaps in a ring, this displacement would allow the manipulation of the positive charge in rotational motion where the concentration of positive charge and its rotation will induce an anti-gravitational field in the manner expected. When this principle is understood more fully, it may be possible to induce antigravity by other manipulations of the space matrix since in a multidimensional construct; induction can be achieved by many different mechanisms. In such a superconducting ring, many different elements can be tried, and perhaps those that work best might have magnetic properties.

### Dimension and Superposition

Every dimension is a superposition of space. Every dimension of space has a matrix of strings. These strings are distance, time, and velocity. Each string comes as positive, negative, and neutral. The strings have spin and appear as spheres. When the lattice of strings overlap or intersect, they form in interference patters or particles that we see. The strings spin at the speed of light, but move slower when the interference patterns create subatomic particles. Nearer to the speed of light, the interference patterns or particles are those such as neutrinos and photons. And in the third dimension and sub dimensions of one and two, the particles of the interference patterns are transparent and pass through matter. The matrices of space or strings themselves are also transparent and allow superposition. Superposition is the compression and overlap of space to higher density and dimension. And in these higher densities, you have more complex rotation of space and interference patterns as well as heavier particles. So all matter and dimension at whatever level is made of the lattice of space. All space has current or the elastic collision of the nodes of the lattice of space, which occurs at the speed of light. The node of the lattice of space for the third dimension is the photon. For a higher dimension, it is the electron. The collision of these nodes or particles of the space lattice is *current*. Current can be harnessed whether it is electrical or photonic. Photonic current can induce electricity via induction. A possible experiment would be to rotate light in a crystal of fiber optics as in a Tesla coil. This rotation, which is also a rotation of the lattice or space, would induce currents in other dimension, such as electrical, since all dimension are related via rotational motion. Rotation means induction. Particles that travel through a space matrix cannot exceed the elastic collision of that space-time, as described in relativity. However, the relation between dimension is related by $C$, so in the fourth dimension, velocity related to the third dimension is $C^2$. But in each dimension, it is still $C$. So relativity changes the relative time between dimensional spaces.

## Current

The elastic collisions through different dimensions of space result in different types of current. The current (light) through three-dimensional spaces is different than the current of superimposed space (electricity) where you find a metal, like iron. So the elastic collision we call electricity is different than the elastic collision we call light. Both are current and wave induced by elastic collisions. In this theory, electricity (electric elastic collisions) and the generation of current via electrons are related to elastic collisions in other dimension: by induction. Hence, it is quite conceivable to harness the power of light from the Sun and by induction to deliver an electric current in a wire. Hence, all elastic collisions, no matter in what dimension, are a source of power when properly harnessed by induction. The elastic collisions of light from the Sun to the Earth can be seen as current ready to be harnessed, like a circuit waiting to be completed. The task is to find out how to induce electric current in a wire from the vacuum of space and establish the relationship between space current of the third dimension and that associated with electrons or electricity at a higher dimension. The only medium that properly manipulates light and space is a crystal and, therefore, must be the medium for experiment and induction of electrical current in a wire. And from there, harnessing a great deal of available and free energy. This might be obtained by rotating space and light and inducing current in a wire by a photonic transformer to higher dimension..

Elastic collisions are a phenomenon of current for every dimension of the space matrix. And ideally, without impurities or imperfections, these collisions occur at the speed of light. So we can speak of dimensional matrixes and mediums of space that have current or properties of elastic collision and that the current is the result of collisions of the nodes or superposition of the dimensional matrix associated with the particular dimension. Hence, we can speak of photonic current and electrical current, and they are nodes of different dimensions of space. Sound is also an elastic collision, and therefore, there must be a matrix of space that will allow sound or molecular elastic collision to occur at the speed of light. Since it is possible for sound or molecular elastic collision to occur at the speed of light, it may be possible via induction or rotational motion to induce antigravity via sound (molecular vibratory motion) occurring in a circular path.

## Space Transparency

Three-dimensional spaces are transparent, which means that the components and particles that constitute three-dimensional spaces pass through more dense space, matter, and subatomic particles. As space superimposes to produce more dense space, these transparencies are replaced partly by particles created by the interference patterns of this superposition. Thus you have more dense space, which is still transparent, and the interference of these superimposed spaces and their overlapping spin that creates the particles that we see. For an example, an electron is the result of the interference pattern of superimposed space. At the level of transparent space of the third dimension and also transparent particles, we are at a level of the speed of light for both spin of spatial dimension string components and the particles created by interference patterns at the same level. Neutrinos are the transparent particles of transparent space. It is my contention that the three neutrinos (electron neutrino $\nu_e$, muon neutrino $\nu_\mu$ and tau neutrino $\nu_\tau$) combine to produce the photon of three-dimensional space, which is the combination of positive, negative, and neutral transparent particles at the speed of light. This then becomes a model for the universe where relationships are related in positive, negative, and neutral. As the interference patterns occur with the result of more dense particles that are not transparent (such as an electron), the speed of particles and supporting space structure slows to a sub light speed. Hence, in an atom, you may have the electron traveling at 75 percent of the speed of light, which is still fast; to complete the millions of orbits required to generate concentric gravity. If neutrinos do have a mass, then they are particles of a higher superposition or dimension,

## Elastic Collision Grid Matrix

maybe dimension 4 or higher. Consequently, the transparent particles that constitute a photon still need to be discovered for the third dimension in addition to those of the second and first dimension.

## Detection of Space or Dark Energy

The detection of space is the detection of a transparent matrix of component strings (distance, time, and velocity) in positive, negative, and neutral. The intersections of these strings are particles or intersection nodes of the matrix of space. Each dimension of space has different nodes, which are heavier at higher dimensions. At lower dimensions, the intersection particles are transparent along with the strings. The elastic collision of the matrix of space is at the speed of light and is called current. Quantum physics measures the particles (intersections of the matrix of space strings), and perhaps it will be much more difficult to measure or detect transparent particles in lower dimensions of space. Even further down the road it will be hard to detect the transparent strings, which are the building blocks of all space and particles. The relationship of strings in various dimensions of space has the effect of what we call dark energy, that motion associated with the perpetual motion machine of space universe, the relationship of all dimensions of space and associated ratios. Each dimension of space exerts a force on associated or adjacent spaces or dimensions. The ratio of mass to space is $\boldsymbol{D^5 = MC^2}$. Hence, with ratios, there is implied force to equilibrium.

Dark energy is space. Space is made of strings with spin (spheres), and the intersections of these spheres are particles. Space and its properties are defined by dimension (*D*), which is the superposition, compression, and density of strings in a geometric space matrix. For example:

**$D^5 = MC^2$ or $E^2$ = gravity = $D^{10}$**

**Energy = Dimension = Space**

**This means that various dimensions are various dimensions of space or energy.**

## Generation of Space Fields:

A black hole is $\boldsymbol{D^{23} = E^4 \times D^3}$. This equation can be reduced to $\boldsymbol{E^2 = G}$ **(gravity)**. From the equation, we see that in a black hole, energy is squared and the field of gravitational space is much larger than photonic space. Before I go on, I want to make clear that each dimension in this theory is a different dimension of space matrix. Hence, you have the photonic field of $D^3$ with its photon monopoles, and in higher dimension you have gravitational fields, electric fields, magnetic fields, and so on. So the space we traverse in outer space has various densities, transparencies, and types. "Outer space" or the ether is not one type of space but a myriad of different and complimentary spaces, geometries, or matrixes that exist together. So $\boldsymbol{E^2 = G}$ suggests that the distance of the field of gravitational space is square the distance of photonic space. Hence, its action on objects is very far away. For the Sun, this action would be far beyond the solar system to planetary objects like Sedna.

Sedna is a planet or planetary body roughly the size of the Earth's moon and very far away from the orbit of the Sun, about $1 \times 10^{11}$ km from the Sun. If you calculate the value *G* for the Earth and Sun using $\boldsymbol{G = E^2}$, you will find that the ratio of gravity between the Earth and the Sun varies by the value $M^2$. And by calculation of *G* for the Sun and the Earth, you will find that *G* for the Sun is about $1 \times 10^{11}$ times that of the Earth. So it appears that the distance of the effect of the gravitational field matrix from the Sun is also about $1 \times 10^{11}$ km from the Sun or the distance between the Sun to Sedna. Black holes have spin and are spheres since everything in the universe is made of compaction and superposition of space matrixes or geometry. The spherical nature of a black hole reinforces the following concepts:

- That gravity is concentric.
- That space is a uniform spherical geometry and that all things are made from this regular geometry in balance.

What makes the gravitational field space of a black hole so strong is $E^2$, which means that the field space of gravity extends far into the universe, generating most of the space that appears black in the universe. And if you calculate the average distance between galaxies, you will find they are based on the equation $\boldsymbol{G = E^2}$, where the distance is squared to the $D$ calculated from $\boldsymbol{D^5 = MC^2}$.
**Gravity = $\boldsymbol{EMC^2}$**

**Higgs Boson, the God Particle, and the Smallest Particle (D)**

Strings ($D$, $T$, and $V$) make up the matrix of space. The strings are transparent and have spin at the speed of light. The monopoles or nodes of the matrix are where the strings intersect or overlap to create interference nodes or monopoles. So the space field is directly related to the existence of matter. In the third dimension, the monopole is the photon, and at the lowest dimension ($D^1$), you have the smallest particle, which is transparent like the field itself. The relationship of $D$ particle is the relationship of string $V$ and $T$. Strings are the mathematical construct of the universe dimension per dimension. The interference patterns created by the space field are the particles that we see. And in the universe there is a ratio and relationship between the entire dimension and superposition, and all the mass and associated space per dimension. If there is a Higgs field, it is the fabric of space, the dark energy. This field contains all the elemental strings by which all matter is made. This field is also the building block of all particles of mass and space dimension per dimension via superposition. Strings are in positive, negative and neutral, and these properties are also seen in particles. Hence, the smallest denominator is in a relationship of three strings in 3 positive, negative, and neutral, or a relationship of 9 and 27 for three dimensions ($D^3$). But to simplify the understanding, $D^1$ is at the lowest level of interference of the matrix.

Since the universe is multidimensional, at the big bang, the existence of these entire dimensions was simultaneous since their existence has a direct relationship to infinity at dimension 23 or the black hole. The entire perpetual motion machine exists by relationship and ratio and, hence, did not progress from a state of low energy to high energy but descended from infinite energy. This universe is a particular qualification of infinity.

There is no particle that gives mass to all other particles, but there are string fields that do, and there is the smallest particle $D^1$ from the string matrix that combines to form all other matter. If this is the Higgs Boson that is being searched for, then it needs requalification. This particle would be transparent much like a neutrino and hard to detect.

Speed of light means elastic collision, which is supported by a matrix. $\boldsymbol{E = MC^2}$ suggests an elastic collision in space. This is now $\boldsymbol{D^5 = MC^2}$, which shows the relationship between mass and space and the construct of strings that result in mass. Light is a current since it is an elastic collision as electricity is a current and elastic collision. The elastic collision of the matrix of space transmits photonic current; hence, as in electricity where the electron does not move at the speed of light, the photon is not moving at the speed of light, but both transmit a wave via elastic collision at the speed of light. The photon, a monopole in the matrix of space, vibrates and transmits a kinetic wave or current. What we see is the wave crashing on the receiving end like a wave crashing on the shore of a beach. Hence, the photon does not travel through space but, as a component of the space matrix, transmits the impulse to use via elastic collision at the speed of light. We see the photon on the receiving end. Yes, to the God particle, you can model it to make two aspects of this particle $V$ and $T$. It is almost like superstring theory where two of opposite spin, $V$ and $T$, combine and then become $D$. $D$ is the smallest subset of superstrings and the element that multiplies dimension per dimension, adding different properties that are also different dimension per dimension. His unites all the equations of physics.

**The String Grid or Matrix of Space, Relativity, and Quantum Particles**

Trans Dimensional Unified Field Theory is a theory that somehow reconciles concepts of string theory, quantum mechanics, and relativity into one conceptual model. In this theory it says that space is comprised of a matrix of strings that makes up all space, particles, and dimensions of space; each space dimension contains specific particles per dimension and associated space dimensions matrix. Mathematics *can* describe all the dynamics of dimension since the strings that comprise a dimensional matrix are mathematical in relationship. The relationship of the string matrix dimension per dimension is related by the multiplier *VT*; hence, a periodic table of dimension from $D^1$ (*VT*) to $D^{23}$ can be established. That is what a unified field theory as described here attempts to do: derive the fundamental mathematical relationship between all dimensions of space that define associated space matrixes and particles.

The order of the theory, however, needs to be developed into various structures of what is seen and unseen, which present-day physics does not have. In this theory, what is unseen (space strings and matrix) is the universal mathematical construct, and what is seen (particles) are the interference patterns of the space matrix or monopoles of the space matrixes and superposition of space dimension per dimension. So the particles are shadows of the larger construct, and our understanding of physics is largely hidden from view. The things unseen are determining what we see, and so our visual understanding is backward to the underlying reality.

**The theory establishes this:**

The structure of the order or universal construction is a matrix of strings: strings are the mathematical structure of the universe, and space is in relationship to three strings that are distance, time, and velocity. Strings form the matrix of space. These strings exist in positive, negative, and neutral, so for three dimensional spaces, you are looking at a geometric matrix of twenty-seven strings. Each string has spin and related together in a rotating active geometry. I say active since all of the structure has rotation and spin. This active geometry is similar to toroids, fractals, and recursive spherical geometries. The intersection or interference patterns of these strings (with spin) form the particles that we see, and the variety of particles is made possible by space superposition in the various dimensions. In the third dimension of space, the intersection node (monopole) is the photon, and the elastic collision of the matrix occurs at the speed of light. The speed of light suggests kinetic elastic collisions. Photons that collide in the matrix are photonic current, similar to electrical current. Current or kinetic motion is a perfect elastic collision of a space dimension and, at the speed of light, is a phenomenon of the three-dimensional space matrixes and perfect elastic collision.

The structure of space dimension per dimension has a mathematical construct. $\boldsymbol{E = MC^2}$ is now $\boldsymbol{D^5 = MC^2}$, which explains matter in terms of space ($D^3$), and *VT* (*D*) is the constant string multiplier dimension per dimension. Hence, we can talk about a periodic table of mathematical dimensional relationships, and the dimensions start with $D^1$ or *VT* and progress to $D^{23}$.

The structure of what is unseen creates the particles (quanta) that we see. Particles are the intersection patterns and superposition that occurs in the various dimensions of space. At dimension 23, there is singularity of a black hole, and the matrix of space converges to a point. By having these three concepts of

- string matrix of space with fundamental mathematical physical relationships,
- the periodicity (*VT* multiplier) of space dimension per dimension, and
- the created quanta particles by interference and superposition of the space matrix,

and these working together in one theory, you have a Trans Dimensional Unified Field Theory, which is the unification of physics and the unified field theory searched for. So you have new ether, and this theory explains all the problems of ether away. First, space is made of strings that make the space matrix; second, space is related dimension per dimension as a periodic table; and third, particles are interference patterns of space superposition.

$\boldsymbol{E = MC^2}$ is now $\boldsymbol{D^5 = MC^2}$. Mass is explained in terms of space and the composing strings of velocity and time. The speed of light is an elastic collision (kinetic energy). Strings are transdimensional. The strings are three—distance, time, and velocity—and compose space. Strings are mathematical in relationship.

Photons do not travel through space like electrons; the elastic collision is the current, and so photons also collide or vibrate as they are in the space matrix. What you get is on the receiving end like a wave crashing on the shore of a beach. The vibration of light is a perfect elastic collision. The photon is a node or monopole in the lattice matrix of space that explains the dilemma of the double-slit experiment. The photon is the effect and not the object of the wave.

Supercolliders are trying to find all particles, but all particles are created by more fundamental strings. Why is that distinction interesting? Because strings are mathematical in relationship, form the greater part of the universe, are the dark matter, and are the Higgs Field. Is the photon the effect of some other root causation? Yes, since particles are interference patterns. In electricity, the wave or current electricity travels at the speed of light, not the electron that moves very slowly. The same is for the photon in the $D^3$ space matrix. How can they interact causally then? They interact through the vibration of the space matrix. You see the photon on the wave receiving end. In light, the same shadows or photons may not push each other around? They push through the vibration of the matrix. Imagine for simplicity people with six arms with jump ropes. The ropes are strings with spin; the people are the photons or monopoles of the matrix. The vibration passes through the lattice of the space matrix at the speed of light. Now, instead of people with six arms, imagine twenty-seven strings (spheres or strings with spin in a spherical and rotating lattice) as in a geometric framework. *D*, *T*, and *V* in positive, negative, and neutral and also in three dimensions are **3 × 3 × 3 = 27**. If these strings have spin and also the geometry is active and rotating, the resulting geometry is radial like a toroid, fractal geometry. It allows movement in all directions, and now you have the simple space of $D^3$. As space moves into the atom, by superposition, it becomes more dense and complex; the geometry is also more complex. The strings and radial geometries slow down and interfere to create heavier particles that move slower than the speed of light. Interference and superposition of strings generates particles. Particles are like two rocks dropped in the water; where the waves overlap, you get a particle at the interference pattern. By analogy, an atom may be a hundred-armed person. I say there are twenty-three dimensions, a black hole ($D^{23}$) being all the lines of force into a point. A black hole is the mega monopole of singularity.

In each dimension, two strings are added—*VT*. Einstein's something that relates energy to matter is *VT*. The strings are *V*, *T*, and *D*—three in all in positive, negative, and neutral that are nine. And the dimensions are as in a periodic table, increasing by the multiplier *VT* dimension per dimension, and the properties associated with each dimension change dimension per dimension. Just like atoms add electrons in the chemical periodic table additively, each space dimension has an increase; however, the dimensions multiply strings. Strings are the mathematical basis of the universal construct. Dimension is the superposition of space and associated patterns of interference. With each dimension come monopoles and properties of space, like gravity and magnetism. Each dimension is another form of space. We live in $D^3$. But atoms are in multiple dimensions that are higher, such as fourteen or fifteen, and black holes are in dimension 23. Higgs boson may very well be space itself and dark matter. And the smallest particle is the monopole of $D^1$. And dark matter is space. Higgs boson is a particle in space that (supposedly) accounts for inertia. That particle may be the monopole of the first dimension *D* or something undiscovered, because *D* is itself a string. So it is very likely, however, strings are more fundamental. Particles are the shadows of the larger string matrix. So you want dimensions to not be directions but just differently integrated strings. I think they are all correct—string theory, quantum, and relativity. Yes. However, I believe by rotation, different dimensions induce others. This rotation is the basis of induction as in electromagnetism. And I think induction has an effect three dimensions over.

Two dimensions would be, e.g., a grid of strings in two dimensions, but you could add a third dimension by laying down more strings into the two *D* but just choosing how they connect to the original lattice. There are three strings—distance, time, and velocity—in positive, negative, and neutral, and in $D^3$, that's twenty-seven. All strings are superposed, and as you add more, the geometry changes. That is the superposition of higher dimensions create more complex geometries, and hence, the particle interactions change, and I think the nature of time even changes in these configurations.

Particles are both the monopoles and interference by superposition and rotation of the space matrix. Particles are the interference patterns of dimensional space. A photon is a monopole or node. Space is highly geometric, and the particles follow the rotation of the space strings. Strings are transparent and are the medium of space. Some particles created at lower dimension are transparent. The strings are the medium of space and also transparent. If there are multiple dimensions, all particles must have a relationship to other dimensions. A particle is related through strings mathematically to all dimensions. Some particles created at lower dimension are transparent. But they are components or more-dense particles at higher dimensions. All particles are separate but are related mathematically by their behavior through dimension. Particles exist in relationship to all dimensions all at once. That's what is meant. Well, that's a motherhood claim. It's true for all physics theories. Yes, but it makes the case here by unification. Space at a lower dimension would be strings and particles and, at higher dimension, atomic. The continuum is space that generates particles, but I like the idea of intersecting strings. It's like a graph, but with edges being the crucial thing. How about intersecting torus-shaped membranes? Well, for a matrix of spheres (strings with spin), which is active and in rotation, you are describing toroid or spherical geometries.

**So to summarize geometric space, I will say the following:**

- Dimensions are not currently conceived correctly.
- Each dimension is a type of (conceptual or mathematical) string matrix. These are just mathematical relationships and don't go off in some real direction, since that is what they are, an attempt to explain.
- A different type of string is attached in a kind of conceptual grid accounting for the second dimension and a third for the third dimension and so on for a while. It's like a graph approach to space, but instead of the nodes being the relevant attribute, it is the edges that are. So a particle, for example, would be an edge. Waves of causation pass *between* nodes and superimpose and are detected as particles or other existents. How do you orient them? Which way is up? They look the same, but they are just connected differently. There is no up; it's all just active and balanced geometry. There is no other space in which they exist.
- The nodes are defined only by their relationship to each other. It gives sense to quantum entanglement. So it's a topological construct. The relationship dimension per dimension is mathematical, and one dimension can induce another by rotation, which they do and relate naturally. It's true graph theory; the "edges" aren't distances. And it's hard to know what distance would mean in such "a lattice, if not "the number of intervening nodes." And if the nodes are not separated by some kind of meta-distance, than how can they vibrate? It's a very tight lattice because the geometries allow radial and linear motion. The wave is a vibration of the space matrix. And what can *radial* actually mean if the only directions are "along the lattice." At times particles that are displaced are replaced by adjacencies. Radial means toroid or like the radial geometry of a sunflower. How big is this lattice compared to a quark? Well, it would be larger since the volume of space to size of particles is also larger. However, strings are the most basic fundamental structure in the physics of building blocks. The matrixes are denser, complex, and compressed in an atom, and the superposition more frequent. Interference patterns and superposition creates particles. Space is smaller in an atom, but larger in space. My view is that the grid of the space matrix expands at lower dimensions and compresses at higher. And geometries of the lattice exist as a rotating geometry.

## Zero-Point Energy and Free Energy

Every three dimensions that is a multiple of 3 represent a dimension of stability in the geometry of space. For example, the third dimension ($D^3$) has characteristics of stability and has characteristics of induction with dimension $D^6$. A spherical geometry (fractal) based on three is stable and represents a toroid geometry. And there is a relationship of "induction" between stable dimensions and otherwise by a factor of 3.Via the rotation of one dimension, a rotated dimension will induce another three dimensions away. For example, the relationships in electromagnetism reveal this. (Andre: Web Link) Now imagine, via a Trans Dimensional Unified Field Theory, electricity is related to just about every dimension somehow mathematically through the sequence of $D^1$ to $D^{23}$. Now imagine that the vacuum of space is a spherical and radial geometry, where the monopoles of $D^3$ space are photons, and that the model described represents the potential for photonic current. When you see light, that light is photonic current. And so the light from the Sun represents a vast amount of photonic current ready to be harvested as free energy. It is conceivable to create a motor that induces electrical current in a wire from the vacuum of space. By rotating space, it may be possible to induce electric current in a wire and harness the vast amount of photonic current. How is it possible to rotate space? Well, if photons are the monopoles of the $D^3$ space matrix, then if you rotate light, you rotate the space. It is possible to rotate space using crystals or fiber optics and induce electrical current in a wire. Perhaps a Tesla coil in crystalline or fiber optic form could be able to do this. It is my belief that maybe in this single step, or an additional combined step, you can create a free energy–generating device that harnesses the free energy from the Sun in vast quantities.

There would be no cost to rotate light. If you rotate light, you rotate space since light is the monopole of the space matrix composed of strings ($D$, $T$, and $V$). Imagine a crystal or fiber optics as in the shape of a Tesla coil to rotate light. Rotation means that light cannot be bounced (reflected) light, but rotated or turned light, meaning light turning as in a magnetic field. This rotation will induce a field of another dimension. By this process, in one or two steps, current may be induced in a wire. All dimensions of space are related by rotation via induction, similar to what Faraday discovered with electricity and magnetism. The idea of light or the photon as a monopole of the matrix of space strings actually resolves the dilemma of the double-slit experiment. Space means elastic medium and elastic collision, and a perfect elastic collision occurs at the speed of light, also a phenomenon of space. The $MC^2$ and its connection to light means elastic collision and space. $\boldsymbol{E = MC^2}$ is an equation that concerns or describes space strings, and $E$ is $D^5$, so energy is space, and when you have atomic explosion, you are creating space, unraveling the space from a higher dimension to a lower one. The strings are the same, hence, the mass, as we understand it, measures up to $MC^2$. The total package $MC^2$ remains the same from a pencil to splitting matter. The energy lost is space.

So the cost of rotating space via crystals or fiber optics requires zero motion since the current or photonic current is already there in the vacuum of space, and light *can* be manipulated via optics, like current in a wire for electrical current. Light needs to be redirected to induce other forces and finally to electrical current. I believe it's all there for the taking, free energy for the whole planet. Space is the Higgs field and dark energy, and free or zero-point energy can be harnessed from the vacuum. The Sun is the generating source.

This theory maintains the principle of induction of one space dimension to another and in a relationship of rotation. Light is a current like electricity is a current and, therefore, must generate a rotational field of some kind. However, I maintain that the induction is from higher dimensions to lower dimensions and involves rotation. The induction from photonic current at $D^3$ may be restricted since three dimensions lower does not exist. Perhaps fields generated by induction are restricted to one or two dimensions. The electricity of a higher dimension induces magnetism three dimensions lower, and the relationship is rotational. Photonic current will generate a field pattern also, but I often thought that because the dimension for photonic current is three, then the current of light must be restricted to generating fields of one or two dimensions as said before and have other restrictive characteristics. Concentrated current (concentrated $D^3$ space) would be a laser, and we know that the field pattern of magnetism around a laser is radial from the laser and varies. This radial or vector pattern tells me that both electricity and magnetism are of a higher dimension and field patterns are restricted, changed, and altered here from what is induced in electrical current. If you can take these vectors and curve them by curving light ($D^3$ space) and hence curving space, you will have artificially created an effect of a higher dimension of space, which will induce the electromagnetic properties that you need to induce electrical current in a wire for generating power. So fiber optics or optics can be used to turn space, not reflect. But what I mean is that the path of light be near to a curve as possible, and that we understand that even though not perfect, what we are after is a circular path that will artificially create or simulate a higher dimension of space in a zero-point free energy creating motor, and that will allow us to draw the current we desire since we have reconfigured space to our use. Reconfiguration requires superposition, concentration, and rotation to give the new field patterns of space the semblance of a higher dimension compatible with generating electromagnetic force.

In the diagram, turquoise lines represent the electric field and the yellow lines represent the magnetic field. This wave is the result of the elastic collision of space. Also involved are the elements that form the superstring space matrix, which also have spin and monopoles (matrix intersections or superstring overlap). When both the wave and its effects are combined with the rotation of space to form a curved coiled line instead of a straight line, the effects produced with regard to time, velocity, electricity, and magnetism will be entirely different. The reconfiguration of space is the key to developing a zero-point energy-generating device. The energy that can be harnessed is already in the space matrix. A zero-point device may not require any moving parts. In a two- or-three-step process, energy can be harvested from the matrix of space. The motion of space or energy is already there and is powered by the Sun. Translating matrix energy to electrical current energy would be a small percentage of the existing available output.

Light in a fiber optic coil strand is similar to electrical current in a wire coil and should have the superposition that would be required. What I mean that light should not be reflected is that the line of the photon or space matrix be as far away from a linear path as possible. It would be like stirring coffee: the walls of the cup keep the motion rotational in a fiber-optic cable as opposed to linear. I think I am right here. The linear path of the photonic elastic collision must already be

## Electric Magnetic Field Wave Drawing

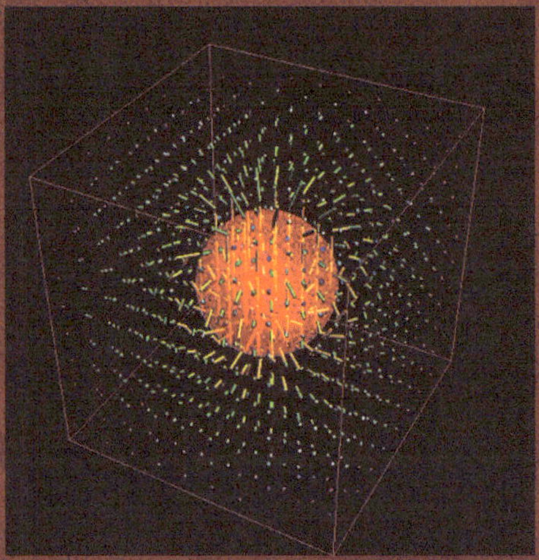

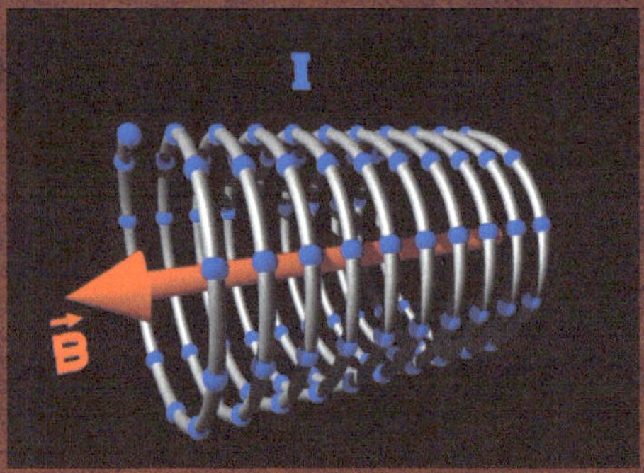


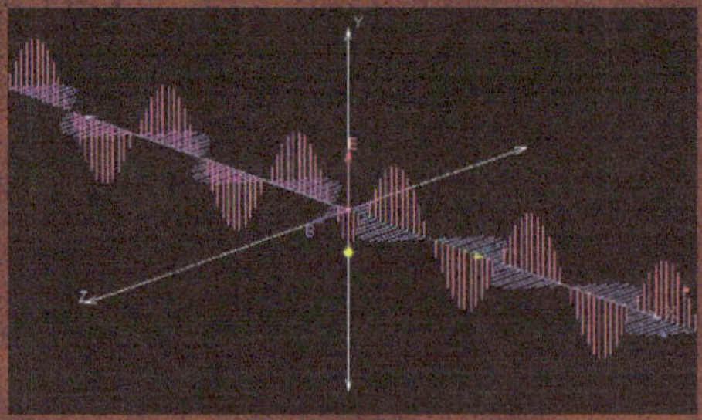

inducing "a field" unknown. However, we know that it is producing radial, electric, and magnetic patterns. The vector quality of these fields have to do perhaps that photonic current is already at a low dimension 3 ($D^3$), and inducing a rotational field becomes restrictive since inducing the lower dimension would be dimension 1 or less or singularity. I do not believe the induction works both up and down, but down from higher dimension to lower.

**Inverse Square Law, Elastic Collision of Space, and Second Law of Thermodynamics**

The inverse square law is proof of the elastic collision of space. All dimensions of space exhibit this characteristic of the inverse square law, whether it is spaces governed by electric, gravitational, or photonic monopoles, where the monopoles of the space define the dimension. The key is that from a "point" source, the elastic field generates a spherical wave field that gets larger and larger, like throwing a rock in the water of a pond. And as the wave gets larger, the energy of the wave spreads laterally since the geometry of space is spherical. The lateral motion of energy also compliments the second law of thermodynamics or the universal law of increasing entropy. The lateral distribution or diffusion (inverse square law) of energy is apparent from a "point" source of light but in the scenario of a flat or broad wave and the ideal situation of elastic space without encumbrances such as

- space density variation,
- expansion or contraction of space, or
- creation or removal of space.

You will detect the apparent inviolate motion of photonic current or light across the universe. Even in such a scenario, small lateral vibration due to spherical space will diminish the forward action. In addition, there is proportion of space field to generator and, hence, limit to path of travel, and hence, the universe appears black. But for the observation or generation object, space is always infinite since you are always at the center.

Space is also infinite by its connection to infinity via fractal arrangement and also by the source of this universe, which is infinity. This universe is a particular qualification of infinity.

**Energy, Space, Strings, and Dimensions**

Science understands energy as $\boldsymbol{E = MC^2}$, but in this theory, $\boldsymbol{E = D^5 = MC^2}$. The reinterpretation of energy or the fact thereof is energy as a mathematical relationship of strings or described as various dimensions of space matrix. The component strings of all universal mathematical and physics relationships that comprise space matrixes are $V$ (velocity) and $T$ (time) and $D$ (distance), expressed together as $VT$. So matter and energy are manifestations of strings ($V$, $T$, and $D$) and manifestations of dimension. And we can also say that the strings or space matrixes or dimensions ($D$) are manifestations of different dimensions of energy. So to summarize

1. Dimensions and powers thereof are mathematical relationship of strings $V$, $T$, and $D$.
2. Energy and powers thereof and mathematical relationship of dimensions ($D$) or strings $V$, $T$, and $D$.
3. Energy (dimension of energy) = dimension (dimensions of space).
4. Energy exists in various dimensions, and relativity deals with the analysis of the various dimensions of energy, associated string construct, which is also the fabric of associated dimensions of space. The building blocks are strings $V$, $T$, and $D$ in positive, negative, and neutral. Strings have spin and are combined in a space matrix different in each dimension.

The $MC^2$ derived from $MV^2$ means that space matrix is a perfect elastic medium, and the monopoles of space matrix are photons that collide elastically within the space matrix. Matter is neither converted to energy or vice versa, but I would like to use the term translated since the same elemental strings are still present.

Various dimensions are various forms of space, and the space of the universe may have various and different characteristics. Space or the ether is not a homogenous vacuum but constructed of different types of space matrixes. The extent of gravitational space is much farther than three dimensional spaces. For example, space of three dimensions extends to just beyond Neptune, while that of gravity extends to Sedna.

One dimension has relationship to another by induction, a relationship first explored by Faraday. A dimension coexists with another by induction, and then that to another and continually repeating as a fractal pattern to infinity. Yet there is a boundary at dimension 23 since infinite energy is not natural for natural mechanism, since the universe is bound by infinity or a particular qualification of infinity and has a mathematical relationship to infinity that actually exists. On one end of dimension 23, it is a fractal move upward in dimension based on energy toward singularity or **infinity**, and in the other direction toward **unity** of the God particle *D* superstring. Between this dynamism is **balance**, a perpetual motion machine.

**The Nature of Space and Energy**

The universe is made up of or completed full of strings. These strings are transparent to atomic matter, yet they have mass. So strings are transparent matter and not the solid matter we associate with their interference patterns of strings that create particles. However, at lower dimensions, some particles are also transparent. You can say that dimension per dimension, space varies in the following:

1. Density: The amount of strings present and their compactness.
2. Number: The number of strings.
3. Structure: The fractal and geometric structure of strings.
4. Superposition: The compression and overlay of string geometry.
5. Information: The amount or quantity of strings and the arrangement of different types.
6. Properties: The apparent nature and relationship of one dimension to another.
7. Magnitude: The magnitude of quality of arranged strings.
8. Mathematics: Strings are pure mathematics.
9. Mass: Strings accumulate to mass; as in $mc^2$, they accumulate to $mc^2$ or space or energy.

Space dimension per dimension changes in properties. The matrix of space in each dimension (twenty-three in all) is made of strings. What we see, the particles are the interference patterns of strings. The strings in lower dimension, such as 3, are of lower density, superposition, number, magnitude, and mass. Space makes up everything, and the space is a perfect mathematical construct, a perfect equation. Particles are the shadows of the larger reality space, and that's what we see. The atoms are shadows. We break up shadows in a supercollider, and we disturb the strings that we do not see. And some particles are transparent portions of these intersecting strings. The very smallest particle is unseen even by smashing in a collider, and the larger portion of strings are also unseen. And even the mathematical hierarchy, the pure math of strings is not seen. Strings represent

**Dimension = Space = Energy**

These are all interchangeable terminology.

Einstein was looking for one simple equation to explain an elegant universe; well, there are twenty-three simple equations. And $\boldsymbol{E = MC^2}$ is really $\boldsymbol{D^5 = MC^2}$. And $\boldsymbol{E^2}$ = **gravity**. And $\boldsymbol{E^4 \times D^3}$ = **black hole**. The unification principle makes one equation impossible since "relationships" are established through the governing dimensions of the universal construct. One equation and its properties is the door to the next, and so on and so forth to infinity. Humans are basically able to see everything since all this can be derived by intuitive logic. Everything can be known by each person truly by thinking about it.

Light is a monopole of the three-dimensional matrix of space. The photon does not travel at the speed of light. Like the electron, it is slow to move. The photon elastically collides with adjacent photons. And what you see is the photon near you. Space is "elastic," a perfect elastic medium. This is apparent with lights connection to $mv^2$ and $mc^2$. But space is not empty; it is transparent and full of strings that form a space matrix. So let's get rid of the terminology *illusion*, and the word *nothing* about the universe and the vacuum of space respectively. I am that I am. Everything that is, is.

The elasticity of the space matrix and dimensional universe is the "perpetual motion" or conservation of all energy in motion. The understanding of energy has evolved and been expanded. The term **Energy = Space = Dimension** refers to energy as a geometric space matrix and powers of dimension thereof. When you explode an atomic bomb, you are generating space. Splitting the atom or fission involves a generation of space. In either a fission or fusion reaction, the generation of energy is the generation of space. There is a relationship to mass and space as there is a relationship to mass and energy. The units are the same.

Space means space dimension, and dimension is another name of powers thereof, and energy is how we define dimension and space currently. Each dimension of space is perfectly elastic with elastic properties at the speed of light. $MV^2$ in the perfect elastic collision of each dimension of space and so becomes $MC^2$. $MC^2$ means or suggests elastic space and elastic collision.

The component or multiplier of difference dimension per dimension is *VT*. *VT* is the "thing" that relates energy to mass. A new way of understanding energy is by redefining it in a broad sense as space and dimension, but yet more specifically as the strings of the space matrix that constructs all that exists, and the strings are *V*, *T*, and *D*.

There is a matrix that forms all space in its various types and dimensions. When you explode an atomic bomb, you create space. The energy is space, the mass is space, and the equivalency is that they are made of the same strings *V*, *T*, And *D*. $MV^2$ means elastic collision; its association to light in $MC^2$ means space is elastic. Elasticity requires a medium. And light and $D^3$ are one and the same. Light or the photon is the monopole of three-dimensional spaces. There are no particles, only particle like relationships of the space matrix. Energy is space? Yes. Space is created in proportion exactly to energy released. The Sun generates the space of this solar system. To make space, I have to release energy. The galaxy generates the space of the galaxy. The release of energy is to expand the space of the universe and to reduce space is to incorporate energy. More space means an enlarged universe. Each center of galaxy generates space, and there are different types of space. Gravitation is one type of space, and there are others such as $D^3$, electrical, and photonic, so on and so forth, each representing a different dimension of space. Space is not something waiting to be filled; in the contrary, space is full of strings or is the construct of strings, the dark energy of the universe. Space is a superstring matrix transparent superconductor.

## Particles

Consider one photon as not being an "independent" particle in the ether of a vacuum space but rather a monopole created by the space matrix (strings) interference patterns of space, and hence, a photon is being a certain geometric position and state of the entire space matrix. A space matrix has a well-defined number of monopole particles in each state or dimension. The matrix of 3-D or $D^3$ space has photons as monopoles that are seen by oscillator excitation or elastic collision. It's not one wave that travels from photon emitter to detector; it's an elastic collision of the space matrix that is perceived as a packet of waves. When the packet of "waves" or excitation hits the detector screen (phosphor), its effect is particle-like, but what hits is an elastic collision of space. It's as though the packet makes up a clump of particle. Particles may not be localized; they are waves (elastic collision) of the monopoles of the space matrix, but they are called particles, and the number of particles is fixed. So *particle* is really just a term used to help one imagine what's going on, but that depends on the context. Are photons considered real things? But then electrons are. It's an electromagnetic field that is moving; or better still, a space matrix that is moving or vibrating, and it's a perfect elastic medium that vibrates on collision, allowing waves or "current." Particles are the interference patterns of the matrix of space by geometry, superposition, and density that effect of this interference is the result in an effect we understand as particle-like. So a photon isn't really a particle; it's just a phenomenon that can be compared to a particle and the interference patterns of a larger mechanism at work. This understanding can be extended to everything we see. A photon is a clump of space matrix, monopole, or a regular interference pattern of the matrix component superstrings in rotation. Space is transparent to higher dimension. By elastic collision, the transparent photon is excited to a higher vibration or frequency and becomes visible or interactive with adjacent dimensions.

## Strings: The Influence of Strings Dimension per Dimension of Space and their Qualities, Both Linear and Inductive (Rotational)

Time is a string, one of which everything is made, and the relationship of strings are "relative" and mathematically precise, and the relationship of strings in the universe are multiplicative in relationship, not additive and subtractive. The relationship of time in relativity is multiplicative. If time is a string, does that make space an associative array? Yes, time is a component of the space matrix along with strings *V* and *D*. And they exist in positive, negative, and neutral. So for $D^3$, you have a matrix of twenty-seven strings. Space is a geometric array of strings. Time is one of the strings; hence, space has the quality of time. It also has the qualities of dimension and velocity, the other component strings.

Each dimension of space has certain properties and monopoles; such dimensions are as magnetism, gravity, photonic, and electric as examples. Gravity exists in a dimension of space; hence, gravity and space gravity experience bending. Light is also in a dimension of space as the monopole photon; hence, photonic space and space experience bending. Space has a quality of time since one of the strings is time. When these strings relate at a higher dimension, their multiplicative effect or property effect generates other properties, such as gravity. In the equation $\boldsymbol{D^5 = MC^2 = E}$, energy is really just a dimension of space on the list of dimensions of space. Yes, energy is dimension 5 in $\boldsymbol{E = MC^2}$. $\boldsymbol{D^5}$ is *D* to the fifth power and also a fifth dimension of space. And so mass is a manifestation of space in dimension and time; $\boldsymbol{M = D^3 \times T^2}$, where $T^2$ is density. Particles are the interference patterns of space matrix in various dimensions of space. There are no particles per se but particle-like effects. Particles are not mass; mass includes everything, strings included. What an observer sees is particle-like but somewhat more space matrix components "in interference" as they move or rotate through each other; but actually there are no particles, only particle-like effects. And there are waves that are

- the elastic collision of the string space matrix, such as what we traditionally define, and also
- interference of the string space matrix itself in rotation, which are particle-like effects or what we know as particles.

And momentum is just a dequeue (dynamically associated array of the space matrix) on the sorted associative array. The matrix or array is in three dimensions. You see the thing about Einstein and relativity is that he skimmed one of the equations for strings in $\boldsymbol{E = MC^2}$. This equation has nothing to do directly with particles, and our understanding of physics was backward. Meaning, the relationship of particle to energy is backward; dimensions of space interacting causes particles, not particles interacting causes energy. Energy creates particles, and not particles create energy. The universe is analogous to a hologram by such a matrix. When you explode an atomic bomb, you create space. The release of energy is the release of dimensional space. The universe enlarges. The matrix of space expands. Expansion happens at every point. Not just one. Also, gravity has different units and comprehension. The big generator of space gravity is the center of the galaxy and each galaxy. The current understanding of gravity is between two objects, but here presented singularly. This new concept is singular. Gravity effect is infinite with its effect decreasing infinitely as well as in a fractal subdivision. $\boldsymbol{E^2}$ = **gravity**; well, the gravitational field of any object $E^2$ relating somehow to dimension, and gravity is a new paradigm. $\boldsymbol{(MC^2)^2}$ = **gravity** and is a gravitational equation. The field of gravity has monopoles. Every dimension of space has monopoles. The elastic collision of the space matrix and monopoles occurs at the speed of light; space is elastic by association of light to $MC^2$ and $MV^2$. A photon is particle-like but really a wavelike clump that is particle-like. Light is merely disturbance in an electromagnetic field. A photon is a monopole on the matrix of space. $MC^2$ means space is elastic by association with light. What you see is the photon on the receiving end, like a wave on the shore or a clump of wave strings. Just as in electric current where electrons don't travel at the speed of light, photons do not travel at the speed of light but elastically collide in their matrix to produce an elastic wave at the speed of light or current. No field, no monopole, means any space. $D^3$ space is the field. Photonic current field is a spatial dimension.

Induction of one field of space happens to another field of space three dimensions over. So if I move my cup, do I induce a field? Induction is the result of rotation at or near the speed of light. Rotational motion near the speed of light induces like magnetic and electric induction. Every space dimension induces a third down. All dimensions of space are connected. Everything is related as in an atom, a multidimensional construct. For one, light needs only magnetic and electric fields, and this simple perception is possible due to induction of dimensions close to $D^3$ because they are three dimension higher. A solid block of gold is a dimension of space. Atoms are the relationship of several dimensions of space. The matrix of space slows down as the interference increases of the matrix due to rotation and superposition. That's why particles are sub light speed since their associative fields are slowing due to special interference patterns. The greater part is space, particles are the fringe. But you see the particles, which are the smaller part, and your physics is backward. So now I'm saying that we live in matrix. Yes, there is no vacuum. You know, insects perhaps see radio waves; you see only a few things too, and if you saw it all, it would be too much information. The particles are as shadows to the real structure of the space matrix that pervades the whole universe. Particles are the smaller part of the matrix and the analogy of the relation of an electron to the proton and the relative size of the intervening space matrix.

## Space, Superconductivity, and Elasticity

When certain materials get to very low temperatures, electric current keeps on going indefinitely with no power source. This is called superconductivity. Current has been classified as an elastic collision, although imperfections in the medium can affect the current. An example would be electric current. At certain states where the medium has no obstruction and the temperature is cold or at the right temperature for superconductivity, elastic collisions become perfectly elastic, and there is no hindrance to the current once started.

The various dimensions of space have monopoles that collide elastically. Space is a perfectly elastic medium. These dimensions may or may not have obstructive constructs that affect the elasticity of space. The dimensions of space vary dimension per dimension such as photonic, electric, magnetic, gravity, and so on. The elastic collisions of space are called current.

Elements such as iron are condensed forms or higher dimensions of space. At temperatures approaching zero Kelvin or lower, some elements exhibit superconductivity. Superconductivity represents perfect elastic space. Outer or deep space ($D^3$) is both elastic and a superconductor. $D^3$ or three-dimensional space has a temperature of 270° Celsius or -445° Fahrenheit or 2.75° Kelvin. $D^3$ space is a superconductor, and photonic current is an effect of that superconductor.

By the above rules, temperature alters gravitational effect and the possibility of an elastic collision in the various dimensions of space. Temperature and density affect gravity, which is another dimension of space. Temperatures indeed affect gravity as it contributes to the total energy momentum density of space. When the temperature allows a dimension to be perfectly elastic or superconductive, the elastic collisions are infinite if there are no obstructions. The spherical and radial geometries would have the effect of $1/X^2$ over the path of transmission; that is why the universe is black.

## Space Is an Elastic Medium

The conservation of energy through distance is an elastic collision, and therefore, the medium of the transfer of energy is elastic. All things are made from space, which is a matrix of superstring velocity ($V$), time ($T$), and dimension ($D$). The properties of $VT$ are multiplicative dimension per dimension and form a periodic table of dimensional energy from $D^1$ to $D^{23}$. Everything is mass (particle and space matrix), and when you have motion, the result is a dimension of energy.

$\boldsymbol{E = MC^2}$ **= space = energy**

Energy exists in various dimensions. Space is an elastic medium and superconductor. There are no particles, only particle-like effects caused by the supposition and interference patterns of the space matrix.

The monopoles of a matrix of space are particles like photons. When the matrix of space vibrates elastically, you have an electromagnetic wave. Light and its effect are the result of elastic collision. The conservation of energy in the matrix is elastic collision. $MV^2$ now $MC^2$ suggests elastic space or the conservation of elastic energy.

Space is a field of superstrings. The collision of the space field creates photonic current. If a crystal can bend light, then it can bend space. Everything we see is a manifestation of energy, even space, the lowest denominator. When you explode an atomic bomb, you create space.

$\boldsymbol{E = MC^2}$ means mass and motion. The difference between $MV$ and $MV^2$ conserved is elastic motion. Energy over distance when conserved is possible only in an elastic medium. Light suggests the matrix of space is elastic. $E = MC^2$ is a superstring equation, giving a relationship of elastic space.

In each dimension of space, the message carrier is the monopole of that space and the matrix of the space. Together we can refer to both as the space matrix. The transmission of message is by elastic collision of the space matrix, which is known as current. Hence, dimension per dimension, the transmission profile should be the same. And the various forces are indicative of various dimensions.

Each dimension of space matrix ($D^n$) has different monopoles or force carriers. For $D^3$, it is the photon, which has a spin of 1 hbar. And we would expect the energy in higher dimensions to be greater and perhaps the spin to be greater. The graviton monopole has a spin of 4 hbar. But the space matrix and monopoles have the quality of transparency. So it is possible for these monopoles to pass through each other as well as the matrix. The interference patterns created by superposition and interference of the space matrix are the creation of particles.

The conservation of energy is the result of an elastic space matrix; this also implies the conservation of mass and charge (spin). The change in the photon is the electromagnetic field and is also the elastic collision of the space matrix. The photon is a monopole or stationary node in the superstring space matrix. What we see is the vibration of the photon within the matrix, which is elastic and allows elastic motion where energy is conserved. Both mass and $spin^2$ show conservation of energy in the matrix and elastic vibration through the matrix. $D = VT$ means that superstrings *VT* (*V* and *T*) are equivalent to *D*, a multiplicative relationship. In positive, negative, and neutral or in relationship to spin, you have three more for a total of nine, and for $D^3$ multiplied by 3 again you have 27. This 27 represents a tetrahedral geometry that is also spherical and radial. Hence, space is allowed motion in all directions. The strings are related via rotational motion; hence the qualification of something that exists is that it exhibits rotational motion of elemental parts and relative motion. Motion becomes energy.

## $C + V$ or $C - V$

It is conceivable that $C + V$ or $C - V$ is possible for an object if within space that is either contracting or expanding respectively. Hence, where there is a high gravity and the tendency is for space to compress, the actual elastic collision of light, which is *C* under general space conditions, might be altered to be $C + V$. The elastic collision of space that produces *C* can be altered slightly due to the fact that space is both a matrix and superconductor. And this measurement depends on the position of the observer. The speed of light depends on the matrix of elasticity and how an impulse can be messaged through space. There are exceptions to *C*, and these are relative to the observer and the motion of the observed object. But the exceptions are minor when comparing motions of the entire system. (Dolphin: Web Link)

## Space Matrix Monopoles

Every dimension of space has a monopole that defines the nodes of geometric space or the intersection—overlap of the string matrix. The monopoles define regular transitions between strings in each dimension of space according to the nature of that space. A monopole is not a particle but has a particle-like effect.

## Multidimensional Universe

In the multidimensional universe proposed, dimensions are from 1 to 23 or from what I call the exoatomic or relativistic world to the subatomic quantum world. And in such an arrangement of periodicity from one dimension to the next, the smallest building block or construct is the matrix of space dimension per dimension made from geometric space or space matrix consisting of elemental superstrings. Relativity and $E = mc^2$ is the discovery of a string equation in the exoatomic world of large bodies.

We have said that each dimension varies to an adjacent dimension by multiplier of strings *V* and *T* or $VT = D$. *VT* is the multiplier dimension per dimension in an ascending periodic table from $D^1$ to $D^{23}$. The basic components of all dimensions are superstrings *V*, *T*, and *D* in positive, negative, and neutral. In each dimension, the multiplication of superstrings are also different properties of that dimension. Strings are the mathematical construct relationship of the universe. $\boldsymbol{E = MC^2}$ is just one string equation in the whole periodic table of string equations. The difference between $E$ (energy) and $F$ (force) is $VT$.

Where $\boldsymbol{E = MV^2}$ and $\boldsymbol{F = MA}$, let us describe the dimension of $F$.

Let us suppose that an unknown dimension $\boldsymbol{D^n = MA = MD / T^2}$, which says that $MA$ is a particular dimension of space.

$\boldsymbol{D^1 = VT}$ is the first dimension of space.

$\boldsymbol{M = D^n \times T^2 / D}$, and from this follows $\boldsymbol{M = (D^n / d)\ T^2}$, which means mass is a certain dimension of space multiplied by $T^2$ (time squared).

Let us figure out $D^n$ and determine the dimension of force ($F$). So force is $\boldsymbol{F = MA = (D^n / d)\ T^2 \times A}$

**Density = mass /** $\boldsymbol{D^3 = (D^n / d)\ T^2 / D^3 = D^nT^2 / D^4}$ **= density**

$\boldsymbol{D^nT^2 / D^4}$ **= density**

$\boldsymbol{D^nT^2}$ **= density ×** $\boldsymbol{D^4}$. By this equation, time rate seems proportional to density and without further influences, $\boldsymbol{D^n = D^4}$.

We know that $\boldsymbol{D^n = D^4 = MA}$

$\boldsymbol{D^5 = MV^2}$ and applied to the matrix of space $\boldsymbol{D^5 = Mc^2}$

**Hence,** $\boldsymbol{D^4}$ **= force, and energy =** $\boldsymbol{D^5}$

There are no particles, only particle-like effects. The particles we see are the intersection or interference patterns of string space matrix. We can think of the third dimension as a matrix of closely packed spheres of three strings $V$, $T$, and $D$ in positive, negative, and neutral. This matrix is the dark matter, and the spherical and radial geometry allows motion in all directions. The monopole of the $D^3$ matrix is the photon. The matrix is also a photonic superconductor. Mass includes all strings (dimensions of energy) or space and interference patterns (particles). The definition of mass is a proportional relationship of volume or $D^3$ and time rate ($T^2$).

From above, $\boldsymbol{M = (D^n / d)\ T^2}$, a certain dimension of space ($D^n / d$) multiplied by $T^2$ is mass. Above we solved the unknown dimension for force so that $\boldsymbol{D^n = D^4}$ **= force =** $\boldsymbol{MA}$, so

**Mass =** $\boldsymbol{D^3T^2}$; volume and the rate of time are inversely proportional in the same manner that time and velocity are inversely proportional in $\boldsymbol{D^1 = VT}$.

Previously it was stated that time rate ($T^2$) is proportional to density. This means that where space has greater density, the time experienced is relatively faster, and where the density is less, time experienced is relatively slower. But when comparing the densities of areas of $D^3$, the variations are there but not as significant as the density of a higher dimension. So the gravity of a larger planet will slow time more in the space around it, perhaps due to the compression of space than would happen by gravity on the adjacent space of a smaller planet. But the ratio should be density proportional to $T^2$. With this relationship all the time, distortion of stars can be calculated because the density of space has an effect of time. And also in addition, mass is made of superstrings that result in $T^2$.

# CHAPTER 2

## The Triad of Being

**Density = $T^2$**

**Mass = $D^3T^2$**

**Velocity and Time are Quantized Functions and express cause and effect. Mass and all space is a Trans Dimensional space matrix and particles are shadows or nodes of the matrix. Velocity is merely the complementary resonance of time.**

**There is no free will without transcendence. Only by a foreign set of mathematical operators to those emergent is free will possible. The universe is a mathematical machine and the results of freedom or randomness are an illusion except those enabled by the transcendence of life forms. There are no emergent principles or observable events that show the beginning of life. Both the universe and life are induced from the outside. Science merely observes emergent principles that coexist with transcendent ones in a harmony that through our perception makes the emergent ones relevant to daily life since they are directly measurable while ignoring what's a prior i. All that is, is, and reality is, and nothing is a myth.**

## Gravity and Time

Gravity has an effect on time. By the equation $\boldsymbol{T^2}$ = **density**, we know that the rate of time is proportional to density. This means that the density of the space matrix in a dimension affects the rate of time. Hence by implication, the density of gravity affects time since gravity is also a dimension of space and has a relationship to space, and also, time is described as a component of mass by such an association. **Mass** = $\boldsymbol{D^3T^2}$ redefines mass and describes time as a component of mass and also justifies the influence of the density of space on time since all mass is both space matrix and particle (interference of the space matrix).So finally in physics, you have closure in elementary components to describe all phenomena, which a unified field theory requires, a description of all physical relationships by common denominators, which I will use the word superstrings $D$, $T$, and $V$.

The effect of time rate is related to the density of the gravitational field or the field of space. The relationship is $T^2$. The magnitude of time from ground to GPS satellite is the same as the difference in magnitude of gravity to satellite by $1 / X^2$. The effect of magnitude is reciprocal, meaning the magnitude of time increases to the satellite while gravity decreases by a similar magnitude. You get the same number for the difference in time by $\boldsymbol{T^2}$ = **density** as $1 / X^2$, which is the difference in gravity. By this solution, we now can relate time to mass, and it would not be possible if gravity had no effect on time. So the rate of time with volume and mass are interchangeable as mass and energy; $\boldsymbol{M = D^3T^2}$ and $\boldsymbol{T^2}$ = **density**.

The difference in time from the Earth to GPS satellite is about forty-five microseconds per day or $4.5 \times 10^{-7}$. The $T^2$ for this difference between ground clock and satellite clock time would be near $2.025 \times 10^{-15}$. With less density of the space matrix, which also includes gravity, less dense space allows time to move faster while greater density allows time to move slower. Whereas as velocity is inversely proportional to time by $\boldsymbol{D = VT}$, gravity is inversely proportional to $T^2$ in the equation $\boldsymbol{T^2}$ **gravity = force** $\boldsymbol{D^3 M}$.

So if $T^2$ is the difference in time from ground to satellite, then the density by proportion is different by the same amount; go back to the equation $\boldsymbol{T^2}$ = **density** mentioned before. If you calculate the difference in the force of gravity from center of the Earth to satellite, its decrease is by known ratio $1 / X^2$ from the distance to the center of the Earth. This difference is at a degree of $10^{-15}$ by proportion for gravity considering the orbit of GPS near 26,400 km from the center of the Earth. Hence, we can see that by ratio, the reduction in gravity is similar to the reduction in the density and time of space implied by the equation $\boldsymbol{T^2}$ = **density**. Gravity affects time due to variation in the density of the space matrix. And for the first time, mass is explained in components of time and volume, which is a necessity to explain the affect of gravity on time.

Gravity affects time implied in the equation $\boldsymbol{T^2}$ = **density**. For the first time, mass is described in terms of time and dimension ($D^3$ or volume) in $\boldsymbol{M = D^3T^2}$, which gets us to $\boldsymbol{D^5 = MC^2}$.

Time is affected by the density of space. Current physics doesn't explain "why" gravity effects time. By the equation $\boldsymbol{T^2}$ = **density**, the rate of time is influenced by the density of space, which also has a relationship to gravity. And that relationship establishes mass as derived from components of time and volume, $\boldsymbol{M = D^3T^2}$, otherwise gravity would not affect time as said before.

The gravity equations in this theory are singular. They are not as in Newton between objects but describe why gravity is by equations that include subcomponent superstrings. And the relationship of superstrings is purely mathematical. $\boldsymbol{E = MC^2}$ is a superstring equation. But for gravity, **gravity** = $\boldsymbol{E^2}$. When you consider the gravity of Saturn and the Earth, they are the same by current measurements. But in reality, Saturn's gravity is much greater. Gravity is a three-dimensional field; weight and associated force is a one-dimensional vector. Gravity has components of line, area, and volume. Gravitational force is not gravitation as a field of space. Saturn's gravity field is much larger than that of the Earth, and so Saturn has many more moons. In the experiment above, we compared values between accepted measurements of current experiments and those that are predicted from this theory. By comparison of tested and predicted numbers, we have determined the relationship between time and space density that has been proved and relationship to gravity by magnitude and proportion.

Time dilation of the gravitational space matrix about the Earth has been calculated and showed that the density of space and time dilation and gravitational dilation are all proportional by the equation $\boldsymbol{T^2}$ = **density**; density of space is proportional to gravity and time dilation. So those relationships justify $M = D^3T^2$ and $D^5 = MC^2$. Time is related to the density of space and density of gravity, which is a dimension of space, and so what of it? All things are connected. The amount of superstrings in a given unit area refers to density. Density is also a function of a dimension reinterpreted by volume ($D^3$). The qualities of density can be revealed through the dimensions that exist. Hence,

**The amount of strings per unit volume = the amount of time squared**

**Density = $T^2$**

**The amount of strings per unit area is proportional to time squared.**

It is the superstring *T* that contributes the quality or property of time. Hence, space with superstrings *V* and *T* have or "experience" those *VT* properties. Space has properties, an unknown precept for current physics, and space is an elastic medium. So time, being related to velocity, is really an anti-velocity, which means time superstring does not exist without reference to velocity superstring. So there is a polarity or anti-velocity established in time. Superstrings exist together and give the quality of space ($D^3$), time, and velocity. Time, by establishing an inverse or polar relationship to velocity, in a sense provides resistance to changes in velocity (like inertia). This implies elastic inertia, similar to the elastic motion of a perpetual motion machine. Mass ($D^3T^2$) is also defined by superstrings in a different order and combination, and mass also includes all strings with associated particles. But in the most basic sense, everything in the universe is a dimension of time and velocity or *VT* or *D*. *D* is the basic particle or the God particle.

**Energy Is Space Made from Superstrings**

When an atomic bomb explodes, you create space. You are releasing space compressed in a higher dimension of time and velocity to a lower dimension of same, like a compressed hologram unfolding like an accordion. There are no particles, only particle-like effects created by the superposition of space superstrings. Particles are the result of the interference patterns of the space matrix. Space and mass have a ration, and space is created by black holes.

Gravity = $E^2 = EMC^2$; $E = G / MC^2$

$G = EMC^2$

Gravity = $(MC^2)^2$

The comparison of the gravity of the Sun and the planets is as follows:

Sun

**Gravity of Sun $D^{10}$**

Mass of the Sun: $1.9891 \times 10^{30}$ kg and **gravity = $EMC^2 = M^2C^4$**

**$1.9891 \times 10^{30}$ squared = $3.956 \times 10^{60}$ $C^4$ = gravity field Sun**

**Gravity density of Sun $D^7 = D^{10} / D^3$**

**$3.956 \times 10^{60}$ $C^4$ / volume Sun = $3.956 \times 10^{60}$ $C^4$ / $1.41 \times 10^{27}$ ($m^3$)**

**$2.80 \times 10^{33}$ $C^4 = D^{10} / D^3 = 2{,}800{,}000.0 \times 10^{27}$ $C^4$**

Earth

**Gravity of Earth $D^{10}$**

Mass of the Earth: $5.9736 \times 10^{24}$ kg

**$5.9736 \times 10^{24}$ $C^4$ squared = $35.68 \times 10^{48}$ $C^4$ = gravity field of Earth**

**Gravity density of Earth $D^7 = D^{10} / D^3$**

**$35.68 \times 10^{48}$ $C^4$ / volume of Earth = $35.68 \times 10^{48}$ $C^4$ / $1.08 \times 10^{21}$ ($m^3$)**

**$33.03 \times 10^{27}$ $C^4 = D^{10} / D^3$**

**Ratio to Establish Difference in Vector of Gravity Sun and Earth**

**Sun density / Earth density = vector$^3$ / 1 (Earth) = 2,800,000.0 / 33.03 = 84,771.41 = vector$^3$**

**Vector = 43.92 greater vector than Earth**

To get to the Sun with a vector twenty-seven times the Earth, the Sun would have to have a mass two times less. The only thing that would modify this is that the core if the Sun is much denser than the fusion reaction surround (atmosphere of Sun). But with such great numbers, it can also be that the mass of the Sun is less by a degree.

**<u>Venus</u>**

**Gravity of Venus $D^{10}$**

Mass of Venus: $4.89 \times 10^{24}$ kg

**$4.89 \times 10^{24}$ $C^4$ squared = $23.91 \times 10^{48}$ $C^4$ = gravity field of Venus**

**Gravity density of Venus $D^7 = D^{10} / D^3$**

**$23.91 \times 10^{48}$ $C^4$ / volume of Venus = $23.91 \times 10^{48}$ $C^4$ / $92.843 \times 10^{19}$ ($m^3$)**

**$25.75 \times 10^{27}$ $C^4 = D^{10} / D^3$**

**Ratio to Establish Difference in Vector of Gravity Venus and Earth**

**Venus density / Earth density = vector$^3$ / 1 (Earth) = 25.75 / 33.03 = 0.779 = vector$^3$**

**Vector = 0.92 times the vector of the Earth, which is close to scientific observation**

**<u>Saturn:</u>**

**Gravity of Saturn $D^{10}$**

Mass of Saturn: $5.6846 \times 10^{26}$ kg

**$5.6846 \times 10^{26}$ $C^4$ squared = $32.31 \times 10^{52}$ $C^4$ = gravity field of Saturn**

**Gravity density of Saturn $D^7 = D^{10} / D^3$**

**$32.3 \times 10^{52}$ $C^4$ / volume of Saturn = $32.3 \times 10^{52}$ $C^4$ / $8.6 \times 10^{23}$ ($m^3$)**

**$3.755 \times 10^{29}$ $C^4 = D^{10} / D^3 = 375.5 \times 10^{27}$ $C^4$**

**Ratio to Establish Difference in Vector of Gravity Saturn and Earth**

**Saturn density / Earth density = vector$^3$ / 1 (Earth) = 375.5 / 33.03 = 11.36 = vector$^3$**

**Vector = 2.24 times the vector of the Earth**

To get to the Sun with a vector 1.08 times the Earth, Saturn would have to have a mass three times less. The distribution of mass within Saturn must be polarized outside to inside, with a heavy center. It is very possible that Saturn has a volume near the size of the Earth and that the atmosphere accounts for the difference. Saturn's atmosphere covers the surface of the planet. Hence, the gravity is near the Earth with an adjustment for volume of Saturn rather than mass. It's very likely that Saturn has a surface and is not entirely gas. These equations will be useful in correctly determining the volumetric size of planets covered by hostile atmospheres that make access to their surfaces impossible.

## Mars

**Gravity of Mars $D^{10}$**

Mass of Mars: $6.4185 \times 10^{23}$ kg

**$6.4185 \times 10^{23}$ $C^4$ squared = $41.197 \times 10^{46}$ $C^4$ = gravity field of Mars**

**Gravity density of Mars $D^7 = D^{10} / D^3$**

**$41.197 \times 10^{46}$ $C^4$ / volume of Mars = $41.197 \times 10^{46}$ $C^4$ / $0.162 \times 10^{21}$ ($m^3$)**

**$254.30 \times 10^{25}$ $C^4 = D^{10} / D^3 = 2.543 \times 10^{27}$ $C^4$**

**Ratio to Establish Difference in Vector of Gravity Mars and Earth**

**Mars density / Earth density = vector$^3$ / 1 (Earth) = 2.543 / 33.03 = 0.077 = vector$^3$**

**Vector = 0.42 the vector of the Earth**

For Mars to be .38, the mass would have to be 1.18 times greater. This value is close to scientific observation.

## Mercury

**Gravity of Mercury $D^{10}$**

Mass of Mercury: $3.3 \times 10^{23}$ kg

**$3.3 \times 10^{23}$ $C^4$ squared = $10.89 \times 10^{46}$ $C^4$ = gravity field of Mercury**

**Gravity density of Mercury $D^7 = D^{10} / D^3$**

**$10.89 \times 10^{46}$ $C^4$ / volume of Mercury = $10.89 \times 10^{46}$ $C^4$ / $0.060827 \times 10^{21}$ ($m^3$)**

**$179.03 \times 10^{25}$ $C^4 = D^{10} / D^3 = 1.7903 \times 10^{27}$ $C^4$**

**Ratio to Establish Difference in Vector of Gravity Mercury and Earth**

**Mercury density/ Earth density = vector$^3$ / 1 (Earth) = 1.7903 / 33.03 = 0.0542 = vector$^3$**

**Vector = 0.378 the vector of the Earth**

Mercury seems to be close to the scientific observation, but slightly denser.

## Pluto

**Gravity of Pluto $D^{10}$**

Mass of Pluto: $1.31 \times 10^{22}$ kg

**$1.31 \times 10^{22}$ $C^4$ squared = $1.716 \times 10^{44}$ $C^4$ = gravity field of Pluto**

**Gravity Density of Pluto $D^7 = D^{10} / D^3$**

**$1.716 \times 10^{44}$ $C^4$ / volume of Pluto = $1.716 \times 10^{44}$ $C^4$ / $0.0072 \times 10^{21}$ ($m^3$)**

**$238.33 \times 10^{23}$ $C^4 = D^{10} / D^3 = 0.023833 \times 10^{27}$ $C^4$**

**Ratio to Establish Difference in Vector of Gravity Pluto and Earth**

**Pluto density / Earth density = vector$^3$ / 1 (Earth) = 0.023833/33.03 = 0.00072 = vector$^3$**

**Vector = 0.08 the vector of the Earth**

For Pluto to be 0.088 seems to be close to scientific observation of 0.08, but slightly denser.

It appears that gravity as a field has a great magnitude. And when this is proportionately converted to vector in $D^7$, then gravity is the proportional relationship of $M^2$ to volume. Hence, a new law is established for the definition of gravity, which is $M^2$ proportional to volume. And also, the current estimation of the mass of bodies larger and smaller than the Earth needs revision, which seems to suggest that larger bodies have less mass than currently defined, and smaller bodies have more mass. Consequently the theories about gravity both classical and relativistic are incorrect; they are more approximations than fact.

It appears that larger-mass bodies (larger than Earth) have a lesser density (less mass) than current science predicts and that smaller planets and planetoids tend to be denser (more mass) than current science predicts, but not always. Hence, Mercury being denser than once claimed and also Pluto are denser as well, although close to scientific observation.

The Sun has two times less mass than believed, and so nearer to being a giant red sun than most would expect, in the latter part of the Sun's evolution. The difference in Newtonian gravity and what is proposed here varies by less than 1 percent but shows the difference more in bodies smaller and larger than the Earth.

## Universe as a Perpetual Motion Machine

When two people fall in love, things heat up, and when love is lost, things cool down, but then a new window opens with another. And when you die, do you really die, or is there life elsewhere? So our notions about "the end" haunt us even to see it in the universe. I like Jim Morrison's view. This is the end, my only friend, the end.

Now I'll pose an idea. Let's say the universe is made up of twenty-three equations or parts. And part 1 exists in relation to part 23. And 23 will always be 23, and 1 will always be 1; since unless this is not so, the entire structure is lost. So then there is no end, since 1 reminds 23 who it is and 23 reminds 5 and 8 and 1 what they are and that they belong. The interdependency keeps it moving like a Calder sculpture. Ever seen one? One of those mobiles that keeps moving? The structure exists because all the parts are always there. One keeps the other present and moving, or else the absence of one is a failure of the entire system.

The universe is a perpetual motion machine. I think of an atom as a perpetual motion machine. The electron moves around and around. This creates gravity; one form of motion translates to another. Does it ever die? I think not. The planet is a machine. The universe is a machine. Life is a biological machine defined by a program, the genetic code, and perhaps in that code evolution is programmed, so death is a necessity for the cycle to progress. Hence, all machines are progressive cycles that are from simplicity to complexity and vice versa. The idea of the end speaks more about your humanity than the reality that is.

For the universe to die, motion would have to stop. I see the opposite; our universe is a perpetual motion machine. According to my theory, when two galaxies come together, the amount of space is cut in half, density increases, the clock begins again. To assume the death of particles means the space is always the static, but if the space is dynamic, then motion never stops and temperatures are never flat. Conceptually, the experiment has to be with natural machines, not man-made machines. As I said before, dimension 23 exists in balance with dimension 1; dimension 23 is high energy, and dimension 1 is space in one dimension. You cannot have one and not the other because one creates or is in balance with the other. And for the two to exist simultaneously in twenty-three steps means motion never stops; perpetual motion keeps the structure intact and balanced. The existence of twenty-three dimensions demands the work of the other twenty-two dimensions. So when the order of the dimensions is in balance, it demands the imbalance of components such as the compression and contraction of matter and space, and the imbalance created brings entropic inertia, which brings those imbalanced components of the order back to its balance and in imbalance once again with the order itself. Hence, the function of the universal order is in disequilibrium with the mass/spatial components, and vice versa as a return to equilibrium of the same components; hence, a perpetual motion machine, like a pendulum swinging back and forth, where one equilibrium of the order causes an imbalances (disequilibrium of another or mass/space) that brings about inertia and motion to a return equilibrium or mass/space offset by a shift from the order (disequilibrium as before). As a perpetual motion machine, the universe is granted eternal motion and eternal change always in a harmonious and dynamic machine. To explain another way, the universal order represents one state of balance where entropy of that universal order forces the compression of mass/space into disequilibrium with its own entropic inertia. So you have two entropic counter movements or multiple movements that shift balance away from equilibrium (of the order) to disequilibrium (of the components) and then back to center or equilibrium. The imbalance between the equilibrium and disequilibrium provides inertia for mass/space to return to equilibrium by entropic principle, and thus counter to the balance progressing by the entropic inertia of the order back from disequilibrium to equilibrium. This is followed by a countermovement once again as in a pendulum or, more correctly, the universe as a perpetual motion machine. Order of one component forces disentropy of a second component and the inertia of disentropy to entropy of the second component becomes in disequilibrium with the first component and brings the first component back to disentropy, which then, by inertia, moves to entropy. One level of organization in the matrix of space dimensions initiates an entropic inertia to begin another level of order. This vibration of perpetual motion is possible because of the perfect elastic medium of the universal space matrix and related dimensions. The creation of the universe is the creation of a perpetual motion machine. Entropy oscillates between infinities of a multidimensional construct by balançing constants to achieve unity. Hence, there is infinity, balance, unity in an algebraic relationship.

## Infinity, God, and Multilevel Hyperspace

There are an infinite number of dimensions possible, but due to the limitations of energy required for infinite energy, dimension 23 is the black hole, which is a limit of natural physics. It is the contention of this theory that the finite is a projection of the infinite. The infinite is represented in a circle or sphere, and the first dimension is a point or center of the circle or sphere.

$\boldsymbol{D^{\infty} = D^{23} \times D^{N}}$, which means that the boundary of this universe is infinite energy and infinite dimensions since to the outside is infinity or the part of God beyond the math of our universe. So then the relationship of a transdimensional universe to infinity is in the following four concepts:

1. Dimension ($D^{23}$) is the highest dimension occurring naturally by the limits of energy and entropy occurring naturally. This dimension is also a black hole.
2. Beyond dimension 23 is that which is traversed to the boundary of existence of the universe via multiple infinite dimensions or potentially infinite energy. We arrive at this point of hyper-dimensions or infinite dimensions just before the boundary of existence or hyperspace infinite (actually infinite hyperspace) as opposed to hyperspace

of a particular dimension.

3. Hyperspace infinite is the boundary between this universe and the infinite part of God not experienced in this universe. It is a place where $\boldsymbol{D = D^{\infty}}$, which is a constant, and there is no time or velocity.
4. **Infinity** = $\boldsymbol{D^{\infty}}$ × **variable**. Here is creation in the making. Like a loop in a program that returns and creates something new.

But what is infinity? It is that which is beyond the infinite of our universe and where God is creation.

**Summary**

1. $\boldsymbol{D^{23} = V^{23} \times T^{23}}$ is a black hole or the highest naturally occurring dimension.
2. $D^{\infty} = \boldsymbol{D^{23} \times D^{n}}$ is infinity between $D^{23}$ and hyperspace of reality.
3. $\boldsymbol{D = D^{\infty}}$ is the hyperspace of reality where $V$ and $T$ is zero or 1.
4. **Infinity** = $\boldsymbol{D^{\infty}}$ × **variable** is God beyond reality or Creation from which all existence of this creation and other creations is derived.

All that we see in this universe is created from space, the apparent nothing. This universe is a specific qualification of the infinity of God that is beyond. This universe is a particular qualification of infinity, which is also revealed in mathematics and the equations that represent the construct of dimensions and superstrings. Every "component" has infinite capacity, but then is qualified by other components, and constants represent unity or equilibrium. Hence, ***X* / infinity = 0 (unity).** In this equation, we find a "definite" component $X$ such as mass divided by or in relation to another component that varies. The limit or qualification of infinity is in relationship of one component to another by a constant. $X$ and infinity are bound in relationship together by the relationship to the constant zero (unity). Unity represents equilibrium and represented by a limiting constant. Hence, our mathematics and physics represent the nature of the universe, which is infinite, variable, and a particular manifestation of infinity. And this infinity is found in all natural phenomenon. And God is an analogy of creation, and hence, qualities of God can be derived by observing creation. The order in the cosmos represents a God of order, which is infinite, both within and without this existence, even as suggested by mathematics.

The universe is an analogy of God. These analogies describe God. Duality, a philosophical construct, is very human. The real nature of the universe is multiplicity, infinity, qualification, balance, proportion, and unity. Man has used myth to cope with what isn't understood, and that was the truth for him in the absence of truth. God is truth and also includes physics. And a small part includes duality. The idea of nothing is a dualism, and if there is nothingness, I think only God knows it. Zero is far from being nothing; it is everything, or better still, "unity" and 1 is balance-proportion. Everything that is, is. It's like with "dual nature of light"; without dualism, one would say, "How would you tell light from darkness?" Without dualism, "how would you tell goodness from bad?" But the real contradiction is that there's no "light" and "darkness"; **everything is LIGHT,** which we do not see because of the limiting nature of our eyes, or information. The duality is more about human nature than reality. And the good is everything, and evil is deficit in various forms, and you can know both without experience intuitively. But in that also has certain truths about God and our limitations. There's just scale from 0 to infinity, but somehow because of our human nature, we break it into pairs and define a human existence.

**Space is light** since at the smallest scale of the space matrix and of the universe, the monopole is the photon, and to see each one, we would see a sea of light. We see the elastic collision of those points of space at the receiving end of our vision, which filters and distills much information. And so our physics sees the shadows rather than the light and cause our view to be backward since particles are the smaller part and the components of space are the larger. Our understanding of physics is backward and is based on a fraction of what is measured and seen. As a species, our eyes see the smaller part of existence.

Before the big bang, there was infinity, and in this infinity, God created the universe. And in the infinities where God dwells and the infinite number of possibilities that are God, God chose **three concepts** established through **harmony or**

**proportion** to capture the whole universe.

These three ideas are the following:

**Balance-proportion** (mathematical relationship or 1 or inertial constants) are connected to

**Infinity** (I) and

**Unity** (0) establish this universe and perpetual motion.

Three concepts established through **proportional constants** (example *C*, 1, or others).
The three ideas were established in mathematics by 1, 0, and infinity. And as concepts for the universe, superstrings *D*, *T*, and *V* were created as an analogy. And as a structure of superstrings, they formed light at the smallest scale. **All space is light.** Our eyes don't detect the dark space, which is exploding in light. So that gives meaning to "let there be light," the relationship of three (tetrahedral geometry) that pervades the whole of mathematics, philosophy, and the universal structure. And God felt it was good. And the big bang was both the creation of this particular qualification of infinity and also a grand unfolding of the space matrix hologram. So you had the large relationships of mass and space established along with mathematical harmonies and the various dimensions or space or energy.

## Mass and Energy

Mass includes all the matrix of space, which also includes all particles of the matrix along with the matrix. Hence, the matrix has mass. When the matrix along with particles has "motion," then you have a dimension of space or dimension of energy. In the relationship of superstrings in the equations or laws in this periodic table, there appears to be a triad of relationship at the most basic level. This triad combines "quantity" with infinity and a stabilizing constant. Hence, this universe is a particular qualification of infinity, defined along with quantity and universal constants. I prefer to call the constant unity, infinity is the variable, and quantity is the relative point of observation. **Energy (*D*) = *CT*** means that infinity (*T*) and unity (0 or *C*) are inversely proportional, or infinity (*E*) is proportional to the sum of unity and quantity. The number 1 is the relationship of polarity and proportionality (equilibrium). The relationship of infinity to unity is represented in the number 1. Equilibrium is established by mathematical relationship of the most mysterious questions that math proposes in numbers.

## Symmetry

Dimensions increase periodically by *VT* (multiplied) dimension per dimension, but the additional properties are multiplied to those that remain the same. This property is known as symmetry. Symmetry allows repeating characteristics through the dimensions. For example, **density = mass / volume**; we can also talk about gravitational density, which is $E^2$ ($D^{10}$) / $D^3$ (volume). Hence, the idea of density is relative to $D^3$ as it is applied through the dimensions. There are qualities and properties that repeat through the dimensions. You can follow gravity all the way to $E^4$, from gravity as energy ($D^{10}$), then density ($D^7$), and then force ($D^4$).

**The principle of symmetry says that phenomena (such as gravity) appear to act differently in different dimensions. And they are also measured differently in different dimensions. But that they have relationships to all dimensions.**

## Tachyons

A tachyon is a theoretical particle that travels faster than the speed of light. This theory presents the concept that space matrixes existing in higher dimensions have spin that are higher than the speed of light to $D^3$, relatively speaking. Hence, motion is relatively faster in a higher dimension, but never greater than *C* in the dimension that you are in. The slight variances above or below *C* (speed of light) in $D^3$ are due to the compression and expansion of the space matrix.

Particles have both higher- and lower-dimension characteristics. Any particle is caused by the interference patterns of the space matrix. When particles are created, the space matrix slows in motion and rotation to sub light speed. The electron is an example. Force carriers are the monopoles related to that dimension of space matrix. When interference of the space matrix occurs in a higher dimension, the interference creates particles, referred to in $D^3$ as tachyons. So tachyons are a relative twin to particle in another dimension. It is possible that this construction or tachyon also has a three-dimensional effect or existence. So we may see a particle that is a three-dimensional manifestation of a higher dimensional activity (tachyon) but can never witness the true nature of that particle tachyon unless in the corresponding dimension where it exists. Do tachyons exist? Yes, and they are in harmony with the construct of the space matrix dimension per dimension, and we can refer to tachyons as particles that exist in other dimensions and have a three-dimensional effect, and so are also three dimensional. When physics evolves to be multidimensional, it may be possible to harness the effects of multidimensions to facilitate effects in our dimension. This may also be possible via anomalies. Anomalies can be created artificially and capture tachyons for our use.

## $E = MC^2$ and the Pythagorean Law

You do not need the Pythagorean law to derive $E = MC^2$. The realization of space as both an elastic medium and a matrix refers to the equation $E = MV^2$. The simple conversion of $V^2$ to $C^2$ for the space matrix because it is an elastic medium is a logical derivation. Hence, $E = MC^2$ means space is an elastic medium bound by the speed of light $C$. However, since space is elastic, you can get C - V and C + V relatively depending on the position of the observer and also effects due to the compression and dilation of space. So it is quite possible to experience this phenomenon in satellite communications to distant planets from the Earth.

The Pythagorean law cannot be used to define exodimension since the properties of exodimensions (outside $D^3$) are multiplicative and not additive. The Pythagorean law can be used to view the dimension you are in, but deriving relationships across dimensions must be multiplicative. Hence, this is a major fallacy in string theory. Superstrings form the matrix of space of which all particles are made by the interference patterns of the superstring space matrix. There are no particles per se, only particle-like effects. And superstrings are not individual notes or packets; they are interconnected matrixes. The monopoles of these matrixes are message carriers. It may be that spin refers to the dimension the particular particle-matrix refers to and also the force carrier of that relationship. There are more force carriers to be discovered to account for the many dimensions.

The idea that Pythagorean law is the basis of understanding $E = MC^2$ and also many mathematical concepts of string theory is misleading. I disagree that the Pythagorean law is needed for $E = MC^2$. If space is an elastic superconductor, then the elasticity means conservation of energy or $MV^2$, and then $MC^2$ is the limit of transmitting information in that medium of space. The idea about string theory is that at the smallest level, there are these strings that spin that make up everything. In my theory, I actually give them the names *D*, *T*, and *V*. In string theory, they are unnamed and, hence, have no real properties that can be ascribed to anything about physics except that they are particles or strings, and the separation of relativity and quantum and now string continues. For me the strings are the building blocks, they form a symmetrical structure, and hence, the symmetry and implications from Noether's theorem about the conservation of energy in symmetrical systems seems a good fit. I call the smallest structure tetrahedral. The relationships of particles of spin are symmetrical and join to form a larger particle. This is somewhat like when *V* and *T* form *D*. So the claim is that small strings are fundamental particles, and maybe one is the fundamental particle. In my theory, I call this one *D* since space is made up of various configurations of *D* dimension per dimension via a symmetric superstructure.

## Density, Orbits, Space Matrix

As we referred to before, planets and masses seek out areas of balance in higher space densities. This phenomenon also includes particles. Let us review this concept from the subatomic world to the exoatomic world.

There are no particles, only particle-like effects. Particles are cause by the interference patterns of the space matrix. These areas of interference of superstrings are areas of higher density. Hence, particles are associated with areas of higher density (mass and volume) in the fabric of space.

Planets are associated with areas of higher density in the fabric of space. The stars are generators of space. As the stars spin, they send waves that extend out and reflect back across the fabric of space. Bands of higher space density accumulate as standing waves and allow areas for planets to orbit in stable areas about the ecliptic. Exceptions occur with smaller bodies and orbits eccentric to the ecliptic that travels through various space densities.

Similarly, the solar systems of galaxies are pulled into spirals of higher space density that orbit the center of the galaxies.

Space or energy or dimension contracts and advances by proportion to a relationship to infinity. Now where are different types of space? For example, gravity is one type of space or energy or dimension. Certain dimensions have a larger size extension than others. For example, gravity extends far out into space, to infinity, except the energy decreases with distance from the source. Hence, the density of space varies. Other space extensions are photonic, magnetic, electric, so on and so forth. We can detect transitions in the matrix and density of space by planetary behavior. We notice that beyond Neptune, planets in their orbits vary from the ecliptic, are eccentric, and generally look different in behaviors, so I expect that there is a transition in the matrix of space through the orbit of Neptune. Transitions in the magnetic field of the Sun and relationship to magnetic fields of the planets also affect the spin and angular position of the magnetic fields of planets. Shifts in the magnetic field in the Sun may also have a harmonic effect, affecting the shift of magnetic fields of planets by inducing complementary relationship and motion. The 2012 shift in the Sun's magnetic field may well induce a shift in the Earth's magnetic field. So as the dimensions of space vary and shift in polarity, this affects similar fields in other planets and relationships in motion that seek a point of equilibrium. Space matrix is not homogeneous. A similar analogy would be that the atmosphere is not homogeneous.

**Energy and Coupling**

As the dimensions become higher, the energy fields are stronger. At lower dimensions, you see fewer dimensional components of the three superstrings *D*, *T*, and *V* in combination. For example, mass is $\boldsymbol{M = D^3T^2}$, and at a higher dimension, **gravity** = $\boldsymbol{D^{10}}$. So as you go up in dimensions, the addition of superstrings and superposition of space increases. Along with this, the binding forces increase. At higher dimensions, the kinetic energy increases, and also the energy required bringing these superstrings together. So we can imagine a great unfolding of the various dimensions of space from higher dimensions to lower dimensions. I like to use the analogy of the unfolding of space dimension in the explosion of an atomic bomb. In a black hole, one would expect one large force combining everything into singularity.

The coupling constant becomes stronger with greater gravity. $\boldsymbol{E^2}$ = **gravity** = $\boldsymbol{D^{10} = EMC^2}$, and the strong force similar to *EM* and weak and gravity as gravity increases. As the dimensions that bind strings move to higher dimensions, the energy of that dimension approaches steps closer to infinity. Our universe is a qualification of infinity, and the higher dimensions are closer to that infinite relationship. The string interaction comes down from infinite energy ($D^{23}$ and beyond) through the lower dimensions ($D^N$). *D* is the lowest dimension and the particle from which everything is made. It may be that *D* has a fractal subdivision of itself, but *D* represents unity in contrast to infinity, and the two are balanced within the outer infinity of God.

**Antiparticles**

When discussing particles, we first must establish which dimension they are created in. It may be that spin refers to the dimension of corresponding space matrix superstrings of the particles in question. Particles and antiparticles behave in similar manner. But what are "natural" particles in a dimension is the natural behavior observed without manipulation, and what occurs as an anomaly in experiment, such as antiparticles, refers to the existence of possible parallel antiparticle dimensions. The matrix of space has the capacity to induce other dimension by rotation of the matrix, rotation (spin) of the superstrings within the matrix, superposition, and interference.

Dimension refers to the matrix of superstrings, superposition, interference pattern, spin of associated particles, and space matrix topology. Every successive dimension higher has more superstrings, whereby you add *VT* for each dimension higher is the multiplier. So you have a periodic table of dimensions and properties. Each dimension has its own particles and antiparticles that behave similarly but reflect a mirror extant.

**Mass and Inertia**

***Mass*** = $\boldsymbol{D^3T^2}$ means that mass has the properties of density (**density** = $\boldsymbol{t^2}$) and volume ($D^3$). Hence, mass includes the following:

- `Particles, which are the interference patterns of the space superstring matrix.
- Space matrix, which are the primary superstrings in a geometric pattern.
- Monopoles of the space matrix, for example, photons.

The dilemma becomes this: does space and photons have mass? Well, yes. However, mass has two areas, and they are

- atomic mass or solid sub light speed of a density greater than zero, which are particles, and
- transparent or energy mass of the density near zero in various dimensions.

The present "definition of mass" is at fault for the understanding mass. The mass of a photon and the mass associated with particles of sub light speed is that they exist in different dimensions ($D^N$). So when they are qualified on the same observable dimension ($D^N$), then there is error. Can the mass of energy be measured? **Energy = space = dimension,** and so the energy of the monopole of space or energy or the photon can be measured. When the Sun burns, it releases energy and loses mass. Rest mass is a hypothetical case for the photon; the error is that photons do not move, but rather they

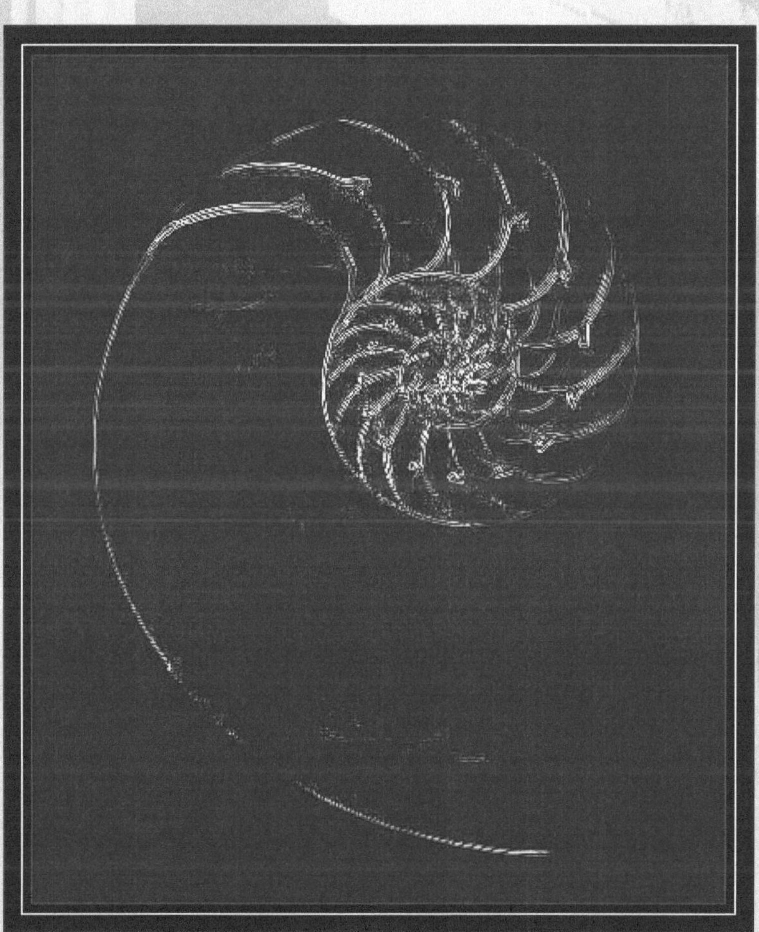

## Fractal Pattern

elastically collide. There is a frame where the location of a photon and the observer are at rest, and hence, a rest mass of near zero. But the dimension (energy) of the object and the photon are different and do not represent the same reference point. Energy releases is transformed through dimension, and hence, the photon released is particle to one reference point and zero to another.

There is transparent mass and particle mass as mentioned before. For particle mass at sub light speed, mass measures the amount of inertia an object has, with inertia defined as the resistance the object offers to a change in its state of motion. Transparent mass is a different dimension of mass, and hence, the laws of inertia are different. There is error to qualify that as the same as particle mass. However, when calculating the total mass of a system, it includes both masses, and the transparent mass is defined as energy. Mass ideally includes both space and particles. Space being energy of the space matrix and photons acting as monopoles of the space matrix can be viewed in current physics with a rest mass of zero. However, everything in the matrix has mass if defined in terms of energy that is defined as mass. The minimum mass of an object is rest mass, space has mass, and the rest mass or total mass can be defined as their accumulation of the matrix unto a point, just as grouping photons unto a point will be affected by gravity since the point acquires mass.

The concept of inertia for a photon and a particle should not be viewed in the same manner, and I think that resolves the confusion about the rest mass of a photon zero, which it isn't really, and also the loss of mass when a photon is lost from or released from a system. **Anything that has volume or density has mass.** The question of a photon is that the density is at near zero, and I don't see it absolutely zero, and also that the "dimension" of that energy is not in the same level as that of a particle. **I think the definition of mass and inertia require redefinition.**

Energy and mass is everything that includes particles and the space matrix. Everything has motion; even the space matrix has superstrings in motion. Motion is associated with mass when we talk about energy. Space exhibits the qualities of the properties of the superstrings, which are distance, time, and velocity.

Particles are never at rest; they are always in motion whether traveling within a system of rotation or a part of larger systems of motion. There is always a frame in which the atom is at rest, relatively speaking, but never in the universe, and that frame is not the same for a photon, which is mass as energy. The superposition of this energy is that higher volume and density will be influenced by gravity and become more mass.

If a photon leaves an object, it takes energy away. And if we take energy as conserved and the photon takes away energy, the resulting atom must have less energy, less mass and less volume. So to prove the loss in volume is to determine the volume of the photon. And thus, the photon has mass, although near zero. Volume defines mass since it is a property of mass. So all space has mass since space has volume, and thus, space is the dark mass. The question is to a much smaller degree, and also that the photon is a dimension of energy, transparent mass, with volume, as a monopole in the space matrix. But you need to redefine mass, and $D^3 T^2$ is the definition of mass. To include volume as a legitimate definition, and what would you gain by such a redefinition? You would allow a photon to have mass since it occupies space, and the ether of space would also have mass, even at rest, however so small, but transparent to higher dimensions of atomic mass at a higher level of energy. A photon is energy since it is a component of the space matrix. What would it mean for a photon to have some volume? Specifically, what experiment would verify the volume? It would mean the photon has mass, although a different type of mass—energy mass and not subatomic mass—which includes energy mass and particles. Well, would angular velocity or spin imply volume? Yes. Would energy imply volume? Yes! An experimental way to measure your definition of volume would be a good next step. An experiment could be the change in energy and volume of a mass. Divide the change in energy by the energy of a photon, and then calculate the number of photons lost. And then by change in volume, calculate the volume of a photon by proportion. The steps in the experiment itself suggest a volume for the photon, and hence, photon has a property of mass and has mass energy, but very small so that in terms of small systems, the rest mass is near zero. However, for large systems such as galaxies, this adds up. And having calculated the volume of a photon, you may be able to calculate the size of superstrings and the total energy of space itself per cubic light-year.

When we say a photon has a rest mass of zero, you need to understand the "current" definition of mass includes solid sub light speed particles, but not energy particles moving at the speed of light. So the definition of mass, when changed to $D^3T^2$, now allows for mass to exist in both terms of energy and particles by the same basis of volume and rate of time.

In current theory, it's the rest mass of the photon that's mass less. Think of it as a photon at 0; unless you give it energy, there is no mass as far as the photons mass goes if it doesn't have momentum, and so it doesn't have mass, or thus the rest mass is 0. However, if the photon has volume, it really does have mass and explains why objects losing photons lose mass.

Atomic objects are a frame of atomic mass. Energy in space or photon is a frame of energy mass. I would say an atom at rest will not lose energy unless acted upon in some manner, internally or externally. An object photon in space is transferring through dimension, such as energy, coming out of an atomic object into space. Find an article on radioactive decay, and if something decays, it's being acted upon; that's what radioactive decay is. It appears obvious that what is decaying is unnatural to its environment, and yes, energy is changing position, so either it's being acted upon externally or by its relationship to surrounding environment. Consider something like the Sun as it gives off light, so something is going inside the Sun that causes photons to be given off so the Sun is losing energy. Is the Sun losing mass because of that? Yes! If the photons given off are mass less, so where's the mass going? So let us redefine and say the Sun gives off photons in the form of space since the photons are monopoles of the space matrix. The volume of photons and the supporting geometric matrix is space surrounding the Sun. This volume has mass since volume is a property of mass. This energy mass we call energy, so space is a form of energy. Mass is converted into energy or space from fusion. Because mass includes both particles and space, space and photons have mass by the new definition. Particles have solid mass, and photons have energy mass. Mass is of several types, such as particle mass and superstring or energy mass, but yes, a photon is a class that becomes particles at higher density. The bounds on photon mass are rather strict; add it all up and you have mass but, of course, still nonzero by current standards of understanding. Zero would be impossible to guarantee. Energy is mass when compressed and exhibits effects from gravity. Since the whole universe has naturally momentum, even space, then all has mass.

There is solid mass and transparent mass; solid mass is sub light. Inertial mass and gravitational mass are pretty much what is currently called energy. But by the new definition here, energy comes in different dimensions and levels of motion (spin). So there are also different levels of inertia since our understanding of inertia is related to motion. Photons have energy, so photons have inertial mass. And also, they have gravitational mass. If you put enough photons in one place, you'd feel the pull of gravity. Rest mass can be confusing since under current thinking, a free photon has no rest mass, but since it has volume, it has an almost zero rest mass, but still it has a mass. Stick a photon in a box, and the box has rest mass. If you fire a photon out of a gravitational well, the photons will red shift. But the same happens for a real mass too. Likewise, if you put a photon in a box and lift it out of a gravitational well, it won't red shift. Same as "real mass"; as you heat an object for example, the particle inside moves faster. Each particle gains relativistic mass, but the object as a whole gains rest mass since each particle inside is now heavier. It may be that in this universe, nothing exists in a point observable from this dimension ($D^3$). So then the matrix of space has volume and mass both in structure and monopoles.

There is the idea that photons travel at the speed of light. This is incorrect. Photons are the monopoles of the space matrix that elastically collide. Hence, the argument that at the speed of light, photons having zero mass is incorrect. Photons have volume; hence, they have mass, and as in this dimension, volume is the general rule.

Inertia can be defined as the resistance the object offers to a change in its state of motion. By this rule, there are different ways of viewing inertia in this theory. For example

- Time, by establishing an inverse or polar relationship to velocity, in a sense provides resistance to changes in velocity (like inertia).
- Forces and motions resolving balance are inertial.
- Resolution of balance between dimensions is inertial.
- The perpetual motion between dimensions of entropic and disentropic action is inertial.
- The motions of separate dimensions seeking balance are inertial.
- The resolution of balance within a dimension is inertial.

Without understanding the variations of dimensional energies and also their effects on inertia, one may be led to think that both classical and modern ideas of inertia are in conflict, but within this theory, the understanding of inertia is resolved, and any conflicts are resolved. I would classify inertia as motion toward balance of interacting constructs. How this balance is achieved describes the inertia.

## Infinite Energy

The universe is a particular qualification of infinity or infinite energy. Each dimension of space is a particular construct or dimension of energy. Energy then is a particular construct of superstrings or space matrix (*D*, *T*, and *V*) dimension per dimension. When we speak of energy as in $\boldsymbol{E = MC^2}$, we are talking about $D^5$ superstrings, which is a particular construct of superstring (energy) as $\boldsymbol{D^5 = D^3T^2C^2}$. Here, **mass** = $\boldsymbol{D^3T^2}$, which means all volume ($D^3$) has mass ($D^3T^2$). Space has volume, has mass, and is energy.

The space surrounding an electric charge is an electric field by its influence and interaction. When the electric field moves, it induces a magnetic field; and when the magnetic field changes, it induces an electric field.

A charge such as an electric charge is a particular interference pattern in the matrix of space made of superstrings. **The "effect"** of that charge **is infinite** as **it affects**

- **through dimensions of energy and**
- **in a dimension of energy.**

The understanding of energy varies in each dimension, and its effect is understood differently in other dimensions. For example, gravity in $D^{10}$ is force in $D^4$, so it is measured differently in different dimensions. Although its effect is infinite by its connection to infinite energy, and its effects are infinite trans-dimensionally.

Through this relationship of trans dimensions, all points of space are related via inertia or balance to an infinite source and infinite construct. Hence, although a charge may be seen as finite, its effect on the construction of transdimensional fields of energy is infinite since the source is infinite. There is finite interference alongside infinite effects. We can speak of a finite portion of energy or space matrix and infinite relationship and transfer of energy trans dimensionally to an infinite source.

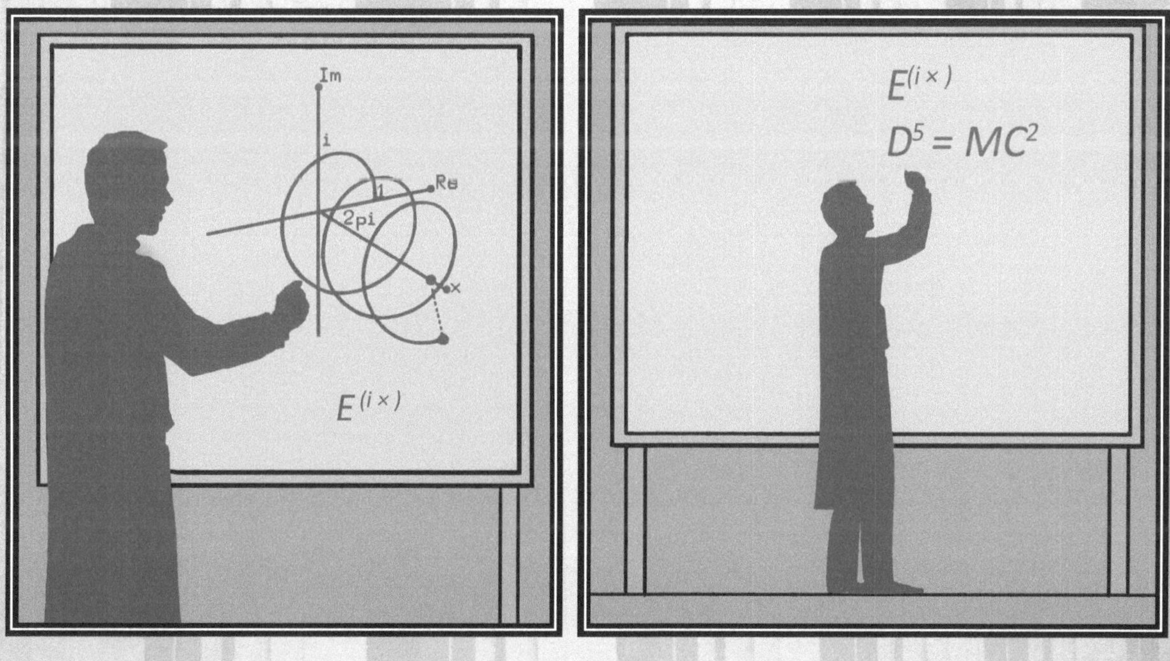

## Non-locality and Special Relativity

Non-locality or action at a distance is incompatible with relativity. Speaking about relativity, an action such as gravity at a distance takes time to accomplish. In this theory, this is explained by space being an elastic medium or communicator. So the elastic communication takes time. And what do I think about non-locality and relativity? Well, in non-locality, there is a connection between objects that allows them to be affected to what is perceived to be instantaneous; this connection happens regardless of distance. So to resolve this disparity, I will say that non-locality depends on your frame of reference. Since space is an elastic field matrix, then the communication between objects is via the elastic collision of the space matrix. Sub atomically, the fields of space are much tighter and form geometric orbits, and so, there seems to be immediate communication because you are dealing with parts of the same construct, smaller distances, and faster velocities. I doubt anything is instantaneous because volume implies mass and time; time, no matter how small. The universe is made from a space matrix, and no matter how small the space, the fundamentals of *D*, *T*, and *V* apply. And also, point of observation determines the relative analysis.

## Noether's Theorem: Invariance of a Langrangian under Time Translation

All space has mass, and everything in space has volume; volume or space means mass. Hence, a photon has mass, and a photon is part of the structure of space. Photons elastically collide, and what you see is on the receiving end of a wave. So photons are not traveling at the speed of light, and the abstraction of zero mass does not exist. Space is not an abstraction and a characteristic of all energy and mass in this universe. Energy is a dimension or construct of superstrings, and particles are their shadows. The type or construct or dimension of energy is defined by the geometric arrangement of superstrings in that dimension. For example, $\boldsymbol{MV^2 = D^5}$, which means that energy as we know it is a construct of five superstrings $D^5$. The actually construction is *D*, *T*, and *V* in five dimensions in positive, negative, and neutral can be expressed mathematically by **5 × 3 × 3 = 45 superstrings**.

The structure of space or energy is symmetrical and, when acted upon, will show the conservation of energy in accordance with the invariance of a Langrangian under time translation. So the most basic structure of space I can imagine is a tetrahedral one of *D*, *T* and *V* superstrings. This expanded into multiple dimensions leads us to spherical, radial, symmetrical, fractal, and tetrahedral geometries that allow for a Cartesian system to be perceived within them. So although space is a spherical, radial, and geometric matrix of superstrings, our perception can also impose a Cartesian system within it. Time is a superstring of the matrix of space; hence, space has that quality. The same is true for *D* and *V*. Energy is defined in various combinations of *D*, *T*, and *V*.

### So how does that tie up to invariance of a Langrangian under time translation?

What does invariance of a Langrangrian under time translation mean? This is about conservation of energy. A system should evolve the same if begun at a different time. So the time does not affect the evolution of a system? Yes. Time translation invariance of the action implies energy conservation. Like a pendulum, I start it now or later, and it's the same. So if you take this as true, you get energy conservation.

Noether's theorem derives the conservation of energy from symmetrical systems. If a physical system behaves the same regardless of how it is oriented in space, its Lagrangian is rotationally symmetric. We can say that the matrix of space behaves in this way, and thus, space is a rotationally symmetric matrix of superstrings. The superstrings are in rotation, and their effect is that space behaves the same no matter where you are. Space is an elastic and symmetrical superconductor, and the angular momentum of the superstrings is conserved since the motion is symmetric. Symmetrical systems of motion account for the conservation of energy and momentum.

So in cases or systems of conservation of energy, the process is always the same no matter when started. It says that any continuous symmetry of the action results in a conserved quantity. Conserved quantity implies symmetry. So for the conservation of energy, the action or model has to exhibit symmetry, not necessarily continuous though.

**So a process or system exhibiting symmetry will, under action, produce the conservation of energy independent of starting time.**

If space is a matrix and waves are transported by elastic collision, the conservation of energy implies a symmetric system. Space behaves the same regardless of it's or the observer's orientation. Its Langrangrian is rotationally symmetric from this symmetry. From this we have angular momentum conservation. Well, the superstrings in the matrix of space have angular motion. And that matrix is also symmetric when acted on directionally. I define space with symmetric structure. By symmetry of the structure of the matrix of space, you have conservation of energy. Space is a symmetric structure, and what more to give credence to the idea than Noether's theorem and the ideas brought out about symmetric systems and the conservation of energy. And really, this space or energy makes up everything. The symmetry of the structure of geometric space leads us to the conservation of energy. That is also presented by $mv^2$, and the concept that space is elastic as well as a superconductor.

The symmetry of the action of a system leads to the conservation of energy. The function that summarizes the dynamics of a system is known as a Langrarian. Space behaves the same regardless of its orientation and position; hence, space is rotationally symmetric. The matrix of space consists of rotating superstring at the speed of light. The structure is complex, symmetric, tetrahedral, fractal, spherical, radial, and alludes to a Cartesian interpretation as well. The monopoles of this structure are photons, and at such a small scale, the geometric grid seems as a sea of light. But this very structure contains the properties of the superstrings that comprise it, such as *D*, *T*, and *V*. And they are composed in more complex geometric arrangement called higher dimensions that result in other properties, such as gravity and other forces and energies. The resultant energies or arrangements are emergent, meaning from basic principle building blocks or primary superstrings, more complex arrangement and properties evolve or develop. The structure of energy is the primary part of all existence while particles are the smaller "shadows" or "interferences." The symmetrical structure allows for the conservation of

## Interference Wave Patterns of Space Create Particles

energy and momentum. The very things physics measures are attribute properties of the space matrix. Both the system or geometric matrix and the motion therein are symmetric.

The elastic collisions of photons are nearly the same despite differences of place or time in deep space; hence, the resultant action (Langrangian) is the same and account for conservation laws of linear momentum and energy. What are the conserved quantities from the observed symmetries of the physical system of symmetric space? The answer is the conservation of momentum and energy. What are the Langrangians that describe the conservation of energy in space? Definitely, it is the structure of space or super space and super symmetry. The smallest space is a tetrahedral arrangement of *D*, *T*, and *V*—a symmetric structure.

Space is transparent because the space exists in a different dimension of energy than the atoms that comprise elements. Lower dimensions of energy pass through higher dimensions of energy.

$\boldsymbol{D^5 = MC^2}$ means that the construct of energy as we know it is made of five primary superstrings *D*. *D* is manifest as dimension; however, as superstrings combine in higher dimensions, the properties of *D* change. $D^5$ is energy *E* or five superstrings of *D*; a superstring is a part of the structure of energy. Does it matter about the density of superstrings in space? Yes! In a dimension they are generally the same distance apart, but through different dimensions, there is compression, superposition, and fractal relationship. The superstrings that combine to form the various structures for different dimensions of energy $D^n$ are *D*, *T*, and *V*. Space is $D^3$; ***D*, *T*, and *V* × 3 in positive, negative, and neutral (× 3) = 27 superstrings**. The structure of superstrings is the structure of space or energy. The nature and properties of a dimension of energy is felt or measured differently in another. For example, **gravity =** $\boldsymbol{E^2 = D^{10}}$ is measured as force as we know it or experience it.

*How do D, T, and V differ?*

**$\boldsymbol{D = VT}$; *T* and *V* are polar and related via *D*, and the model is a tetrahedral arrangement of three strings**.

**Why not two or four strings?**

**The basic algebra of this tetrahedral matrix is a relationship of balance, infinity, and unity. Zero is unity, 1 is balance, and then infinity is qualified. It is implied in $\boldsymbol{D = VT}$.**

$D^5 = MC^2$?

Yes. That means space is an elastic superconductor. Photons propagate by elastic collision. The density of space in a dimension is relatively constant and changes through dimensions (such as an atom). There is a slight adjustment due to elasticity that can happen. For example, near a large planet or star the speed of light may be $C + V$ or $C - V$. At the center of the tetrahedral arrangement, a photon is formed by overlapping superstring interference. The superstrings rotate. Space is any dimension of superstrings; particles are their shadows or interference patterns. The dimension refers to the configuration of superstrings that form a dimension of space. The symmetry of the space matrix implies the conservation of energy and is confirmed by $MV^2$.

In this theory, the strings are like spheres with spin. They form a structure and overlap. Where they overlap, you get a monopole. I think in three dimensions, you have twenty-seven strings that form a photon at center. So I imagine an elastic structure, like a bowl of marbles; you hit one in a perfect elastic relationship, and the wave moves through at the speed of light. So the photon does not move, but the wave does like a wave on a beach. Elasticity means conservation of energy and vice versa. There is motion in the structure components and motion through the structure and the structure can also bend.

## Experience of Multiple Dimensions

Our universe is multidimensional, but we really don't experience all the aspects of this multidimension. For example, in the third dimension, gravity is experienced as a force from the tenth dimension, so to approach gravity by scientific observation alone will lead you to the abstraction of Newton's conception, but not the truth of the mechanisms at work. This reality even plays with the units. The units or one dimension cannot be understood exactly the same from the

# Atom and Matrix Field

measurements in another. So then this theory understands the properties of superstrings, and the equations reflect the arrangement of superstrings dimension per dimension.

To summarize, multidimensions affects several things dimension per dimension:

- **Units.**
- **Perception from the experienced dimension of human perception.**
- **Transparency of one dimension on another.**
- **Expression of one dimension as experienced in another.**

The space of the human existence is transparent to the higher constructions of atomic mass. These relationships are occurring between dimensions but also have relationship.

In the theory, I describe a photon as the monopole of the matrix of space. This particle is made of the superposition and interference of space superstrings, the things that make the matrix of space. The same superstrings combine differently to make space of higher dimension and density in an electron. So to answer your question, the elemental superstrings *D*, *T*, and *V* interact in a higher dimension of space or energy to create the particle electron. Electron has volume, and photon has volume, and all volume has mass by the equation **mass** = $D^3T^2$.

The difference between photon and electron is in the geometry of the dimension of space and energy where they exist, the superstrings that compose that dimension, and the interference patterns of the rotating superstrings that result in an electron. The spin of superstrings in different dimensions shows differences in associated particle (interference pattern) spin. These motions are associated with different frequency. But frequency is the effect and not the cause of the difference. Differences in all particles are the composition, superposition, and compression of the matrix of space, and an arrangement is a dimension of space or energy.

After the derivation of $D^5 = MC^2$ in this paper, then it's followed by a description of density; **density** = $T^2$. I realized or had the idea that space changes when you add more strings. A periodic table, if you will. The change of space was related to measured changes in physical properties or units. So the idea came about a periodic table dimension per dimension.

**How far does that that go, you ask?**

Well, even superstring theory would have limits. I felt a black hole was the limit. So then I redefined the idea of energy, and dimension was a different dimension of energy and also of space. Our understanding of energy was only $E = mc^2$ by current understanding. Also, any dimension of energy was a dimension of space. Space or the matrix of it in any dimension had associated particles. And then of course, there is much more space than energy. And then back to Einstein, if matter was energy, energy was everything, and matter included not only particle but the surrounding structure of space. **Mass** = $D^3 T^2$ means all volume has mass and that the variation of density of space changes the value of time.

I think that relativity is describing the properties of the space matrix superstrings, and quantum is describing the associated particles of strings.

With regard to **gravity** = $EMC^2$ and the force of gravity at $D^4 = ma$ **or force**, what you have is the realization that what we measure is a narrow band of dimensions. If you calculate gravity for the planets at $EMC^2$ and then bring the magnitude down to $D^4$, you get the gravity force of all the planets without using any constants.

All forces are "induced" by the presence of the structure of space in other dimensions. So that force of gravity induced from its field space in dimension 10 is a phenomenon that says the effect is different than the source. And the source cannot be measured directly, only the effect. With regard to **gravity** = $EMC^2$ and the force of gravity at $D^4 = MA$ **or force**, what you have is the realization that what we measure is a narrow band of dimensions. If you calculate gravity for the planets as mentioned before at $Emc^2$ and then bring the magnitude as I did to $D^4$, you get the gravity force of all the planets without using any constants.

$D^5$ represents a matrix of superstrings. The idea that things are linear is from the perception of scientific observation in the dimensions that we measure. Hence, this theory I propose for gravity is **gravity**= $\boldsymbol{D^{10}} = \boldsymbol{EMC^2} = \boldsymbol{E^2}$ acts as a force in $D^4$. So what exists in one dimension is measured differently in another. This understanding requires a requalification of units. Our science is basically a narrow band dimension of observation, and much of it is approximate abstraction. Here we see things as they are as a matrix dimension per dimension. So when I speak of a dimension, I am speaking of an arrangement of unseen structures, what I refer to as superstrings.

There are elementary units in rotation that make up everything. In the theory proposed here, I call them *D*, *T*, and *V*, and *D* would be that smallest one or God particle. The thing is these combine into twenty-seven for space in a complex tetrahedral geometry or three-dimensional spaces. That geometry allows for a Cartesian interpretation. Now the structure is so small that space could be simplified to be viewed as a bowl of marbles with the photon as a monopole or center of each unit structure. Space has a structure of spinning units that form a space matrix. I show what the relationship of these units are by equations or laws. Energy is defined dimension per dimension as space matrix. So then we see how these structures change from space to space gravity. The problem with scientific observation is that human perception and experimentation is from a narrow band of dimensions, so much of the time we are measuring the effects rather than the cause. For example, I propose gravity is $\boldsymbol{D^{10}} = \boldsymbol{E^2} = \boldsymbol{EMC^2}$, ten dimensions of superstring structure. And this is experienced by measure as force or $D^4$ or fourth dimension. So the force of gravity is not the structure of gravity. This being the case, most of physics is approximate abstraction focusing on a four-dimensional effect of gravity as it relates to human perception. However, by intuitive models such as you propose, many questions arise and suggest remarkably different solutions.

The ultimate goal of a unified field theory is to present a model with mathematical equations that relate all phenomena or suggests a structure (mathematical) that relates everything. A periodic table or the structure of the dimensions of the universe allows for differentiating the various dimensions. There has to be mathematical laws that actually suggest a model and interpret solutions.

It is understood by current science that there are three dimensions of distance and one of time. And perhaps some might even say that dimensions have infinitesimal steps. However, current science does not consider time, distance, and velocity to be component superstrings of a space matrix. This manner of viewing components of space as within a structure and in discrete packets eliminates the myths of chaos, irrationality, and vacuum. Everything we experience has order. We observe within narrow limits or dimensions, so then how can we ever understand the whole in ascending and descending scales by observation? We cannot. Even made more complicated when you consider some dimensions are invisible or immeasurable from one dimension to another, and what we think we are measuring is the effect induced by a cause. And some dimensions are inaccessible to measure in a conventional way. And finally, it will take an intuitive leap of the infinite conscious mind to model something beyond itself. So we stand as scientists looking at the following:

1. Dimensions that are immeasurable directly.
2. Effects that come from causes in other dimensions.
3. Ascending and descending scales.
4. Infinity.
5. Multiple dimensions.

But even with this dilemma, it is possible to mathematically model this entire system because in the end, math and this universe are emergent principals, one reflecting the other in this reality.

Some will say there is only distance and time, and then perhaps only distance or that time is not real. But to say this makes the mistake again of perceiving the universe from the fixed point of traditional scientific observation. Like a horse with blinders, you see only what's in front of you; that is your reality and is the very thing preventing you from finding unifying principles of the truth of the real universal constructs. But this is a larger problem than finding out the world is round; it is realizing that the dimension that you experience may be 10 percent of the whole reality, and this needs to be realized in advance.

To see more on the scale of dimensions, we need to first realize our view is fixed, and then model something that includes the entire construct descending and ascending. Well, some might say that the scale is infinite in either direction. I think not. This universe being a particular qualification of infinity means that there is periodicity like the periodic table. This means that there is infinity, but that the energy required reaching that higher level is held back by entropy. There is motion toward balance in a universe that is perpetually moving from infinity to infinity. Hence, although we can say the dimensions are infinite, there are contrary and multiple forces that pull it back. Infinity, unity (constants like 0 or *C*), and balance (like polarity of 1) qualify the infinite. And this universe is the conscious of God means that it exists within infinity in a certain way as a structure balance from its creation to its destination or existence. There exists a boundary between the conscious of this reality and the infinity of conscious, so there is a connection.

Infinity, unity, and balance are related by ratios, and the three are separate and also related and also united as a unit. The three are not ratios but related via a relationship of three.

Velocity is not a ratio of distance over time. Velocity is real property of space in an entropic and disentropic relationship. The ratio represents the order or relationship between infinity, unity, and balance. The ratio is representative of a multidimensional perpetual motion machine. Time and distance and velocity are only relative by this rule of understanding.

This theory presents a model for all scales of dimension, which is required in a unified field theory and to understand each dimension and scale, much like the periodic table in chemistry. Science is limited by the current models.

The structure of space means space has component form. Is space infinite, in that it is connected to infinite energy? Yes, but in that it has limits and ratios to other dimensions of energies, no. To say there are only three dimensions is a huge flaw. Three dimensions are but one product of a tetrahedral; radial geometry within a relationship of infinity, unity, and balance comprises as much as twenty-three dimensions. This theory presents all objects and details.

**Emergence and Transcendence**

The expansion of the universe is entropic and the contraction disentropic. The balance between these is perpetual motion. The disentropic action includes structures that build up such as mechanisms associated with life; hence, life is not an anomaly except for its transcendence or origin (a priori) from outside this universe. Processes within this universe are emergent or show emergent behavior. Such examples are the evolutionary cycles of galaxies and the development of elements of the periodic chart from simple structure, such as hydrogen and helium, to the more complex elements. With regard to life, there is no emergent behavior that leads to the beginning of life, emergent being processes observable as emergent within creation, from inanimate matter to the beginning of life. Hence, it is designed and transcendent. Truths (God) are an analogy of life, and universe and sciences will prove this. This universe is an analogy of God. The dependency of things in this universe is an analogy of the dependence of creation to an external source and transcendence of that dependency. Darwin did not define emergent processes that result in the beginning of life. And emergent processes of this universe will not prove the creation of the universe and also the beginning of life, both transcendent principles. However, by our transcendent nature, these truths can be perceived intuitively and described.

There is no difference between discovering science and God; they are the same. And it may seem that I use the word *God* quite liberally, but God is transcendent as life is; you will no more understand creation than the beginning of life from emergent processes. Natural processes in this universe are emergent. Life is transcendental. I'm not talking religion; I'm talking science. Religion is historical. You think you have to be religious to understand God, that's not true; you just have to think God transcends this universe like your soul in that body of yours. The infinite is qualified here in this experience.

***D / T* = *V*** means infinity, unity, and balance are expressed in relationship together.

**The word God has no inherent meaning without defining it, you say; your statements have no meaning, you say.**

God is that which the universe depends; dependence is a quality of all things, and this universe does not exist and/or for itself, and that dependency puts creation dependent on something—God.

**Well, who created God? On what is it dependent?**

That infinite beginning has no boundaries. Intuitively, I would say that this universe is qualification of infinity, and therefore in infinity, the argument of beginning or end no longer exists. The wild card is infinity, but here that is controlled or qualified. And you call it science in this reality, and here there is a beginning and time.

**I don't follow as to how that proves God, you say?**

We have all had the thought of God, and in that passing of your intuition, may the greater truth lay that transcendental experience.

## Infinity, Balance, and Unity: The Universe Is Conscious Energy

Realizing that the universe is a particular qualification of infinity and that in this universe it coexists with balance and unity and also represented in mathematical relationships of three, the primary tetrahedral structure of the universe, questions as to the clarification of these follow. What are infinity, unity, and balance?

Infinity is expressed as God, the beyond containing this universe from which the creation of this universe proceeds. Many would ask then, what created God; however, that question is meaningless where there is no time in a framed work of infinite constructs. Hence, the beginning and the end is really an emergent principle within this universe and not transcendental like infinity from outside the universe. The entire universe is an analogy of God. An experience that is infinity and also an analogy of God is conscious. The intuitive mind is infinite and without bounds, without beginning or end. It is the rational mind guided by emergent principle that particularly defines the conscious experience as limited, but the true source of the conscious mind is the transcendental infinite. And because the infinite is conscious, the infinite is also the infinite conscious of God, and our experience shows connection to the same source and may even be a part of it. The whole universe may be a particular manifestation of God's conscious of which we are participating and sharing. So the physics or science of truth is also a physics of faith, and science can lead us to God as well. The connection to this source of infinite or God is by complete harmony of the conscious and subconscious mind to that source where conscious dwells: God. And hence, the vehicle to this may very well be prayer, faith (biblical), meditation (Buddhist), and those things identified by various religions and spiritual groups saying the same things but in a different way. All express a harmony to the source, and that union can enable just and transcendent events such as miracles, healing, and possibly a spiritual science yet to be discovered. Miracles are transcendent. These different ways, whether it is religious or spiritual, suggest a different scientific view in each case. All suggest a conscious harmonic connection to the infinite source or God, the conscious from which everything is made—from the infinite of consciousness, to the creation of this universe where that consciousness is qualified, and then the transcendental of the infinite into this universe. The source of the universe and all life is in the conscious of God, the infinite. The conscious of life shares the conscious of God; hence, intuitively, we all have the question about God. Life is transcendental from the infinite matrix source. (Amoroso: 1-9)

To qualify this infinite, the universe has balance represented in the number 1 from which the perpetual motion of the universe is derived, and also unity or zero, the multiplier of quantity and fractal relationship. The structure of the three suggests a nature of God: conscious, unity, and balance. And these may be reinterpreted as a trinity metaphorically or philosophically as love, truth, and order; and in various forms it means that structure is the emergent rule of this creation and reflecting a nature of the source within the infinite.

Love is God's conscious from which everything comes. This universe is God conscious. The infinite that God is, is like your mind. Your conscious is shared with God's.

**So is there an equation for God?**

Equation is an emergent principle of this creation. God is transcendent. What value is transcendence? Infinity. It is that which is not emergent.

Transcendence has its origin outside the universe, and here proposes transcendence as a possibility that can only be accessed intuitively. Intuitive concepts are seized through consciousness. Like yoga when you are in harmony with God or praying. (Jahn: 1-52)

**Can I prove that?**

Yes, because that source created this emergent reality. Measured is an emergent principle. I didn't say infinite is measurable, explainable, yes. Transcendent rules are different. The infinite conscious is defined in rational thinking of an emergent world. Love is transcendent, and also the quality varies according to emergent reality. You think everything can be rationalized on the same playing field; not so. Even this universe with multiple dimensions, scientific observation only sees a narrow band of dimensions. Gravity in dimension 10 is experienced as force in the fourth dimension, so what is gravity? You will never know by measurement; a great philosophical and scientific lesson here. Even for questions about love and God.

**The universe is chaos, you say.**

No, everything in the universe is ordered, a perpetual motion machine; it is also "connected to" infinity. Chaos is a myth, human knowledge and semantics are full of myths. The early myths were real, like sea monsters, but today we still have logical and semantic myths. Duality is a myth; it has its place sometimes. Chaos and nothing are myths. I'm saying everything is consciousness, and in this creation it has order, a particular order, and there is no chaos. As Einstein said, "God does no play dice with the universe." The problem with current science is that the "math concepts" to current understanding are approximate abstractions, not really representing the true reality. For example, Newton's gravity equations are observational and abstract approximations where concepts of radius are included in gravity equations, but the true equations are about volume. Everything that has volume has mass, so photons have mass. Space has mass.

Infinity is consciousness, like your mind and those things that are a part of it, such as love. If infinity didn't exist, then we wouldn't have a consciousness. Everything is consciousness and always was. Beginning or end is an emergent rule only of this machine universe. You are transcendent since that is the rule of life. You are not here, only looking in, but you are everywhere; your body is your window into an emergent experience. God created this universe and said it was good.

**You might as well just say God created us and leave it at that, you say.**

Perfect. But you get all these questions like science and God are separate, who created God; mixing emergent and transcendent truths, that's chaotic. We assume that a mathematical abstraction represents true science—not true; we observe from a point and the reality changes as we move through it. There is much to know and revise. When we can say about our science that it is exact, we will know that science leads us to God, but we are already there because our conscious is infinite. Our intuition speaks to us about the infinite while our rational mind argues the difference because it's from a point, and the point doesn't know that reality changes through the creation. So you get approximations and myths. Most of science is riddled with myths from scientific concepts to rational constructs of cultural semantics. So where does it end? This is the human struggle for truth.

**Myths**

There are no determinate processes that result in the beginning of life; life is transcendental. God created the universe and then created the life in it. There's no determinate process that we know of that directly results in creation of life. So there were two big bangs, the creation of the universe and the creation of life. God created life by imagining it as he did the universe. Just as yoga. You can see God and move mountains when you realize that your conscious is infinite. You see, a human or scientific observation only measures a very narrow band of reality or dimensions. Reality changes as you move through its dimensions of energy. You can call them fields for simplicity.

So those equations of Newton are scientific abstractions and approximations. Gravity in dimension 10 appears as force in dimension 4. Newton's gravitational equations are inaccurate to describe the true nature of gravity; it shows but one effect. Hence, science limited in its observation to a narrow band of dimension defines more the effects than the causes of physical interactions. Science needs a broader view; a unified theory will do that. And such a theory will ridge to a complete view of the true nature of physics and what properties are induced by a transdimensional relationship. $D^5 = mc^2$, **density** = $T^2$, and **mass** = $D^3T^2$ are equations that describe the true nature of physical relationships as we move through dimensions or reality of this universe. Where we live and measure via scientific observation will always possess an abstracted view from the source. Myths need to be dispelled, such as a photon not having mass. Some dimensions are transparent to others and hence hidden, even the mass. All volume has mass, and a photon has volume and has mass. What does that do to the concept of rest mass for a photon? There is no such thing. Major revisions of the myths are required. A photon has no mass, there is no God, and concepts like chaos and nothing are myths.

So many errors and myths exist in science, language, logical structures, and cultures. This theory is the key to uniting relativity and quantum and in that theory is also a metaphysical proof of God without referring to scripture, and it also suggests a future for physics that is far more fantastic than you ever imagined. (Anderton:Web Link)

The idea that this reality is conscious is major, but there are rules; there is no chaos—another myth. The only chaos is when science mixes emergent principles with transcendent ones, and when science comes to an understanding of a universe via the perception of one or two dimensions when there are multiple, the rules change through reality. So the idea that the supercollider is going to understand the universe via an experiment in one dimension is false. Only the intuitive mind making a quantum leap can understand the reality of a universe born from a transcendent structure. Science can go only so far with experiment grounded in what's in front of you to construct a unified model. Models of multiple realities require a mind that assumes things change and are related by truths that are experienced differently depending on where you are: an intuitive approach. We experience gravity as force, but gravity exists in another dimension. Experiment from a narrow band of dimension, and the results do not project to realities beyond, and so many proposed models are wrong. And Newton's equation for gravity is the shadow of the truth, just like a particle is the shadow of the matrix of space. You see only a percent of the whole reality. Your eyes don't realize that space has a structure and that there is a structure of photons, which you only see when excited. The universe is a sea of light, and only God sees that; you see darkness and the shadow particles. And say, well, I understand the universe, the first big band was when this universe was created in the conscious of God's mind, and then he had another with the creation of life. Life is not an emergent principle like the evolution of matter from simple to complex. There are no emergent processes we can identify now that lead to the beginning of life, and so Darwin is wrong with regard to evolution being natural in its origin; everything shows dependence, even this reality to a transcendental source. When you pass away, you will rejoin that source, and it is prequalified again in another reality, and all this physics will mean nothing. What you take with you is your character or conscious, and the idea that the infinite conscious is truth, love, and God. But in your quest to know truth, even in science, it will lead you to God once all the myths are dispelled.

In ancient times, myths were sea monsters; today they are logical structures that are incorrect and outdated semantics. We battle these myths throughout our history, in the classroom, in science, in politics, and everywhere. There is transcendental experience and emergent experience. Energy is only one qualification of infinity. This universe has access to infinity. Infinity is conscious; God's mind is qualified in this creation. He is here. This universe is an analogy of God. Your mind is infinite but qualified by determinate experience. That infinite conscious is shared with God and transcends this universe. Infinity and conscious are the same and God.

**Dimension and Angular Momentum**

Every dimension of energy is referenced to a different spin of the monopoles of the space matrix associated with that dimension. Hence, to say that an effect of a physical property or performance has to do with spin is partly correct, since objects having properties of different dimensions interact a certain way, and when these are altered or forced to relate in unnatural or different ways, they will cause associated effects or induced anomalies. So the effects can be said to be related to spin or changes in the relationship of spin or altering the effect of adjacent fields of energy or dimension. The issue with respect to gravity and its effect on attraction or antigravity has many inroads. And I suspect in a situation where antigravity is induced strongly or slightly, in some way positive flow is being induced, a reverse in polarity from the natural gravitational field. We also must consider other fields that are also present when experimenting on the Earth since the Earth has a magnetic field. Hence, an experiment that produces a powerful effect rather than a weak one would be more conclusive. The Boyd Bushman experiment requires further investigation but appears encouraging.

## What Set the Universe in Motion?

Infinity and conscious are the same and God.

### Infinite is quantity, not an object, you say?

Infinite is not quantity, there is the error, although infinity is qualified in this universe. Particles are interference patterns of space. For example, drop two rocks in a lake; where they overlap is your particle by analogy. There is more space in an atom than particle, space is the substance, and the particle is on the edge, like sparks or a shadow.

### And what are the rocks, you say?

The rocks are the balance in motion from infinity to unity, so it's a motion already that space has in its components: perpetual motion. So the spin of the components of the matrix are already there; that motion is natural and was there from the big bang. But the rock is an analogy. You are God, so to speak. But the interference illustrates the small percentage, which is particle. You are basically asking what set the universe in motion.

## *D*: God Superstring (Particle) and Spin

If you start with a $D$ God particle, you can model it to make the two aspects of the fundamental particle and call this the motion side $V$ and the time side $T$. When you square $D$, you are adding spin direction. Spin refers to the dimension of a space matrix. When I square $D$, I am adding another $D$ to the matrix. The effect of adding a $D$ is similar to adding a $V$ and a $T$. $D$ exists in a tetrahedral arrangement to $V$ and $T$ in a stable matrix structure. These superstrings have spin and charge in positive, negative, and neutral. Adding a $V$ and a $T$ changes the dimension and properties of that matrix and also the associated particles created. Two opposite spins added in $V$ and $T$ are a balance and has matter extrinsic properties. To add spin is to add a superstring. I believe the compression and size by addition may be altered as in a fractal arrangement as the dimensions increase.

## $VT = D$: Polarity of Time and Velocity

By polarity, it includes opposite spin of the superstring. So velocity is inversely proportional to time and is like time. The superstrings exist together, and since one does not exist without the other, I say polarity. When you talk about frequency, you are talking about the elastic collision of the space matrix. Here, I am talking about the structure of the space matrix and that these two superstrings exist together, spinning and balancing each other—polarity. These same properties are a part of space. Space is a tetrahedral structure of superstrings. Space is energy is space is dimension; dimension per dimension, the structure changes.

Now the way I use polarity for antigravity is that I say to reverse the structure model for space in an atom; positive around negative in orbit. Time and velocity have opposite spin, and thus, I mean polarity there.

$D^1 = CT$: Time no longer changes in the hyperspace of a dimension. $C$ is used to calculate the elastic collision time for a dimension of space or the maximum transmission time information is communicated in a dimension. In a similar fashion to $D^5 = mc^2$, we are really looking at $D^5 = mv^2$, but to calculate the hyperspace for that dimension, you are dealing with total space or energy or dimension.

The constant $c$ is used to explain the elastic collision of space. The structure of space being definite can be proved and is so in the theory. You can't say $E = MC^2$ without saying it means conservation of energy, elastic collision, and transfer of information. Energy is space is dimension. The structure of space is qualified, definite, and has associations to the infinite. But it has mass and structure and order just like a structure. Photons or light are transmitted by elastic collision, or space vibrates, and the monopoles or photons elastically collide at *c*, and the wave carries to your eye. You get a photonic current to your eye. *C* means elastic collision, the max transfer velocity of an elastic collision in space. **Mass** = $\boldsymbol{D^3T^2}$ means all space has mass. Photons have mass has volume; that revises physics. **Gravity** = $\boldsymbol{EMC^2}$ revises Newton; force is induced as a fourth-dimensional effect. And Newton's equation is just an approximation. The value *C* expresses the energy for that dimension or configuration just like $\boldsymbol{D^5 = mc^2}$, so yes, using *C* has a reason for a particular calculation, defining the maximum of energy and hyperspace for a dimension. Time is inversely proportional to velocity. With regard to units, the relationships are rather best expressed as magnitudes since concepts like density can be expressed in a new way through the dimensions. So do not interpret time inversely proportional to time. And at this point, units are best expressed as magnitudes.

Certain dimensions induce others, and you can also have dimensional fields of space that are having polarity.

**Cause and Effect**

The wave associated with light is an elastic collision of a matrix that vibrates. The matrix itself is a fixed source, and so are the photons. What is received by your eye is the effect, not the photonic source of the current.

We view reality in a narrow band of dimension. Hence, we experience gravitational effect in

$$\boldsymbol{D^4 = MA}$$

rather than gravity at the source in **gravity** = $\boldsymbol{EMC^2}$. Hence, our gravity equations of newton are approximate abstractions or an effect of a source. So they do not represent the source gravity.

**Explanation of Energy**

Energy is redefined here as different dimensions. So the typical manner of viewing energy is gone. Energy is a dimension of space. So there are different structures of energy or different dimensions. The components of energy are *D*, *T*, and *V* in different configurations dimensions per dimension. A periodic table, if you will. Time is a component of energy in every dimension structure. Motion is a component of energy in every dimension structure. Dimension is a component of energy in every dimension structure. Energy as $\boldsymbol{E = MC^2}$ is but one dimension. $D^1$, $D^2$, and $D^3$ are all dimensions of energy. So goes the expression **dimension = space = energy.**

So that *D*, *T*, and *V* are components or superstrings of energy, they make energy in its various states.

Energy and space has a structure of superstrings. Now, implied in every relationship in *D*, *T*, and *V* is the concept of infinity. *T* and *V* achieve balance within a measure bounded by infinity. You can never go too far either way naturally. Implied in this construct is that this reality exists within infinity and has a certain form.

**Energy and Mass of Space, Volume and Mass of a Photon**

**Mass** = $\boldsymbol{D^3T^2}$ means all space as we experience it has mass. Space includes particles and space superstrings. It means that all volume has mass. Therefore, space has mass and a photon has mass since they both occupy volume. Calculating the mass of space and the mass of a photon is possible with this theory. Let us take an atom. First establish the mass of an atom and the gravitational pull. Then take the equations here to establish the expanse of space projected by the nucleus. This should be the outer limit of the electron cloud or orbit in the same manner that Neptune is around the Sun in what I call within the ecliptic without the types of eccentricity of Pluto. So now you will have three figures:

1. Mass of atom.
2. Volume to outer shell of atom and space generated by the atom (this can be calculated from $D^1$ from $\boldsymbol{D^5 = MC^2}$.
3. Gravitational pull using Newton's equations as an approximate (scientific experiment for force).

Now run the real gravitational equations for $D^{10}$ and bring it down to $D^4$. You should expect a difference; that will be because the nucleus size needs to be revised lower than the outer orbit.

Now rework the mass of the entire atom with new figures, which are the volume of the nucleus and the volume of the outer shell; having the mass of the nucleus, you can calculate the mass of the outer shell, which includes space, and therefore subtract the mass of the electron and have the mass for space. This would be the mass for **space in an atom**.

For outer space, I would use the Sun in a similar experiment. The Sun generates the space for this solar system, and it also coexists with the space from the center of the solar system. Then you can calculate the mass of space by a similar exercise. The amount of space surrounding a nucleus is proportional to the mass and is a basic principle in this theory. Both the nucleus and space around are both mass and are both spaces.

For determining the volume of a photon, it may be possible to use radioactive decay experiments to measure the amount of photons released and the effective reduction in mass and volume, and hence, determine the volume of a photon and mass of a photon. Already knowing the mass of space, then one can determine what percentage of the density of space would have how many photons.

**Dimensions of Energy and Energy of Space**

Each dimension of space is a different dimension of energy, and these would be $D^1$, $D^2$, $D^3$, and so on. These dimensions are dimensions of energy. So this theory has many equations for the energy of space dimension per dimension.

**The Value of *C* is Relative Depending on Observing Dimension**

The value $C$ is constant for each dimension, but $C^2$ is relative to the adjacent dimension. So in the fourth dimension, the value is still $C$, but coming back to $D^3$, I experienced $C^2$. I think transdimensional travel is the way to navigate space, and you can travel to the center of the galaxy at $C^2$ in a couple of seconds. So $C$ can be manipulated on how you navigate through dimensions whenever that becomes possible. I see space as a lake, and you are a speedboat. The matrix elastically collides at $C$, but you will bounce off the surface going fast enough into a higher dimension, and then fall back down, but for that short time that you bounce up, you are at $C^2$ to where you were coming from.

**The Generation of Space and Outer Space**

In this theory, space matrix mass exists in a ratio to particulate mass, which is saying that the interference patterns of space produce particles and atoms. When space is compressed and in tighter relationships, you have the construction of atoms and denser constructs.

Each dimension has a hyperspace for that dimension or the calculation of maximum space for that dimension. Definitely as we move toward a black hole, toward singularity, space is in a tighter configuration, but there is never chaos.

So for the solar system, the most of the mass is centered in the Sun, and it generates the space for this solar system. This solar system is also contained within the space of the center of the galaxy. However, the gravity at the center of the galaxy or a black hole is squared ($E^4$), so the space generated around the center of a galaxy in ratio with the center of mass is also squared. So there is much more space extending from the center of the galaxy by ration than the center of the solar system by a factor squared. Hence, there are vast stretches of space generated by the center of the galaxies, so their adjacencies are large voids of space that you have to traverse to their nucleus centers.

Atoms also contain matrixes of space that determine their structure, and this space also is in ratio to the nucleus, but better stated, the nucleus is in ratio to the matrixes of space.

So then space becomes varying and more complex than once assumed since everything is made from different dimensions of it, and these different dimensions here are called dimensions of energy.

At the most minute level, space is made of three superstrings *D*, *T*, and *V*. *T* and *V* are balanced between infinity, and the center of that balance is called unity. And that relationship is made possible by *D*. The tetrahedral arrangement establishes stability in the structure of a perpetual motion machine. The universe is a perpetual motion machine existing and qualified within infinity. Metaphysically, that infinity is conscious. And the emergent principle of mathematics that describes all this relates infinity, balance, and unity. And there are constants that define this perpetual motion and establish a certain order or natural order of emergent activity.

In a universe of multiple dimensions, $D^1$ is the point of balance, and then the scale swings in dimensions of positive to a parallel construct. So the very relationship that is the model for space at the simplest level to complex formations is also the transdimensional model of the entire perpetual motion construct. The behavior of the fractal or small also is represented in the behavior of the entire system. The idea that when you reach dimension 23 or higher, entropy or disentropy is pulling you back, and so you see naturally the scale to the upside is limited. That is not to say that it can't be engineered, but naturally, the structure is being held up. Just as in chemistry, you can't have elements getting heavier and heavier on their own.

## Space and Time and Frequency

A photon is made of superstrings that overlap—the monopole of space. They are transparent to the atoms in your body. Space and its matrix and monopoles are transparent to higher more-dense dimensions. When excited, however, via an elastic wave, the additional motion or kinetic vibration causes them to be felt. Photons are all around you and the structure associated with twenty-seven superstrings. Now, we should think of strings as space and particles as particles as the interference patterns of space. The equations derived describe the construct of space, not particles. However, when we think of strings, there are associated particles dimension per dimension. And also in each dimension, strings have an associated spin. Photons do not move since they are monopoles of the space matrix. They vibrate. Photons are composed of the overlap of strings, which I identify near twenty-seven. This is a stable part of space. So the *D* particle is really a *D* superstring. But there may be an associated particle with this level of space.

Time is a superstring and really exists. Time and velocity are bounded by infinity and have an inverse relationship. Different dimensions of space induce different effects of time. A structure dimension of space could be said to exist at a different frequency due to component superstrings, composition, structure, and compression. But time within the matrix exists. With regard to frequency dimension, some dimensions are transparent to others because they exist at a different structure, compression, and motion. You can call this frequency, but it needs to include structure and compression and components of the space matrix.

Time is real, and time travel is possible. Time is a superstring.

*D* does not exist by itself; it is related to *V* and *T* to form a tetrahedral framework and the relationship between unity, balance, and infinity. With the three, you have a periodicity between dimensions. It also gives space its structure: spherical, tetrahedral, and radial, and interpreted as a Cartesian where we are.

Since space is a matrix and *C* is the maximum transmission time and energy is conserved by elastic collision, space is the classic example of a superconductor.

I said twenty-three dimensions in this theory, but think it might be twenty-one. $Dimension^{20}$ is gravity squared as a black hole, and the additional three might be the most compressed space. I have gravity at ten and much more work needed for other forces.

By frequency transition, I imagine traversing from one dimension into the other. This has great implications for time travel, I agree, especially in a relative context, which I describe. I also believe time can be induced in a lab.

I think black holes at gravity squared are not stuck. And through the dimensions or frequencies, there are ratios from high energy to outer space at low energy. Black holes are the generators of mass and space. Unlike a star, the ratio of space is doubled. The thing about this theory is that certain concepts repeat through the dimensions. Gravity is squared in a black hole.

To summarize, I see frequency concept is a summation of the various things that come together to form a dimension. So then, yes, transdimensional travel leads to time travel, and different dimensions exist at different frequencies. On the structure of space and a photon, I agree it has components, but here; time is real and photons transparent and only is not when excited. Because at that point, space becomes more compressed in an elastic collision, and a shift in frequency allows them to be felt.

It is quite probable that dimensions exceed a black hole, but at the same time, at some point they are naturally restricted by the entropic effect of perpetual motion bringing energy back into balance.

## Infinity

Like Buddha learned, extreme behaviour wasn't the way to Nirvana. After seven years of eating bugs as an ascetic, he found that moderation was the way. Everything follows order in the natural world of emergent processes, no exceptions; exceptions are the anomaly, in your free will, which is transcendent activity. How do you know we have free will? Because in that anomaly, you have extremes; it is not the emergent behaviours of natural law limited by balancing equations. You are allowed behaviour in harmony or disharmony of the natural laws of the universe and the analogies implied by such natural processes about the truth of God. God allows free will.

Slowly we will dispel the myths and discover God, and largely the way to understanding God is to dispel the myths about reality. Myths surround us at every step from logical constructs, culture, to semantics of our vocabularies. Myths block the truth about reality, God, and life. And I should say transcendent truths are free will and conscious. Be careful of scientific observation, that archaic system based on observing a narrow band of dimension, and assuming 1 percent of discovery can tell us about the remaining 99 percent of reality. How do you know we have free will, you ask again? I will repeat. That archaic system developed as scientific observation is flawed, this universe has multiple dimensions, and we observe from a narrow band of dimension. Gravity is a ten-dimensional phenomenon and expresses itself as force or four-dimensional, so we see the shadows of the truth; and unlike science, your behaviour is not predictable, hence, you have free will. It is not emergent as the natural and predictable processes of this universe.

This universe has multiple dimensions, and we observe from a narrow band of dimension. All things emergent are predictable. There is a story about God where he says everything created is owned by God. So what can you give someone that owns everything, and only free will allows this, and it's "thank you." The principle of love reflected to the source and to others is an analogy of the creation of the universe. A perfect harmony that is in sync with the conscious from which everything is made, the infinite that makes up everything. In scripture, it says let your sacrifice be thanksgiving. But then it says there is one thing greater than the sacrifice of thanksgiving, and that is mercy, because when you forgive, you make an exception, and that anomaly is free will and love that returns in balance with the infinite conscious, and the natural rule has none of those options, emergent processes are fixed; that's why we call it science. But you say why can't forgiveness be part of the natural machinery? It can't unless it is programmed, and then it's imitation, not real free will. Life is transcendent. The universe is emergent. Conscious is the infinity that makes up everything and that we are a part of.

Infinity is conscious or God, and you have that too as your conscious infinity and share in it, and that transcends this reality. What does it mean for your mind to be infinite? You are not a single being fixed in this reality of an emergent universe but exist on more than one level; it means that your conscious exists outside this universe, but how does it affect a mind, and what would a finite and infinite mind be? Consciousness is the same stuff that makes this universe, but also it is that part that we call infinity, only here it is qualified; all my mind knows is this universe, just because it's not material doesn't mean it's not part of this universe. It only knows this because it occupies that brain, but not forever; your intuition allows you to have great insight into transcendent reality and truth. Hence, the analogies become meaningful; this universe is an analogy about the truth of God.

**Oh, you're talking about the soul, you say!**

Soul, conscious, and infinity are the same.

## Space and the Hierarchy of Nature

In nonmathematical terms, why is it necessary to assume the existence of up to several dimensions to explain the laws of nature?

There are many ways to answer this question. Let us look at the nature of space. Here it is presented as a structure, and elastic structure and particles exist by interference patterns created by the motion within the structure. Space is also defined as energy, which is a certain configuration of space. Space is what makes everything, since particles are the smaller portion of it. Space has mass both in the space and in the particles, which is also a form of space. Now look at you, you have a body, and that space is configured differently to outer space. And to continue; what about electricity and magnetism and gravity? How are the different forms of space here related, all space but in different forms? So now we have a theory that says everything has space made from a structure that exists in different forms in nature. What makes one form of space different from the other? That similarity will actually unite and explain the differences in these forms. So each form of space is related to others by some adjustment. It so happens that this adjustment is like a periodic table in chemistry. Except instead of adding electrons, you are adding space. And the elements are superstrings *V*, *T*, and *D*. And instead of being additive, the properties are multiplicative. So the math involves multiplication.

So now in nature, everything is made from space and varies in a periodic chart. It so happens that this variation is by *D* or dimension, and each dimension is a different form of energy and a different form of space. Sometimes one form of space will induce another through rotational motion as in electromagnetism. Now you as a man, you experience only a narrow band of dimension and call it nature. So your scientific observation is limited to that, and your equations are also limiting the truth. They become approximate abstractions. Take, for example, gravity in this theory. Gravity is ten dimensional and you experience it as force or four dimensional. So these activities in other dimensions are experienced differently where you are. But understand that by understanding the various dimensions, you can understand the relationship between everything.

Multiple dimensions and their interaction allow for balance between motion and energy so that nature exhibits balance, unity, infinity, and hierarchy. This principle also is found in the tetrahedral arrangement of the smallest components of space *D*, *T*, and *V*. We mentioned before that in these three superstrings you see are relationships between infinity, unity, and balance, and the relationships exhibit hierarchy and symmetry or equality. This has profound meaning as a metaphysical concept that hierarchy, harmony, and equality can coexist as a model for nature. The universe is an analogy of God. It enables the idea that God allows equal access and privileges to his love, while at the same time, with this equality; allowing hierarchy means that both exist harmoniously together. Even in nature, it is true that capabilities are not naturally equal in all beings; however, their consideration as having the rights to fairness, justice, and love participating in a harmonic coexistence exists alongside the necessary hierarchies that also allow diversity. In a political sense and sociologically and philosophically, we are not equal in capability and diversity, but we all have equal rights or the right for equal consideration under God's code of harmonic truth. So hierarchy extends to the arts and also incorporates symmetry and rhythm and equality of elements and these exists as a recipe for a masterpiece work. Hierarchy and equality exist together. Equality means equal access for love and consideration and harmony with truth, wisdom, and life. Even in the Christian definition of God, there exists hierarchy, equality, and unity; diversity and unity, equality or symmetry, with order along with infinity, so the model of infinity, unity, and balance remains.

## Unified Field Theory

Science is developed from observation and intuition. Intuition needs to be verified by math, or then you have no theory. Math is the emergent characteristic represented in the universal order of this creation. There is a need to define reality correctly, to know the truth. To accept something because of convenience is not correct, especially when there are so many questions and none are answered and also so many contradictions.

Observation is not the complete story if there are multiple dimensions.

The defining of other dimensions will allow us to travel to the center of the galaxy in sixteen seconds travelling at $C^2$. Stay with a four-dimensional model, and you'll never leave the solar system.

The purpose of a unified field theory is to have the model for the complete picture of reality of physics. Settling for something with exceptions is not the goal of such an endeavor. All right then, you have a theory, but not a unified theory. The theory presented in a transdimensional model is a unified theory because there is a model for everything. The bottom line is that any theory should propose laws, and the one presented here has many equations. You could say that in the realm of human perception, there are few dimensions, but I wouldn't want to say what number exactly. Also, the problem with many current theories is that they still rely on current terminology and ideas. Einstein proposed four dimensions. Quantum theorists talk about particles, string theorists are looking at math, and relativists think nature has waves. So everyone is trapped in a box. When you talk about space, I think structure (particles, strings, and waves that elastically pass through the space matrix structure). When you talk dimension, I think twenty-three for *D*, but many for time and velocity as well. So the vocabulary and options are so much greater. I would like to say everyone has a contribution, but I think what transdimensional theory does is tie it together. Energy is presented as space in multiple dimensions. Outer space is an elastic superconductor.

When you talk about four dimensions, force is a four-dimensional phenomenon. $\boldsymbol{D^4 = ma}$, **density squared** $\boldsymbol{= T^4}$. All the conversation in the world will not get you a unified field theory without equations. And then once mathematically expressed, the math will do the talking, present the ideas.

Transdimensional physics involves multiplicity. What is time? Well, it's multidimensional. What is space? Well, it's multidimensional. In the theory, I propose time squared or $\boldsymbol{T^2 =}$ **density**. That explains why the change in the density of space affects time. Now here you already have two dimensions of time. And without it and relationship to density, you cannot explain why gravity affects time. And in this theory, gravity is really $D^{10}$, and using the equations in the paper, one can calculate the gravity of anybody without constants. What you "experience" is gravity as a force or $D^4$ from its real place in $D^{10}$. So there you are, basing current science on your perception alone, but more is happening. Why not base chemistry on four elements? Well, that was good for a number of centuries, but not now.

History has shown that most theories are wrong. Acceptance and truth are two different things. We are not scientists for acceptance; politics is good for that, and many empires have crumbled on it. Look at Galileo and Copernicus, not so highly accepted.

Every exception is a doorway to a new truth. When you see an exception, it means your theory could be wrong.

Just get something that works for everything, and you will have something. Approximations like Einstein and Newton are good for winging it. In construction, there are many such tools; they use them, knowing that they are in a safe area.

It's naturally human to want two or three. But that is more about human nature than the reality that is.

It's not a question of complicating dimensional structure; actually, as presented, the structure is amazingly simple, but space has a structure. And why are there only four in modern thinking? Are we like the fourteenth century people who thought there were four elements fire, water, earth, and wood? Space and dimensional diversity are what they are, and what appears complex to one generation is simple to another.

Emergent has to do with processes emerging from simplicity to complexity. For example, it is true that atoms change in the periodic chart by adding electrons, and as star systems recycle, more complex elements are created. So in the universe, natural processes and structures from simplicity to complexity are emergent realities. Those things are characteristic of the natural rule of this universe.

In the theory proposed, $\boldsymbol{E = mc^2}$ is really $\boldsymbol{D^5 = mc^2}$; there's five dimensions right there, and we haven't even described gravity. There is a periodic chart waiting to be developed. The processes in this universe are emergent and, hence, require a number of dimensions; this does not rule out infinity. And the source of the universe, such as creation, is transcendent. So the entire reality has a transcendent source, even life, but once in place, exhibit emergent rules.

**Experiment to Discover *D, T,* and *V* Matrix of Space**

From the previous discussion, we have the equation

**Density =** $\boldsymbol{T^2}$

An experiment that shows density is proportional to $T^2$ in such a manner is revealed by the variation of the density of gravity around the planet Earth and the effect on time by GPS satellites. Having this relationship is mathematically possible to get $\boldsymbol{D^5 = mv^2}$ and $\boldsymbol{D^4 = ma}$; it is also possible to derive these relationships mathematically without experiment as achieved in the paper.

Also in this paper, I have an experiment to calculate the volume and mass of a photon, and then calculate the mass of space. So then it would be possible to determine the relative size of the matrix components.

To actually see the matrix is going to take superior technology, but determining and confirming the math of *D*, *T*, and *V* is attainable. But when you get to the point that you realize that light is an elastic collision of the space matrix, that space is a superconductor with a structure and photon as a monopole, that the double-slit experiment is resolved by explaining space as a matrix structure, and Noether's theorem suggests the symmetry of space matrix structure and the conservation of energy; we are led in this direction of structure to discover components. Then it's a task of mathematically confirming complimentary geometries. The geometry that will allow the effect of point to point throughout space needs to be radial, circular (spin on the strings), allows Cartesian interpretation; this leads to several tetrahedral forms, and that subunit is three, and that leads us to *D*, *T*, and *V* superstring structure suggested by the mathematics.

**The Origin of Mass**

If you look through this paper

**Mass =** $\boldsymbol{D^3T^2}$

is explained, and **density** = $\boldsymbol{T^2}$. Everything is made from space, which has a space structure or matrix. Space is made from superstrings *D*, *T*, and *V*. Space structure varies when another *VT* is added into the structure; this variation is called dimension or $D^n$. Each dimension of space is another dimension *D* or dimension of energy. So now energy is expressed differently as $D^n$ in various dimensions and is also explained as space in various constructions. So to simplify the concept,

**Space$^n$ = Energy$^n$ = Dimension$^n$**

The origin of mass in $\boldsymbol{D^3 T^2}$ means that

- **all volume has mass,**
- **all mass has volume and time, and**
- **mass also exists within the construct of various dimensions of space.**

Outer space, which we call a vacuum, is a dimension of space, and an atom contains or is made from multiple dimensions of space, and each dimension of space has a relationship to mass. In some areas, outer space may be a combination of several dimensions of space since other fields or dimensional spaces mingle. Mass is an arrangement of superstrings that has volume and the real phenomenon of time and also the phenomenon of density. A change in density affects a change in time in space. That explains why the change in the density of a gravitational field affects time.

**Similarities and Characteristics of Different Dimensions of Space**

Everything is made from space, and space is made from a matrix of superstrings (*D*, *T*, and *V*). There are different types of space, which are also called energies or dimensions. The matrix of a dimension of space energy has monopoles like gravitons for gravity space or photons for photonic space, depending on the dimension of space. Gravity is $\boldsymbol{D^{10}}$ **= gravity** = $\boldsymbol{EMC^2}$; gravity appears at $D^4$ as it is experienced by our perception or narrow band of dimensions by humans. Each dimension of space has varying densities from its source. Hence, gravity follows the rule of the inverse square law as well as light.

Something needs to be said for the geometry of space, which is spherical, tetrahedral, and radial; although allowing a Cartesian interpretation, the radial nature of space also follows the inverse square law due to this radial phenomenon. Photons do not travel since they are the monopoles of $D^3$ spaces; they elastically collide and transmit the wave to our eyes so we see the photons at the receiving end like a wave on a beach. The geometry of space solves the inverse square law as well as the duality of the double-slit experiment.

Space is an elastic superconductor. Space is transparent to higher dimensions, except when excited with a wave, which alters the vibration frequency, which allows the wave to be felt. Different dimensions of space are at different vibration frequencies.

There is no coincidence in the apparent relationship of gravity and light and the similarity in the inverse square law to each. It is because all space has structure and made from basic elementary superstrings *D*, *T*, and *V* and vary dimension per dimension by *VT*. According to the equation **density =** $\boldsymbol{T^2}$, we know that density of space varies according to $T^2$, which is also a relationship to the inverse square law. That is why the experience of time also follows the effect of the inverse square law with regard to the change in gravitational density. It also affects the measure of light and gravity. The change in density is a change in time, which is also a measure in the change in the apparent field of space. This is measured and results in the inverse square law due to geometry and change in density.

**Geometry of Space**

In this paper, it has been said that there are from twenty-one to twenty-three dimensions of energy or space. When we discuss the gravity in a black hole as twenty, we go to say there may be one, two, or three dimensions beyond that. This is whether the entropy of the system in balance with unity of the first dimension slows the natural progression to a higher form of dimension. Dimension 21 would be a force; 22 would complete twenty-one steps of three from one as a geometric progression of induction, and 23 would complete steps of three bounded by two vectors of force while claiming the last as a volumetric component. Various numerical progressions imply different repeating properties of space energy as in a periodic table. For example, we can talk about density throughout the progression of space and also explain how the form of gravity is altered in a black hole to $E^4$.

Geometric space needs to provide the following characteristics:

- **Basic unit of three (tetrahedral geometry) from the three basic superstring *D*, *T*, and *V*.**
- **Radial geometry suggesting close-packed spheres at a very minute scale. This will allow point-to-point relationships since the scale is very small.**
- **Spherical units for superstrings with spin and organized in a structured matrix will allow three geometries in one such as a tetrahedral form, a spherical form, and radial form.**

When I put these criteria together to form a geometric model for space, we get a geometry similar to that in a sunflower (spherical, radial, and tetrahedral) as a two-dimensional abstraction, but in reality and in a volumetric form, it's quite complex with increasing complexity as you add dimensions of space energy or geometry by adding *VT* superstrings. To symbolize this form simply would be three intersecting spheres. This abstract form has roots in ancient symbolism. Space needs to be a geometric model and, as such, resolves many dilemmas and explains many phenomena as mentioned in previous chapters in this paper. For example, geometry explains the double-slit experiment or space as an elastic medium. The smallest *D*, *T*, and *V* are a simple tetrahedral form, and as the dimensions increase, the forms become more complex. The tetrahedral forms are transparent to some dimensions; however, where these energies interfere, particles are created. These superstrings are like stationary waves, meaning they spin on themselves within a structural matrix. When these structures within a dimension and between dimensions overlap to create a noticeable interference, the motion of the matrix slows down because of the interference or particles that result. I would imagine an electron to be the result of geometry in motion; the circular motion of an electron indicates the circular motion of the space matrix geometry and subcomponent superstrings. The sub light speed of the electron shows a shift in frequency of the matrix, allowing particle formation. The matrix without heavier particles is moving at the speed of light. In outer space, the photon is a monopole of that space. All space is light since the scale is so small to us. We only see the photon monopole when the frequency is shifted due to the additional energy of an elastic collision travelling through the space matrix.

To simplify the concept of space, imagine at the most minute level a bowl of marbles. When so small, you get radial, tetrahedral, and spherical arrangement that can allow an elastic collision to travel in a linear path from any point.

The number 27 was used to refer to twenty-seven spheres or superstrings in space matrix geometry for $D^3$ or the third dimension. So you can see as we approach an atom, the matrix has more information, but it's still space or energy or dimension. In the case of an atom, we see an arrangement of multiple dimensions at higher density of the space matrix or ether matrix. Also, the matrix itself is more complex in arrangement.

## Superconductivity of Space

Outer space is an elastic superconductor geometric space matrix $D^3$. Space has form and geometry of a symmetrical structure. This is suggested by Noether's theorem. The monopoles of space matrix are photons, and the elastic vibration of space allows a wave to pass through. Photons do not travel through space, but what you see is the photon on the receiving end of an elastic collision, like a wave crashing on the shore of a beach. Space monopole photons are transparent until excited by vibration, increasing their natural frequency. Frequency can be used to adjust the dimensional state of space, increasing its interaction with adjacent dimensions.

All dimension and constructs of space have a state whereby the matrix of space is a perfect elastic collision. This state is called superconductivity. At this state, monopoles exhibit superconductive properties as elastic collisions of the monopoles or un-resistive "current," such as electric or photonic. The speed of space matrix elastic collisions is the speed of light. The nature of superconductivity is very much tied into the geometry of the space matrix. Elastic collisions, whether they are elastic for a dimension of space matrix or for the commonly understood meaning of object collisions, can produce waves at the speed of light. Therefore, it is possible to produce a medium whereby sound travels at the speed of light and also geometries of space that result in high temperature superconductors. (Dume: Web Link)

When you speak of space, you speak of the geometry of the space matrix and the temperature allowing a perfect elastic collision for superconductivity at the speed of light. When you think of metallic compounds, it may be that the superconductivity may be enhanced at higher temperatures due to the complementary symbiosis of multiple dimensional geometries in a compound to produce a high temperature superconductor. Where a metallic compound may not exhibit superconductivity at a high temperature, that same compound combined with another in denser space matrix geometry may produce the special and special mechanism for a super elastic collision at higher temperature, since superconductivity is a function of geometry as well. Hence, superconductivity is supported by the following:

- The dimension of the space matrix.
- The geometry of a space matrix.
- The temperature and vibration frequency of a space matrix.
- The combination of geometries of a space matrix whether in one dimension or multiple dimensions.

Understanding that higher density matrix geometries may enable superconductivity at higher temperatures will lead to development of materials that can exhibit superconductivity at room temperature.

## Objective View of the Universe and God

Is there an objective good or truth coming from a deity or something? Or is it good that it minimizes the loss and maximizes the gain? This universe is an analogy of God, and that analogy is a source of truth since it refers to the source, God or good or truth. All truth comes from God and is the structure of reality, so all good things or a truth described, comes from God and describes God—one and the same. Good is a universal and intuitive (spiritual) transcendent truth experience that can only be partly explained by the emergent rational mind of man but more fully by the transcendent nature of man. The rational man qualifies its explanation (from an infinite source) in this universe life experience. These truths are qualified in this existence. Semantics and semantic structures speak more about humanity and its myths rather than truth, so then the understanding of truth requires the dispelling of many myths created in the absence of knowing truth. Truth and life are substantive where myths are false. Hence, the human interpretation of God can be very misleading, even though almost everyone desires it and thinks about it suggested by human intuition. The conception of good is wrong on purely a rational base. Good is transcendent like free will and not an emergent apparent predictable component of the universal machine; however, an intuitive mind can perceive the analogies. In much the same manner, people are better understood by intent than semantic explanation.

The universal example in science as in this theory teaches moderation and balance to infinite options; balance, unity, and infinity are entropic and perpetual motions. This refers to a truth about God that allows motion from the point of balance. This is also an analogy of God whereby moderation is the path to Nirvana. The anomalies in this universe are those that are transcendent and allowed to experience the truths alluded to in this universe. Life is pain in so much as that harmony is not understood. The universal truths and harmonies are suggested. God has qualified his infinite to espouse a harmony of nature. Our transcendent nature, also being infinite, is also qualified by our actions, and this allowed can be an exception to defer from nature's example or the harmonic structure of infinite consciousness or reality. We have the option of harmonic cooperation or deference. And even as entropic action pulls this universe into perpetual motion and balance, so too our actions are pulled toward the source of harmonic creation by suggestion through life. By our free will, we are led back to balance and embrace of harmony. The prodigal son has returned. And this is so evident in science since it too is a qualification of truth about God.

Truth to be known is not obvious, but can be understood intuitively as the transcendental experience that is. And this can also, by analogy, be referred from nature. You don't have to experience bad to know good or bad intuitively. The key word is *experience*. By human rational thought of limited perceptions and its narrow view of emergent processes, good is sometimes augmenting the chance of survival of what you value the most. How does this differ from transcendental good of infinite truth? Transcendental good refers to precepts that invoke a participation in the truth of God. These have nothing to do with the rational empirical mind as a source but can be rationalized after they are discovered intuitively. For example, a harmonic understanding of truth as the organization of all reality is not necessarily the same as the anomaly of free will accessing such understanding. There is the nature of truth, the nature of intuition or the conscious infinite shared with God, and also the limited perceptions of rational scientific man with limited physical perceptions, and they are not the same. And so the rational man limits its intuitive link to truth by myths created by semantics, logical structures, and culture. Man is further limited in experiencing this universe by a limited number of dimensions via perception. The device of physical man is limit, and the device of intuitive man is infinite. It would be easier to understand many actions more by intent rather than explanation. But this intent, without referring to truths about the harmonies existing, fails to achieve harmonic coexistence with creation.

Because the knowing of truth is transcendent spiritual, without intuitive knowledge, we can't say WHICH is the right conception of truth, but we still can spot problems with existing conceptions. So I can't say what is objectively good without understanding truth, but I can in some cases say that what someone thinks to be good, in fact, will harm him, which is more about human nature than truth.

Life is transcendent, so I have a reason to believe that man's conception of good is misguided when limited by myths and a purely rational approach. God transcends this universe like intuition in a spiritual sense, like precreation to creation, like life to evolution. We do not know what is good positively, but we can have a pretty good idea of what IS NOT good. Good and truth are intuitive and represent the matrix of infinite consciousness by which all is created and exists. Science leads us in this direction as the discovery finally is the discovery about God.

**Transcendent Origins of the Universe**

God for most is a personal experience, and you can describe it; and unless it is observed, it becomes questionable. Even scientific observation is questionable if the universe is multidimensional, and we perceive this reality through but a few dimensions; science becomes abstract approximation. So then we are left with man's perceptions through his physical senses that are very limited, which includes his semantics and the many myths or falsehoods developed through these senses involving semantics, semantic structures, culture, and logical structures. But as a social society, we rely on cumulative experiences, those that are shared. And then we leave out our transcendent nature, intuition, our real existence, as a laughable notion. And we confuse reason and intuition, discount intuition, and validate reason. I don't think that most of us are like Columbus, braving out into the "unknown" (laughs) to claim meaning in our lives beyond our physical perceptions; to know thyself is not a mandate. God is the drugstore down the street for a primitive man, but for us down the street, he is known. So what once was a myth is dispelled. So in the end, God is real if he is known and cumulative and historical, and our understanding of God and science needs to transcend the human limitations of myth and the current understandings in science, which are false. But you are more than that, aren't you? You know the known (unknown, laughs) by your intuition, that gift that says all is known and there is nothing to know because there is no unknown. Didn't you see that in intuition? You can guess at some things correctly, and the odds are a thousand to one against it, and truth is found. Don't be so square about life; go to the museum and understand art, poetry, and that spiritual side of being human, which you are discovering, because you too are an artist. The idea of God has crossed your mind by nothing more than intuition, and intuition is the source of all truth. This universe is an analogy of God, and the infinite consciousness of God is qualified to make this universe, just as your infinite consciousness is qualified by your experience in this reality. To know a truth is to know God, and it is absolute, part of the matrix of the universe. Sometimes our reasoning mind will then analyze the idea of truth by imposing typological myths by the unlearned or untrained reasoning mind. Truth becomes distorted and covered over by untruths. Structures like duality, which tells us more about human nature than truth, will result in questions like "How does one differentiate a truth from a lie?" The premise is incorrect and flawed.

How were you able to distinguish what you said as true intuition and conscience? The conscience in you has access to what is right since it is part of the universal consciousness space matrix or infinity. So untarnished these intuition and conscience are not wrong, so to you, neither of these can be wrong, you say? Well, yes, unless corrupted by what must be unlearned, myths that diffuse our awareness of our inner self. What you believe to be right is so in the ideal sense, not unless you willingly alter it. And as the Toltec's claim, it takes a lifetime to unlearn the myths. You should know what action by "intent" is, and the completion is the action and explains the true nature of the action. But we use words to explain other things in our actions, sometimes far from intent. So language is in conflict with intuition of the untrained mind. Intent reveals all when your mind is receptive to creativity and multiplicity; it is akin to the nature of God. When someone says he is a philosopher, let him espouse a truth. I say that because if I haven't heard a truth, then you are not a philosopher. The same with science: give me a law, not an approximate abstraction, which is most of science.

So someone tells you a phrase, and it sounds limited. Most ideas are like boundaries that are measured and limited. And you see the phrase as a circle created. You are in it when you believe it; it is diffused when rephrased intuitively in light of an infinite construct. How people use language is more fundamental than what they say and reveals how they are purely intuitive or rational or both. Truth enables. A truth does not make assumptions that cannot be explained with intuitive reason. You don't have to fail to be wise; the idea that failure brings wisdom is a rational myth. Wisdom speaking to us from creation is an intuitive truth. Wisdom is infinite. But better said is that all truth is available, and this eliminates the semantics limits and interpretation of human limits.

What do you mean by intuition? How does intuition tell you what is right? Intuition is a transcendent process and, by its very nature, is a part of the infinite consciousness of God, whereby all that is good and truth comes. It is a shared source. As the matrix of space from which everything is made, so the analogy is that so from God, everything is made. The matrix in part refers about the whole. And so, life transcendent into this universe is an anomaly as free will is an anomaly, your ability to defer or, better stated, your capacity to harmonize and create.

Every statement has an exception, and that is a door to new knowledge. This being the case, most of science is in errs; our reason also.

One might say, if my intuition tells me you're wrong, then according to your logic, you must be. But if you ARE wrong, then the premise of you being right (my intuition being correct in the first place) is also wrong; you are playing with dualities. You must examine closer. But then since truth is absolute, one of us is definitely wrong or right. What must be discussed are ideas; your notion of me is intuition, what you presume is not intuition, but my notion of you is my intuition, rational is not intuition, reason is not intuition, duality is polar opposites and the biggest trap for the truth because in many instances, it is a myth.

Truth is truly "intuition," as I would say. The universe is multiplicity not duality; thinking or presumption is NOT intuition. Everyone can know truth, even liars—it's just not what everyone else would describe as the truth. Presumption is accepting as true; like a myth there is truth assumed correct, falsely developed by perception. Perception is an excuse to deny responsibility derived by intuition. Truth is qualified by experience, logically explained if something is "true," it must be absolute. Truth is absolute. Truth is not situation; that is what makes it true. Love is a truth, absolute, and your intuition knows it, speaks to the heart without introduction. Because a truth is unknowable doesn't mean it's not true or untrue.

Is knowledge an object of intuitive logic or rational emergent logic mind? Both. Those things of this world are particularly here, and those things of intuition are transcendent. How is knowledge approached differently by rational and intuitive mind? Truth is the object of intuition, and rational emergent reality is but one model of it. The problem with reason is that human perception is finite while intuition is infinite, and in the absence of correct guidance, the reasoning mind is led by many myths or falsehoods. This is very true in science where a universe of multiple dimensions is assumed to be the narrow band of dimensions, which is human perception making scientific observation also, flawed.

So why would we assume there are two faculties if one can explain everything? There are many faculties; here discussed is the infinite nature of intuition by its transcendental source. I think that most of philosophy presumes ideas derived were described rationally on its own terms when intuition is a source.

The universe is hierarchical. Hierarchy means equality is justified through diversity. Ask why do we perceive and animals not? If we are "evolved" from animals, why is it that we are the only animals that can experience the better gifts of God? We participate equally in diversity. This theory speaks of hierarchy of a periodic dimensional construct and the equality of parts. A synergy, if you will.

Through science, there is no excuse for not knowing God because of transcendent truths and analogous truths expressed in an emergent construct.

The variability of cause and effect is an analogy of a truth about God, free will—variability in relation to infinite possibilities. Cause has the flexibility for an infinite variability in infinite number of possibilities. God is infinity. We are a particular manifestation of infinity or consciousness, but God is infinity and is consciousness that includes our consciousness or transcendental nature. Variability in infinity is creative.

Just like your creative mind is analogous to a creative God.

Intent is perfected in love. I had a description of wisdom; at first I thought it knew how to live, but then I expanded it to include awareness of truth is wisdom, and by that you know how to live, and that awareness includes an awareness of God. Awareness of your transcendental nature is the first true step to science and the truth of reality.

So in this universe, we have its harmonic reality; there is no chaos. Chaos is a myth. And the qualification of infinity is this reality and the balance of perpetual motion and relationships from infinity to balance and to infinity. Without intention or action in contradiction to established harmonies or truths, there is no sin. It's one thing to speak about free will working within a system modeled for creatively exploring its truths, and it's another to act with intent outside of this process where that is the intention to alter God truths from consciousness. So the sin is not the knowledge of good and evil, which intuitively we all have, but the **action of intent** away from God. Knowledge of evil then is carried to the **practice of intent** in disobedience (sin) to the universal love in our conscience. So a good story to illustrate is the Garden of Eden. We knew well, we knew evil but did not practice it; we did not have the action of intent to rebel. And so Satan claimed to know both good and evil; he knew rebellion and the intent to remove God from his life. When religion refers to sin, it is in the first motion by intent. It justifies what Christ says about evil in the heart; it means evil begins with intent, and the knowledge of good and evil in the garden was intent to do evil, not a philosophical or intuitive knowledge of it, but intent. By intent, actions are known. In science, by equations, actions are known. There are exceptions due to flaws in the understanding of human nature, and then there are exceptions by absolute truth, and free will is evidence of exception. It is also true that this is also one of God's characteristics: the capacity to act freely.

What is not by the story in the Garden of Eden, and in a sense, let's include the idea of karma. What is not known are the consequences, the entropic action of the universal consciousness to correct and maintain its perpetual motion to balance and is an analogy by observing the perpetual motion of the universe. Good is preserved. That God preserves good. There are general rules, and there are exceptions.

That the philosophy of infinite consciousness of space and mind is free will and entrapment of this motion is contrary to established law.

That God is love and infinity, and that is analogous to the universe and creation and positive multiplicity, and that unconditional love is beyond reason to surmise. It allows diversity being comprehension.

**Scientific Observation**

You can come to an idea of God without religion. After getting to this point, then you become religious. All science leads to God. But of course, if you are one of those who say "scientific observation" is the rule for science, remember that you see maybe three dimensions by your physical perception, and I don't mean $D^3$, but $D^3$, $D^4$, and $D^5$, and in a multidimensional universe of twenty-three plus, then how accurate is your scientific observation if your perception is limited to three? It's not. Your math deals with reality as an approximate abstraction, so almost all of it falls short of being truth. And there are so many exceptions in your theories that identify new directions for understanding of truth. Being a transcendent being, you need to transcend this experience via your intuition and infinite mind and be careful of relying totally on physical perception, which is an emergent logical universal form. Yet even emergent truths are an analogy of God, but this is only realized by transcendence awareness because a transcendental source qualified this experience, existence, or universe. Infinity is qualified in this universe, and infinity is consciousness and God. And this transcendental infinite qualified was the big bang. So then scientific observation for the future is very limited for the future man and becomes an archaic system, like the abacus. So those of you that make scientific observation the Holy Grail for the discovery of cosmological truth are in error. And to the atheists suffering from the dogma of scientists entrenched in primitive beliefs, no wonder we have such an unenlightening populace. What we measure is the shadow of the truth. For example, **gravity = $D^{10}$ = $EMC^2$**, but we experience it as $D^4$ or force. So we actually think gravity is a force. What we see is actually the resultant effect of the dimension we are in coming from another location and is not the gravity at all. So after this error, how much more can we measure or come to a conclusion of God without a transcendent awareness? So you see the blind lead the blind. But there is a God and very easily understood when science adopts a new method and finally reaches to laws, which actually do represent truth. This is the future of the human race. We will travel to the center of the galaxy in sixteen seconds and to others galaxies in similar time.

## Geometry of Space and Pythagoras

## Conservation of Energy and Momentum

Space is a symmetric structure or matrix of superstrings. The unit of superstrings *D*, *T*, and *V* is the sphere, so strings that are components of the matrix have spin. Now how are there spheres attached or related in a matrix? According to Pythagorean theories, circles overlapping where the centers are shared (as in a vesica piscis) are a form from which all numbers and geometries can be derived. This is the model appropriate for a matrix of space, one that also represents flexibility in math and geometry. And a symmetric structure adds credence to the conservation of energy and momentum as it relates to Noether's theorem. This flexibility is the basis for the structure of space. The circle and its symmetrical repetition through space to form everything combine concepts of unity (0), balance (1), and infinity in a hierarchical transdimensional relationship. It is also the basis of tetrahedral geometries that constitute the space matrix.

Irrational numbers tell us that this universe is a particular qualification of infinity that the solutions are not confined and static. We have both whole numbers and irrational; a relationship of infinity, unity, and balance is demonstrated hierarchically.

The nature of a rotating sphere of a superstring means that the matrix of space exhibits angular momentum. Unless acted upon by torque or force, the angular position remains the same. So space has the conservation of momentum because of the rotational symmetry of the structure of the space matrix and also the conservation of energy, which means space, is elastic. Symmetry and conservation of angular momentum are related in Noether's theorem. In an elastic body of space, since the matrix is elastic, you have the conservation of energy. So there is both conservation of energy and momentum. It is the angular momentum of the superstrings that add stability to the matrix of space, associated particles, and transdimensional constructs that result in atomic form and other dimensional energies.

## Missing Mass, Dark Energy, and Quantum Foam

The Planck-Wheeler length (1.62 × 10 - 33 centimeters) is the length scale below, which space as we know it ceases to exist. At this scale, space is said to become quantum foam and an organization of singularities. It is also the level where superstrings are assumed to exist. Virtual particles are created temporarily. They are created in pairs using energy borrowed from a nearby region of space; they are temporary since they have to give the energy back. And according to current understanding, dark energy is a hidden energy that is causing the expanse of the universe.

According to this theory, all space has mass, and that mass can be measured in the dimensions you perceive and also hidden on other dimensions that are transparent to dimensions we perceive. Also, energy is redefined as dimension ($D^N$), which is also a dimension of space. Space creates everything, is a structure matrix, elastic, and the vacuum is not a vacuum at all. The basic components of the space matrixes dimension per dimension are superstrings *D*, *T*, and *V*. So everything is explained; all phenomena in terms of *D*, *T*, and *V*. Particles are associated with different dimensions by spin and are the

Vesica Piscis

smaller percentage of space energy, the interference patterns of the matrix. Space itself is in ratio of total space mass to particulate mass. These ratios expand space because they are transdimensional inertias or entropies that swing from infinity to infinity through balance. So this universe exhibits perpetual motion existing in infinity, and the motion is toward infinity and back. This balance exists in multiple forms such as fractal arrangement, motion, proportion, and other inertial and entropic constructs.

Is dark matter in any way analogous to Wheeler's view of virtual particles erupting continuously out of quantum foam? There is no chaos, only structure. And there are no singularities, only structures. Nothing is virtual about the universe. If the universe is anything, it is fractal, and the structures become smaller via similar patterns and structures, but at some point, entropic inertia requires a relationship to balance, and balancing effect of this universe restricts movement to infinity. So these singularities never happen. Things happen according to structures. The concept of singularities and foam are due to lack of understanding about what we observe, and we probably observe three dimensions ($D^3$, $D^4$, and $D^5$) by our perceptions, and it's impossible with current technologies to either go up or down through dimensions since these dimensions are transparent to our perception. For example, gravity in this theory is $\boldsymbol{D^{10}}$ **= gravity =** $\boldsymbol{Emc^2}$, and so this ten-dimensional phenomenon is experienced as force where you are. So how then, using current scientific observation and models for physics, are you able to determine clearly what is beyond? You can't. The notion of singularity and foam is the dilemma of not being able to observe. There is hidden mass across multiple dimensions and are transparent to where we are.

In this theory, it is proposed that all matter is made from the matrix of space composed of superstrings *D*, *T*, and *V*. Particles are caused by certain interference patterns of space. Energy is redefined dimension per dimension, so *E* as it is known is but one dimension $D^5$. So energy is a dimension of space, or simply, it is dimension. Some dimensions are transparent to others, so it is impossible to perceive the mass. Particles are the smaller construct of mass in the matrix of space that also has mass. All volume has mass. Space can be quantized, but not in the manner presently understood. A photon is a monopole of space or a part of the matrix of space. Photons within the structure or geometry of space (tetrahedral) elastically collide. So the elastic collision does not describe the photon. Photon has volume and has mass. You see the photon on the end of an elastic collision. $\boldsymbol{E = mc^2}$ needs to be rewritten as $\boldsymbol{D^5 = MC^2}$, which means that energy comes in various types, and also in this dimension, it is a certain construct of space. The photon is only a percentage of the space matrix and, therefore, is not the total amount in *E*. The photon being a monopole of space is a small percentage of the superstring matrix having mass as well. To understand, energy (*E*) is the total of all particles and superstrings dimension per dimension. Dark matter then becomes that portion of the space matrix or energy that is not identified, and that portion includes space and both particles and space in other dimensions. This also includes what I termed earlier as transparent mass, that mass that is not detected from point of observation from the dimensions of perception.

All dimensional states have a resonant frequency and, therefore, interact differently with adjacent dimensions depending on this. The deep dark space has a resonant frequency that is natural for that state of naturally spinning superstrings that make both the matrix and containing photonic monopoles of the matrix geometry. So when the matrix collides via elastic collision or an elastic wave passes through, that portion of the matrix frequency is slightly altered, and those excited photons we see. They shift slightly from transparent photons to those temporarily seen and experience a slightly increased mass due to frequency and motion. That we see. If you could see space as it really is, it would be a sea of light, but your perception is limited to certain states of the space matrix. You can say that the matrix of energy dimension per dimension is part of a continuous geometry or form of mass and that the quantum characteristics are part of a resonant system of mass. But this is further expanded via multiple dimensions.

Now why would a shift in frequency shift the quantity of the mass? Because **mass =** $\boldsymbol{D^3T^2}$, where $D^3$ is volume and $\boldsymbol{T^2}$ = **density**. So a change in the frequency or motion alters the density of space. Through an elastic collision, at the center of the elastic wave space is of a higher density, and the shift in photonic mass is up.

The energy of a system is determined by *D*, *T*, and *V*. This would be affected both by motion within the system and also when acted upon externally. Hence, it is correct to say frequency affects the energy of a resonant system. The equivalency between superstring parts of a dimensional system, let's say $\boldsymbol{D^5 = MC^2}$, is the expression of the superstring structure of that dimension. As the values of *D*, *T*, and *V* increase in magnitude with a shift in frequency of that system, so does the mass. However, this is within limits of that dimension.

Changes in the frequency in motion do alter the density of space, hence the mass and volume of space.

**Black Holes**

Black holes in the theory are $D^{23}$. So a black hole is still a particular construct of space; there is no singularity there. In a black hole, gravity is squared or $E^4$. With this concentration of energy, there is also a balancing effect of the vacuum or, better said, matrix of space. Black holes are the generators of the space of the universe. I have entertained this notion with the construct of an atom and generation of space around the atom, and also for the Sun in our solar system. So even in an atom, there is a proportion of particle to space or, better put, energy mass space to energy particle mass space. The distinction is the particle or interference and what human perception sees. So what is beyond a black hole is infinity—the door to consciousness or conscious reality, God, creativity, eternity, whatever you want to call that. There is a boundary, what I call the hyperspace of the whole universe where there is no time and then where the emergent rules of our universe no longer apply. The best relation to this transcendent truth is your own mind or spiritual being. Also, take note that what reality is in another dimension is not what we observe where we are. So trans-dimensionally, there is a lot of heat moving to $D^3$ from $D^{23}$, but in a spaceship at that vibration or dimension of $D^{23}$, there may be no heat at all, and your relatives in $D^3$ are vanished from your sight. So at a higher dimension and in other vibration frequencies, you may be able to live on planets you would not be capable of in $D^3$. You may be able to sail into a black hole with no problems. If you could build a transdimensional spaceship that could operate within a higher dimension, all those effects of heat and distortion would not happen. They happen because from where we are, the transition from a higher to lower dimensional construct involves a lot of heat. So at a higher vibration frequency, you might be able to fly by the Sun or live closer to it without the effect of heat. I think if there was a craft that could travel into a black hole, I think it would be so sophisticated as to actually travel through time, and as such, we would also need to work out the relativistic effects of transdimensional travel; the effects on time and location or distance would be monumental. Assuming that at these higher dimensions that we eventually reach close to realms of conciseness, perhaps at the point, the spaceship itself becomes an extension of our consciousness. And so at some point, as an evolved spirit or consciousness, you might be able to do it with your mind alone. And it might be not far removed from when Buddha meditated and saw God. So having all these resources, I would have no problem traveling anywhere. But as it stands in current technology, I think not.

**Time and Density**

It is correct to say that frequency affects the energy of a resonant system. However, more accurately I would say the change in density because we need to acknowledge that space has the following:

1. Structure
2. Elastic medium
3. Superconductor
4. Photonic monopoles
5. Variations in density

And knowing that density varies due to elastic waves affects the mass or energy since they are partly interchangeable. **Mass** = $\boldsymbol{D^3T^2}$, so time is involved. The change in density also affects time.

There are two types of frequency to be observed.

- The first is internal to the system of the space matrix, which includes motion present within the matrix itself.
- The second is external to the system as an elastic wave passing through.

The frequency of an external system does not define the particle and mass but contributes to it. The frequency natural and internal to the system matrix does in fact define the mass.

The frequency acting upon the system externally is a measure of the elastic collision of the space matrix structure that includes energy superstrings and particle energy mass. Note that I used energy twice there to differentiate between particle and superstring. The frequency is a distortion in the space matrix fabric due to an elastic wave. There's also the frequency of the matrix itself without external disturbance. This has its own characteristics and could define a geometric arrangement and mass. The minimum number of dimensions is debatable to support a resonant system. Each dimension of the space matrix has a resonant system. $D^3$ has one system of geometric structure.

Transdimensional systems are contiguous, meaning joined in space and time. So there is mass of space, which is layered in multiple dimensions, and then there is atomic mass, which is multidimensional. These systems exist together; hence, with dark space, you also have gravitational space existing together in adjacent dimensions. If you want to focus on $D^3$, it is but one dimension of energy ($D^3$). Energy has been redefined dimension per dimension. **Energy = Space = Dimension. Mass = $D^3T^2$.** So space or energy has many definitions that include superstring and particles within the volume. Now with regard to $D^3$, the change in density of volumetric space can be defined in terms of the spatial movement of the matrix system (mass) along one axis. A higher spatial dimension can affect movement in a lower spatial dimension by one axis or vice versa. The example I use is that gravity is $D^{10}$ and is affected as force ($D^4$) as we see it. A particle is the result of movements in space of higher spatial dimension and reflects interference patterns. In some cases, it may be one axis of a higher dimension, and in other, more than one. This needs to be reviewed on a case-by-case basis.

Time does contribute to the energy of a system since a change in density affects a change in time. In **$E = Mc^2$**, let's rewrite **$E = D^3T^2C^2$**, so time is a component of energy and mass within the structure of space. An elastic collision will affect the magnitudes of these by change in density. Time is real to describe energy, mass, and resonant systems.

Time ($T$) is multidimensional, as is dimension ($D$) and motion ($V$). Time is legitimately balanced superstring in a tetrahedral relationship to $D$ and $V$. It is one of three components of the periodic transdimensional table that changes the nature of energy dimension per dimension. It expresses the nature of energy in a new way and thus explains why the change in the density of gravity affects time. It is the link that ties quantum mechanics to relativity. The idea of one unit of measurement is truly a myth created by a philosophy of scientific observation, which has its limits in a transdimensional reality. The exceptions in current theory cannot be resolved by limiting time to one dimension; just as gravity has a true nature of ten dimensions, explaining it as force is what we experience. But the real structure extends beyond, and we see the shadow.

The definitions of mass and energy in this theory include time. The idea of a particle in terms of a higher spatial dimension has its place, but time is also there in the formulas. Time is the reciprocal of velocity, and these exist within infinity in a balance. There is inertia toward balance.

The periodic table does not stop at 4, just as the periodic table in chemistry does not. The ladder here is multiplicative. **$D^5 = mc^2$** is already five dimensions, and we haven't even described gravity, magnetism, so on and so forth. The idea of space-time as four seems like the idea of fire, water, wood, and air. Things are much more diverse than what we see at first. We are just beginning. There are multiple dimensions, and the properties change vastly as additional superstrings are added dimension per dimension. All dimensions of energy have $D$, $T$, and $V$ and component superstrings of the space energy structure. This structure is symmetrical as implied by Noether's theorem.

## Particles and Periodicity of Dimensions

The theory also points to properties that repeat through a periodic table. The concept of density, gravity, force, volume, and others are a repeating characteristic quality through the dimensions. I have outlined some of these, such as induction and relationships of field and force through the dimensions. It may be that when associating particles with dimensions, you may have milestones like spin to identify particles with certain dimensions, and also the transdimensional relationship that results in a particle may in fact involve a minimum value such as three or four dimensions, but remember that transdimensional means through all dimensions, so any effect can eventually be described and connected to all other dimensions simultaneously. However, as in a periodic table, there are characteristics of periodicity, and these may have relationship to the creation of particles and described in minimums and or maximums as numbers.

The triad form refers to the nature of triad space and all things made from velocity, time and displacement; you only need three units. The commonality of all forces from diverse origins expressed in the numerator refers to the multidimensionality of operators. That which infinitely enables through the freedom of numbers is by nature love. Which means that the freedom of numbers is only really free through that which enables infinite, and hence love and the first cause are one and the same and all truth which proceeds from and to all possible realities can be comprehended through various and different matrices of creation. The universes imply unbounded aspect of creation and this dream is but one type of physics or emergent form. All maths is representational and represents functions of time and velocity related to consciousness through thought memory and will. Dimension is the resonant radius of action, which is also geometric, and also reflects possible degrees of action. In this theory you have three units, equivalency of all things in the most basic steps. As we thought mass to be mass, it's now broken down further into components of space and time squared. Look at the traditional mass in E, you can break down mass further. All properties are various degrees or products of time and velocity, almost like a periodic chart, time and velocity are complementary, inverse, and basically existing together. Everything is made from time and velocity. Time and velocity emerge from unity, basically made from the same substance but differentiated by symmetry, hierarchy, polarity, and other features; one not existing without the other. So in the triad system of "form" you have this tetrahedral hierarchy, below that I conjecture through the math is a binary code of 1 and 0 through which 1 is divided where space becomes infinite and this universe rests within a matrix of infinity. I doubt that we can experience that. We exist in perception in dimension 3 - 5 where energy is 5 and space 3. For the science of observation there is limit. The product of time and velocity are resonant effects. And time attributed to charge, and time and velocity attributed to spin. From the arrangements of these we get residual nodes in the matrix of the ether which are particles, which we observe. And so our science is based on what we see rather than that which is invisible being the primary cause.

The antigravity force is reverse polarity in the ether localized or as a field due to inverse resonance in the field. Space is a product of D which is also the radius of action of a resonant field; a field which has polarity. The inverse of the D field or 1/D is polarity for the inverse resonance that is required for an anti-gravitational effect. According to the Euler function where $E^{(ix)}$ as a graphic plot generates a resonant wave, the cross section of which can be a sine wave, the inverse function would be the cosecant. It is in the nature of the cosecant matrix of 1/D that antigravity field of opposite polarity will induce counter force to the polarity of the D field. D is the radius of action of resonant space. 1/D is reverse polarity or counter the natural polarity of the either of space. The antigravity force is reverse polarity in the ether localized or as a field due to inverse resonance in the field; it has to do with the generation of positive flow. Antigravity relates to cosecant and arcsine functions of other spatial resonant quadrants.

# Chapter 3

## Cause and Effect

## Hierarchy, Symmetry and Asymmetry

$$D^4 = F = \text{Volume1} \times \text{Volume2} / R^2$$

$$G = 1/T^4 = 1/\text{Density1} \times \text{Density2}$$

All constants with units are functions. Gravity is represented as a function of volume of the inertial system of localized space.

Truth is like harmonics, it should resonate to every heart that is in harmony with the matrix. Awareness of truth leads to wisdom, and by that you know how to live. Divergence with truth leads to pain and suffering, and when conflict is with self, the nature of being is evil. Life is self-transforming. All I know about truth is because I love. Truth is a standard unto itself, before myths, before religion, before men and incorrect axioms. What I discover about truth, it always was and is.

## Time and Velocity (*C*)

## Four Phases of Space Matrix Construct

## Hyperspace, Transdimensional Superstrings, Pythagorean Constructs, and Duality

In this theory we have described time as inversely proportional to velocity. Both time and velocity are superstrings of the matrix of space. We also have said this universe is a particular qualification of infinity, where concepts are related by infinity, balance, and unity, best described symbolically in the circle or sphere. So in $D = VT$, $V$ and $T$ are inversely proportional and related in infinity by the constant $C$. So to illustrate this relationship, the following equation is presented:

$2V - 1 / T = C$

$V$ represents velocity, and $1 / T$ expresses the inverse relationship of time, which is an inverse velocity. They appear as dualities, time balancing in an opposing motion or direction; in a sense, time is a velocity. The constant $C$ is the qualification of infinity bringing unity to this duality, and 1 is the balance. The unity expressed is symbolic of the motion of the universe. So an adjustment in velocity upward is an adjustment in time downward. We don't experience time in the sense of these equations since our perception is different; however, we do note that time exists. Time is also real. Time and velocity are related in infinity as a pair; both are transformative in relation to $D$. We can take this model of polarity and use it to discover other polarities or polar superstring energies trans-dimensionally.

**$V - D / T = 0$; in the hyperspace of calculating $C$, we drop off $D$ and so the following:**

**$2V - 1 / T = C$, where $C$ becomes a point of unity like the number 0**

$2V = C + 1 / T$

$T = 1 / 2V - C$

**Summary Equations**

***Phi + Phibat = Velocity + Time = 1 = D***

***Phi × Phibat = Velocity × Time = -1 = -D***

$VD = V$

$V(V + T) = V$

$V(V + T) - V = V - V = 0$

$V([V + T] - 1) = C - C = 0$

$(V + T) - 1 + C = C$

$V + T - 1 + V = C$

$2V + T - (V + T) = C$

$V = C$

$2V + 1 / V - (1 / T + T) = C$

$2V + T - (1 / T + T) = C$

$2V - 1 / T = C$

$V - 1 / T = 0$

$V + T = 1$

$C^2 = 4V^2 - 4V / T + 1 / T^2$

$0 = 2V^2 - 2V / T - 1 / T^2$

**There is also a condition of $2V - 1 / T = C$, which allows for the compression and dilation of space. Time and velocity form a polar relationship within infinity like a pendulum within a circle. It can be described in the hyperspace of the circle and also the area within the circle. A sphere as an analogy would also work.**
So $C^2$ becomes

$C^2 = (2V - 1 / T) \times - (-2V + 1 / T) = C^2$

$C^2 = 4V^2 - 4V / T + 1 / T^2$ **in the hyperspace of energy** $C^2 = V^2$

$V^2 = C^2$**, and we can go another direction toward an equation already known in current physics:**

$V^2 = C^2$

$1 - V^2 = 1 - C^2$

$(1 - V^2) / C^2 = (1 - C^2) / C^2$

$(1 - V^2) / C^2 = (1 - V^2) / V^2$

$(1 - V^2) / C^2 = 1 / V^2 - 1 / 1$

$(1 - V^2) / C^2 = T^2 - 1$

$(1 - V^2) / C^2 + 1 = T^2$

$T = SQRT\ (1 - V^2 / C^2) + 1$

**$1 = 1 - V^2 / C^2$, which can get you to $T'\ SQRT\ (1 - V^2 / C^2) = T$ where $T'$ is the observer and $T$ is the object, and in hyperspace, we have balance of polarities time and velocity.**

**"Within" hyperspace the function becomes**

$C^2 = 4V^2 - 4V / T + 1 / T^2$ **and an expression of a Pythagorean relationship as**

$4V^2 - C^2 = + 4V / T - 1 / T^2$

$4V^2 - C^2 = + 4\text{Acceleration} - 1 / T^2$

**So we see a relationship between velocity, acceleration, and time "within" hyperspace in a Pythagorean relationship. This area is the one of our perception.**

Now let's take this equation and redefine $E = MC^2 = D^5 = MV^2 = (D^3T^2)\ V^2$

$D^5 = E = MC^2$**; mass =** $(D^3T^2)$**, and so**

**Since all transdimensional relationship within hyperspace involving addition are established through infinity, balance (1), and unity (0), the following:**

$C^2 = 4(D^3T^2)\ V^2 - 4(D^3T^2)\ V / T + (D^3T^2) / T^2$

$C^2 = 4\underline{D^5} - 4\underline{D^4} + \underline{D^3}$**, hence by using 0**

$0 = 2D^5 - 2D^4 + D^3$

$2D^5 = + 2D^4 - D^3$

$2D^5 - 2D^4 = - D^3$

**And so we can find in the Pythagorean relationship that $D^5$ or energy as we understand it is a relationship of force and volume.**

The negative force can be identified as the Casimir effect acting within volume.

$2D^5 = + 2D^4 - D^3$

**2*E* = + 2Force - Volume**

**Which the force being in two parts can be expressed from two force to force**

**$0 = 2D^5 - 2D^4 + D^3$ = energy - force + volume, where 0 represents balance of all dimensions. Here you see energy, force, and volume where all three are in equilibrium.**

Dimensions $D^3$, $D^4$, and $D^5$ represent the dimensional densities that we live in. Sure enough, this is represented in the Pythagorean Theorem. It is possible to think that other alien beings may also inhabit other higher-dimensional densities.

We have said before that multiplication represents the periodic transdimensional table of the space matrix dimension per dimensions. Here in these equations, we have established the relationship of polarity, unity, balance, and infinity, or more specifically, the relationship to infinity, unity, and balance. These terms now have meaning in math. We know that by multiplying $VT$ to $D^3$, we get to $D^4$ in the definition of energy as well of the structure or geometry of the space matrix for that dimension $D^4$ and similarly for others. When we relate $D^3$ to $D^4$ through a Pythagorean relationship, we get $D^5$.

The circle or sphere is the perfect model or analogy for illustrating the space matrix and relation of energy, entropy, infinity, balance, unity, and perpetual motion of the universe. The superstring shell or sphere is the superstring matrix whose properties are the periodic chart of a transdimensional system. This matrix has natural motion and energy or zero-point energy. This motion is also tied to density and forces such as Casimir that hold it together. Hence, there are ratios that are entropic and disentropic that keeps the relationship of various multidimensions in proportion through their adjacencies. For example, there is always an extension of empty space associated with the higher dimensions of atomic mass. Hence, around something like the center of the galaxy is the generation of a proportional amount of empty space. So there are forces that allow expansion and those that allow contraction. And with the matrix itself, there are forces that maintain its structure that include vibratory, rotational, and motion. A superstring is a sphere that has perpetual motion originating from the big bang, zero-point energy. The hyperspace of a dimensions would be given, for example, as in an equation such as $E = MC^2$, and the area just below, which is characteristic of the dimension of space, would be $E = MV^2$. And at the area or volume of this structure that provides a relationship of unifying balance such as $E = + F -$ **volume,** which represent the Pythagorean relation of **forces binding** the space matrix. And at center, finally we can explore the dualities of energy that exist within infinity such as time and velocity. This is all supported by infinite source black hole, supporting fractal or scalar infinities, and transdimensional infinites. So let us define five zones dimension per dimension within the transdimensional framework that is

- hyperspace defined by $C$ such as in $D^5 = MC^2$;
- transdimensional matrix below hyperspace defined by superstring space matrix relationships and superstring equations such as $D^5 = MV^2$;
- the Pythagorean relationship is the **binding** structure of space and tetrahedral framework presented here in $D^3$, $D^4$, and $D^5$;
- the dualities of energy relationships such as velocity and time, the binding force defined by balance via $C$; and the infinite source black hole.

Hence, our perception of the universe is as in a Pythagorean relationship of the 3, 4, 5 triangle or $D^3$, $D^4$, and $D^5$ matrix densities, tetrahedral qualification of infinities. These are the densities of our perception. Around that we have polarities in a structure of balance, hierarchy in the tetrahedral form, the periodic transdimensional relationship, and the exploration of hyperspaces with the constant $C$. As an analogy, we are essentially a pendulum swinging in infinity; infinite consciousness is qualified in various densities in a multidimensional framework.

## Reality Implied by Science

Every revolution in science creates new problems, but in a sense, everything is already known through conscious intuition because it already exists in the consciousness of the universe or God. There is life all over the universe; who can deny the multiplicity and divergences of creation? Life in this theory is transcendent, meaning of spiritual origin or spiritual consciousness. Life is transcendent to this universe. Your spirit perceives intuitively and your body through the senses, hence, transcendence and emergence respectively. Consciousness transcends this universe, and this universe is but "one" qualification of consciousness or infinity. God's consciousness is qualified in this creation and transcends, and in the same manner, so do we representing our own bodies and lives. There is not emergent behavior that goes from inanimate matter to the beginning of life in this emergent reality in the universe. There is an absolute infinity, and it's qualified in various creations, and this is but one. That infinity is consciousness, God, which includes your consciousness or spirit.

There are multiple dimensions here in this theory, so there could be aliens living on a higher dimensional frequency we never see, although it exists, and God willed it. If you want to know about the Creator, this universe is an analogy of God, and that analogy represents truth. God is imagining, like when you do, except his is creation. We exist in God's imagination with free will. There are three superstrings *D*, *T*, and *V* that make everything so by analogy. I imagine this is close to the three essences of God and resembles the symbolism of the trinity. God is also infinity, much like in the Hindu *Bhagavad Gita*. Think of a pendulum on a clock: one way is time, the other velocity inside infinity; it swings toward infinity and is pulled back. This universe is a clock in infinity, you are in infinity, your mind or spirit is already looking from or the whole of reality of consciousness. So consciousness has different levels and forms. The space matrix, which is a form of consciousness, has structure; reality is consciousness. The infinite mind and spirit is intuitive, but your body has five senses. By intuition, you have access to everything and all knowledge; by perception, consciousness or infinity is qualified, and so your perceptive reasoning needs intuition to get to the real truth. Think of it like this. Let's say you are a being in heaven with fifty senses, and then someone said you are going to sleep and wake up in another form of consciousness, and when you wake up, instead of fifty senses, you have five to work with, but you still have the fifty, but you are going to have trouble because you will put more importance on the five. Human nature is limiting in that way.

Emergence is a rule within the universe after the big bang; you are within and without, you are emergent with your body and transcendent with your spirit.

**So this spirit thing, it needs some kind of way of transcending between dimensions?**

It exists, yes; it uses the body to transcend.

**So it has to have some kind of vessel in the other dimension too?**

Yes, whatever consciousness is has a vessel, but not in the manner we think of vessel in an emergent reality. The idea of beginning and end is entirely emergent or human.

So space has structure, say tetrahedral; imagine a bowl of marbles. You don't see these marbles since they are invisible to your perception. Then the Sun hits them, a wave goes through them, excites them to a higher frequency, and you see them. So you say I saw light. This could be a holographic universe, but the whole of dark space is all light. It's just that your eyes are designed to see it only in its excited state. So the geometry is tetrahedral and looks like six billiards around one if planar, and more when spherical as it truly is.

**Why the need to transcend the universe?**

The universe was created as a place for the infinite consciousness to experience and understand God. The universe is about God's nature or truth by analogy. Why God has a certain nature is the mystery. The awareness of God's truths and tenants is the source of wisdom, and by that you know how to live, merged into the infinite matrix of harmonic consciousness. But we can observe it and, by experience, discover its truth, meaning about God. And all the virtues that are derived through this experience tell us about order, truth, and love of God. For all the analysis, the most infinite is love, creative and evolving.

**Do I still have free choice in this universe?**

God is hierarchy and equality existing together, and in that, diversity and multiplicity are enabled. All things were designed in harmony of motion; call it love if you wish.

**But it's confusing. Haven't you effectively created a finite universe that contains infinite universes?**

This universe I call it a qualified universe, within infinity. It's a certain type of infinity, a transition, so even though it seems finite, it's not really. Think of your own mind as infinite. Best example is when you dream, anything goes. Now in reality, that's God dreaming, but it's real for you, and within that, there is free will that is allowed—the one big anomaly of emergent reality and representing transcendent reality.

## Constant Pi and Phi and the Energy of a Superstring

So thus far, we have defined superstrings as a spherical components combined in various tetrahedral geometries dimension per dimension in this Trans Dimensional Unified Field Theory. These dimensions are $D^n$ and represent various dimensions of space energy. These dimensions are not necessarily scalar but represent change in space matrix geometric construction, density, and compression (scalable), superstring count, resonance, and more. So a dimension has a difference that goes far beyond merely to say scale, even though it has a fractal nature within a dimension and between dimensions. The geometries become complex. And so the basic building block of the space matrix geometry is the sphere, and the emergent geometries are spherical in every dimension. We know from geometry that the area and volume of a sphere is as follows:

**Area of sphere = Pi $R^2$**

**Volume of sphere = 4/3 Pi $R^3$**

So now let's explore the area of a sphere applied to the construct of this unified field theory as we did for the Pythagorean Theorem. Let's reinterpret the meaning for the volume of the sphere in this theory. All mathematical equations and their constants become representations of physical truths about the cosmology of the universe.

**Volume of a sphere = 4/3 Pi $R^3$**

Similarly,

$D^3 = V^3 T^3$; hyperspace solution of that is $V^3 = C^3$, so

$D^3 = T^3 C^3$

So now we take the volume of sphere equation for superstring time ($T$) and substitute

Resonance

**$D^3$ volume of sphere = $T^3$ $C^3$; $C^3$ = 4/3 Pi**

**$D^3$ volume of a sphere = 4/3 Pi $T^3$**

**In this equation, the superstring is time (*T*)**

**Volume of superstring *T* at $D^3$ = 4/3 Pi $T^3$**

So let us revaluate constant *C* in terms of geometric constant Pi and Phi (**Pi = 3.142** and **Phi = 1.618**). Both constants naturally occurring in nature, they represent the ratios of energy between superstrings that make up the space matrix.

**$C^3$ (constant) = 4/3 Pi $r^3$ = $Phi^3$; $D^3$ = $T^3$ $Phi^3$**. From this we get a sequence through dimensions as follows:

- **$C^1$ = $Phi^1$; $D^1$ = $T^1 Phi^1$**
- **$C^2$ = $Phi^2$; $D^2$ = $T^2 Phi^2$**
- **$C^3$ = $Phi^3$; $D^3$ = $T^3 Phi^3$**
- **$C^4$ = $Phi^4$; $D^4$ = $T^4 Phi^4$**
- **$C^5$ = $Phi^5$; $D^5$ = $T^5 Phi^5$, etc.**

Now the true expression that connects phi and pi for the example above is an ellipsoid that includes a circle.
**Volume of an ellipsoid = 4/3 pi *abc* = $phi^3$ for a circle = 4/3 pi $r^3$; *a* = *b* = *c*** or spherical coordinates for a circle.

- ***a = b = c: sphere***
- ***a = b > c: oblate spheroid (disk shaped)***
- ***a = b < c: prolate spheroid (like a* football)**
- ***a > b > c: scalene ellipsoid (three unequal sides)***

**Phi = 1 - 2 cos (3 Pi / 5)**

**6/5 × $Phi^2$ = 3.1416**

Here we have shown how proportion defines the energy of superstrings dimension per dimension and how the value *C* becomes phi in defining the hyperspace of a superstring or the total energy of a superstring. So then the energy of superstring $T^3$ in $D^3$ is

**$D^3$ = $T^3$ $C^3$; which is also related to $D^3$ = $T^3$ $Phi^3$**

What this means is that the magnitude of energy between the superstring dimension per dimension is ordered according to the golden ratio or phi and varies by power of the dimension in each dimension. This is amazing and explains why the golden ratio is so common in nature. It also refers to the value of *C* of the entire superstring matrix and also varies according to the power of the dimension in each dimension, which means that the relative value of *C* to an adjacent dimension varies by a value of $C^2$. Hence, a spaceship travelling through a four-dimensional space in hyper-dimensional travel will be at $C^2$ in the fourth dimension relative to this third dimension.

**The relative value between adjacent dimensions is $C^2$**

Because of the unit phi and pi, the space matrix is a symmetrical composition that incorporates asymmetries. Hierarchy and symmetry exist together. *C* is a limiting proportion of the entire space matrix, and the constants pi and phi are superstring proportions of components within the space matrix. So energy dimension per dimension has overall constants particular to transdimensional space and subspace constants particular to transdimensional subspace. Therefore, the

# Phi: The Golden Mean

constants related to spheres and those related to natural proportion actually reflect the proportions of energies of space distributions by geometry in space and subspace. These proportions tell us that space is an arrangement of energy superstrings in geometric composition based on a sphere. Noether's theorem suggests a symmetrical space matrix, but phi suggests that the matrix also has internal superstring asymmetries that are symmetrically balanced.

**Phi Equation**

**Phi = *a*/*b* = *a* + *b*/*a*; 1/phi = phihat; 1/phihat = phi; phi$^2$ = phi/phihat**

Now let's convert phi into a superstring expression:

$V = D / T$

$V = D / T = D + T / D$ = phi = 1.618

$T = D / V = DV - D^2$ = phihat = 0.618

$D^2 - DV + T = 0$

$D^2 - (a + b) D + ab = 0$

$1 - 1.618 + 0.618 = 0$

$X^2 - X + 1 = 0$ **if** $a = 1$ **and** $b$ = **phihat**

**If** $a$ = **phi and** $b$ = **phihat, then the following is the result, which will be reviewed later:**

$X^2 - X - 1 = 0$

Here we have the expression of time as an anti-velocity. This expression can also be used to identify equations for antigravity. Anti means the inverse of the standard model, and hence, reversal of polarity. The equation expressing phi also expresses the distribution of energy with the subspace matrix.

We can continue exploring all mathematical equations related to geometries and constants to illustrate the myriad of subspace energy relationships that superstrings are capable of manifesting. These have applications trans-dimensionally. What makes this possible is that common denominators superstrings *D*, *T*, and *V* have been established. And these have various relationships within the universe trans-dimensionally. The anti of a relationship expresses the reversal of polarity of a natural system. Hence, mathematics that expresses the reverse in polarity for gravity would be an understanding of antigravity.

**The Physical Nature of Superstrings**

Light does not travel at the speed of light; the elastic collision of the matrix of space does. The vacuum is geometry of superstrings in *D*, *T*, and *V*, so the experience of the qualities of space has nothing to do with the elastic vibration of space in a direct way. The elastic vibration of the space through a superstring matrix from an external source is one thing, and the experience of the qualities and properties in space internally are another. The elastic vibration of space from an external source occurs at the speed of light, and it is also a photonic superconductor, meaning that the monopoles of this matrix are photons. The photons are the shadows or points in the matrix. The larger reality is the matrix. The photon we see is excited to a higher vibration, and we see that as a wave crashing on the beach. As a person, you could never be within the matrix at the speed of light because you would bounce off like a speedboat over the surface of the water into a higher dimension, and then be relatively to this dimension at $C^2$. **Mass =** $D^3 T^2$ means all volume has mass. Photons have volume and have mass. There are no photons at the speed of light; there is no vacuum, just as there are no electrons at the speed of light in current. It is conceivable that even sound can travel at the speed of light if its matrix is elastic. Even high temperature superconductors are possible, by this understanding.

In this theory, the matrix of space is composed of a geometric array of superstrings. So the notion that there is a multitude of different shapes that alters in motion is incorrect. Their vibration patterns remain intact due to that fact that they are a part of a geometric array. The basic form is a sphere and is altered by the warping of space or matrix. So the types of forms within the matrix may distort from a sphere to an ellipse and possibly some extreme extensions in atypical situations. So let's answer or address the issue of vibration pattern: the pattern remains near the same in a natural state and when in motion. Excited states include the following:

- **The bending of space.**
- **The compression of space.**
- **Exciting space via an elastic wave passing through external to the system.**
- **Relationship with adjacent spaces and dimensions.**
- **Internal distortions.**
- **Transdimensional effects.**

In this theory, energy is redefined, so the matrix of outer space is different than, let's say, the matrix in a copper wire. These same strings are forming different geometries or arrays for each dimension of space, which is also a dimension of energy. Energy as we know is but one dimension, but there are many. So the superstrings have different geometric configurations dimension per dimension. This is the manner superstring is presented here. It is the subspace or subcomponent of space. Particles are the interference patterns or nodes of space configuration. So they are the smaller part of space that we see, the visible part, so when you see a chair, you are looking at the shadow of space. Hence, objects retain their form well.

## Consciousness and Infinity

Consciousness is a nonphysical and infinite energy. In this theory, infinite energy is qualified in this universe or to make this universe, its laws, and emergent reality. So consciousness is transcendent, meaning an infinite source or God from which this universe is dependent. A similar relationship would be our physical selves to our own consciousness. And thus, the universe is an extension of consciousness. The universe is an extension of God's conscious imagination. Consciousness and the universe are from the same source, connected and also separate or distinct. The theory describes the relationship of emergent laws to infinity, but the mathematics of infinity is out of reach since this presentation of reality is but one aspect of that infinity. We can, however, describe some of the nature of that infinite truth by analogy since this universe is an analogy of the truth of infinity or God. But since infinity is the source, we may have greater control on this universe than we realize through the conscious mind; and what is allowed and not depends on the harmonic union to this infinity and its precepts, which now leads us into a spiritual discussion. And between the two, infinity consciousness and this universe is the mind. And it may, in fact, have a scientific bridge between infinity and reality as we experience scientifically. This theory describes energy as existing within infinity, and that infinity is consciousness. This theory attempts to present the math of this universe consciousness, how that is mathematically connected to infinity or consciousness, but not describing infinity, although describing infinity sounds like an interesting proposition. And perhaps once we understand this universe well, that would be the future of physics.

Distance, time, and velocity may, in fact by analogy, represent three infinite essences such as awareness, thought, and will respectively. And these together represent love, and within them, hierarchy and equality. How one gets from infinity to the universe is in the will of creation that established order or what I call infinity, unity, and balance in association. The approach to the solution of making this a science has its resource in intuition, your infinite consciousness.

## Traditional Superstring Theory and Trans Dimensional Unified Field Theory

Each dimension of space energy has an associated array of strings and resonance. So we can associate certain physical properties with the array, such as force or gravity. The problem with string theory is that it has no laws and equations such as this theory proposes. So for all its use, it's pure conjecture. It keeps alive the argument for a unified field, but nothing more. Atoms are a multidimensional construction of space. And since it is multidimensional involving the matrix of space superstrings of various dimensions, the supercollider will never get to the bottom of the truth of it. You are essential, a three-dimensional being shooting at a multidimensional object and trying to explain the model of the atom as three dimensional when it isn't. The attraction of all particles is due to the attraction of subspace matrix array. Particles are attached to the matrix, and the matrix is held together by subspace forces of their natural arrangements. No photons are exchanged. Don't confuse elastic collision of space with the internal motion of space components. Electromagnetic wave has nothing to do with internal wave motion of the space matrix. These are two events, separate, related only by temporary phenomenon.

We are not proposing conventional string theory here. Here the idea is that the subcomponent of all energy is space and that space has structure and components that lend to the physical properties that we see. We call those subcomponents of space here superstrings. The ideas presented here are all original, so there is no other source. The components are superstrings in the sense that they exist at the Planck length and are basic components of all space and matter. These components are also presented by physical properties that string theory has avoided; hence, string theory presents no laws, but here laws are presented. Laws are presented trans-dimensionally.

Strings as defined here are made from three primary energies (distance, time, and velocity) and everything—every physics law, every part of space, and every particle—is constructed from them trans-dimensionally. All physics and phenomena can be understood from these three strings. These also form a tetrad, which is the basis for all tetrahedral structure of space and also seen in nature, the significance of both pi and phi (the golden) ratio in nature, and relate to create all physical phenomena. All energy in the subspace is organized according to the proportion phi. Each dimension of energy is related in a periodic table by adding $D$, $T$, and $V$.

There is no basic relationship between an electromagnetic wave, which is an elastic collision, and the rotational motion and inner forces holding the space matrix together. By that I mean they are totally different constructs. This misunderstanding is also the misunderstanding that separates particle and wave understanding of photons. Photons are fixed and are not travelling at the speed of light. Geometry resolves the dilemma of the double-slit experiment.

After all its years, string theory has very few proponents and has yet to become physics in the true sense of proposing physics laws. All of nature has structure, including space. There are no random personalities that move and change as strings are proposed traditionally. As Einstein said, "God does not play dice with the universe." Space has order and structure.

Let's try a simple analogy. Imagine if space was the lake or ocean. When you drop two rocks in, you get waves; and where they cross, at the points of interference a mark, this would be a particle. Two rocks are not two strings. I am showing only that string wave interference results in particles. How many strings are needed to create particles depends on dimension of space or energy and the particle we are looking at. In this model, I propose twenty-seven strings for one photon, and the density and geometries rise as we go into an atom. Atoms are multidimensional. So you do not see the ocean, which is space; only where there is interference pattern is there a particle, which appears visible. Now imagine an atom and you are the electrons, and far away is the proton, about the distance of a soccer field, so the ratio of space to particle is large. You see the particle but not the space creating it. Now imagine space as a bowl of marbles, and where each marble touches is a photon. We do not see the marbles of space or the photons. This space is elastic and a geometric array. It means if you send a wave through it, there is conservation of energy. So I send a wave through, and that is an electromagnetic wave. When the wave is at a crest, the matrix is compressed, and the photon becomes visible. The photon and space have its natural resonance, much like anything, and then what happens when acted upon by an outside source such as a wave, the natural resonance is slightly adjusted. If space has a structure, that means the vibration patterns behave in a predictable manner; the vibration patterns do not change but may vary within limits depending on outside influence.

The particle is caused by string interference. Strings are a part of a larger string matrix or space. The vibration of string is not enough to cause interference; they have a natural resonance, but the spin within a matrix or array does create interference. There is no such thing as a curled-up dimension. Dimensions are determined by the arrangement of strings. It takes twenty-seven strings to give rise to one photon in this theory. The problem with string theory is that all the equations do not represent physical interactions. None of the math is connected to "physics" or the description of physical phenomenon, such as mass, distance, time, velocity, etc. And then you are assuming the ether or space has no substance, and in this theory it does; space is in fact the substance that creates particles. There are no random string particles. The current physics model is backward, proposing that particles create mass or strings create particles. Stings create space, and the interaction of the strings in that space creates particles and are mass. A dimension of space is a dimension of energy and represents a configuration of strings. An atom is a multidimensional construct of rotating space and is more complex than to say an electron is an independent random entity orbiting a nucleus. Photograph an atom and look at the negative, and you will see a complex geometry of static and dynamic strings with spin. One string to create one particle assumes that the universe is particle interaction and space has no structure or space is a vacuum. String theory becomes, in a sense, a reinterpretation of quantum mechanics at a smaller level although still keeping unification idea alive. A part of the particle zoo. The reality is that the universe is multidimensional, and so what you see as a particle is a manifestation of processes induced in higher dimensional configuration. We can walk through space, a certain dimension of energy, but we cannot walk into an atom and observe and take notes. What we see in reality is not the reality or true nature of an atom. You cannot reduce a higher-dimensional physics to our present dimensional units. This theory presents a multidimensional framework to understand all physical phenomenons. Gravity as a ten-dimensional phenomenon appears as a force in this theory as we perceive it. The distortion of physics is that we perceive in dimension 3, 4, and 5 and base our entire math on scientific observation and believe that is all there is, when it's less than 1 percent. We are dimensionally limited beings tying to assess a multidimensional structure.

The basic idea of string theory is correct, that there are basic constituents of reality at the plank length that unify all physical phenomenons, and there is no singularity. But they do not create the fundamental forces singularly. They are space or energy and combine in patterns that produce more complex space or energy. The particle is a sideshow. And in this theory, the strings have been given physical attributes of distance, time, and velocity, and their combination produce all the phenomenon that we know. A string in this theory is a single-wave sphere. Its relationship to an adjacent sphere is perhaps one particle. It is not producing infinite particles.

Let's not confuse the internal motion of the string matrix or space with elastic waves passing through space or electromagnetic wave. There are two different events. The electromagnetic wave and interferences are not creating particles since it is an elastic vibration.

## Density of Space and Planetary Orbits, Accretion of Mass and Expanding Earth, Toroid and Vortex Geometries

As has been discussed in this theory, there is a ratio of the space matrix to mass and gravity generated around objects. This is a ratio that also contributes to the expansion and contraction of space or dark mass-energy in the universe. The gravity of such space can be described according to varying density from the source. Generally, by $T^2$ = **density**, the density of gravity space is affecting time. The planetary source, star source, or object source can be dynamic, such as the Sun or the core of the Earth, and it can be static, such as a solid body in space. Around a dynamic body, space is organized into areas of standing density like standing waves, since the source both generates and transfers pulses to the space that it supports. As a planet evolves, the extension of space is at a maximum at its birth and extends to its farthermost, and as it evolves and energy is translated to surface mass and possibly a larger volume of the object, a portion of the space is retracted. Also noted would be a lesser gravity. There may be some merit to the expanding Earth theory where the enlargement of the surface area also involves some portion of mass accretion from the ether or space matrix. The pulsing of space may create higher density toroid forms orbiting the planet or Sun star that may be more natural for planets to orbit; we have said that areas of higher density along the ecliptic are ideal for planetary orbits. Also, the spin or the planetary-generating body also creates vortexes at the poles. When dealing with the vast gravity of a black hole at gravity$^2$ and also the patterns of space across long distances, the bands of density form helical patterns, resulting spiral galaxies.

## Science of Consciousness and Free Will

There are no emergent balancing equations found in nature to control free will. The process of creation or destruction of the human mind does not follow patterns of entropy, disentropy, or balance that we see in nature. You are allowed to act in this universe without equation; however, in the spiritual sense of consciousness, acting freely in relationship to the universe, although allowed, involves pain since the infinite harmony is pulling all existence back to balance. So although the spiritual existence is free of this created existence, the moral truths remind us those extremes from the balancing harmony of the creator result in pain, in much the same way as balancing constants in the equations of the physical world call us back to harmony. So although we are free in this universe, the analogies suggested by the rules of things in this universe refer to transcendent truths as well. We see a beautiful piece of art with its hierarchy, equality, relationship, and balance and see a reflection of the truth of the creator and creation, and also of the precepts of our spiritual origin or transcendent consciousness. God preserves the balance and also allows free will, much like time and velocity, balancing energies, allowed to move in extremes, existing within infinity but never touching infinity; like a pendulum swinging back and forth in a clock, so too are our lives in the embrace of the Creator. We are free to explore all the extremes of existence that is possible, but to be free to exit from the matrix of reality and its harmonies, no. All that is really is, and we are a part of all. I am that I am. And motion is allowed, like motion of the universe, but are constantly referred to center, and this is no more a richer experience when we ourselves are transcend experiences into another reality of truth and learn the many riches of diversity and infinity that exist.

So we act freely, and it may seem in this balancing back to harmony that freedom is conditioned, but yet in our consciousness, there still exists even greater freedom in all this motion, a subset higher where free will really exists, and that is in the intent of our consciousness. To say thank you and to have mercy reflect an action that is truly free of any preservation of the matrix. This intent, as is the imagination of our consciousness, is akin to the creation of the universe that God made, with no limits, and hence, in our intent and intuition is free will to capture or release love as the Creator imagines the world to be, and this extends to our actions with consequences. And in this, all knowledge is available since it has the same source.

From the theory, it is evident that transdimensional relationships are established through induction. One set of precepts creates the existence of another and that moves on to the next. One has its existence in relationships to the other. In a sense, this can be described mathematically, even as we see in electromagnetism and other phenomena. This becomes an analogy for both consciousness and the creation of the universe. Mathematical induction is representative of the creation of one set in relation and from another implied by its boundaries. Therefore, when you have a thought, you create an existence, and that is induced from your infinite consciousness; in much the same way, it becomes an analogy of the creation of the universe. Hence, all reality is induced from higher source and represented by a higher math, and then it's a math that allows induction onto another state of being. I ask how is gravity a force and then describe how from $D^{10}$, gravity is induced as a force $D^4$ where we exist in our perception. How does one establish $D^{10}$? By intuitively establishing a framework and logically explaining the steps that get you through all the levels, and in relationships to consciousness, perhaps there are multiple or infinite levels also, and all the levels together are "consciousness" and what we see is just a manner of manifestation of that infinite construct. So I have an idea in my mind. I have induced a reality, by my free will, so my options are multiple and not seen as one example, as in this creation. But by that there must be many states of being or creations. But to keep the conversation simple, I have induced an experience. But here, if there is a creator, what is induced is shared, where multiple free wills collide. And the outcome of free will is an unpredictable future. There are options, and wisdom has value, wisdom being the awareness of truth and knowing how to live. So then when the will of God or of that which induced this universe and the will of your consciousness are in harmony, there is active will into this dream joined, such as miracles and faith would show.

We have described the universe as emergent, meaning processes and procedures of nature originate naturally and are observable, so what are the exceptions? Exceptions would include any external interference or transcendence. Those things related to free will and human consciousness are definitely not emergent since we have stated that life is not emergent. Free will is an anomaly in the universe machine animated by transcendent activity. We are visitors in this place, as is all life, to play out an experience and discover truths about creation and the Creator. The infinite has represented the nature of its essence and qualified a portion of infinity in this reality dream; infinity is consciousness as our own consciousness is infinite from the same source. Love is extended in diversity and discovers and is discovered. Love loves and wants to be loved. Hierarchy and equality exist together and manifest in a hierarchical diversity.

Consciousness is transcendent infinite. The whole of math is representative of all the variables of consciousness; math is the qualification of consciousness representing this reality. Science proves the existence of God. Atheists exist because the science is so incomplete, and we follow those quantum physicists like they were druids. Creation does not exist and/or for itself. Just like your own thoughts, dependence is a rule. You create in your mind. And God does it for real; you are a mirage of that with one anomaly: free will.

The math of consciousness exists outside this emergent world. Many think consciousness is an emergent truth, like other world mechanisms. I don't think so. Math explains this world; consciousness is transcendent, so assuming present mathematics can explain it, something totally different might. If life is emergent, then there is a math for it. Let's take exception to this idea since there are no emergent processes that go from inanimate matter to the beginning of life. Emergent means natural to physical world processes. If your math cannot predict that, it cannot predict consciousness, since life and consciousness are transcendent. You are trying to explain consciousness while your math has not yet even explained the universe and would need to go much further. Think of this as a hypothesis. If reality exists in the imagination of God, so how does that imagination of this creation differ from his total consciousness of what absolute infinity is really? God is immense for sure. I would ascribe infinity to consciousness and a thought to a qualification of it. God creates from an infinite consciousness; in that is an order, a math. Infinity suggests vastness beyond a formal set of equations akin to consciousness, and because we can't grasp either infinity or consciousness, they must be one and the same. The equations and formal sets are unlimited, and this is free will. Awareness (self and beyond), thought, and will are free and is consciousness.

Everything needs a set of rules to "be," and by analogy, mathematical induction is the key.

Metaphysics and science are closely related. Science suggests a discourse whereby the reality of God is said to exist. Metaphysics is a study of being and the world and includes cosmology. Science began with philosophy and metaphysics. At first, Einstein did not believe in a God, but through his theories and scientific investigation, he realized there was a beginning to the universe. He tried to get rid of the beginning with his cosmological constant and later said it was the worst mistake of his life. And so then he believed in God through science. Our science is primitive, but as it becomes more accurate, the same sort of conclusion shall be reached in science via multiple paths that God exists.

> "In the view of such harmony in the cosmos which I, with my limited human mind, am able to recognize, there are yet people who say there is no God. But what makes me really angry is that they quote me for support for such views". (Einstein)

In this theory, we talk about a matrix of space that exists trans-dimensionally. Space represents energy in various dimensional states. The energy approaches toward infinity at ascending levels of vibration or dimension. We suggest here that the infinity is akin to consciousness or God, and that in this reality that infinite energy is represented in a way of this universe, or as I said, "qualified." So can one ascend to higher levels of consciousness as a scientific exploration? Yes, because science is also an objective exploration of shared consciousness and represented in math. The math of emergent and transcendent processes is different, but since they represent consciousness, they do exist. Hence, it is conceivable to know the source, which we can refer to as God, that infinite consciousness, which is the source of all things. Dependency is a rule, and one state of consciousness or level of energy is related to the next, similar to adjacencies and induction. The idea of the beginning is an emergent truth of this reality, and for God and infinity, it always was without beginning. The nature of this world is an analogy of truth, and path to the truth of the nature of existence and of the source is possible in a myriad of ways. Levels of energy are akin to levels of consciousness; we see a set of rules in all things, and hence, a set of rules needs to be present for something to "be," so on and so forth, such that science is constantly reminding us by analogy of the truths of existence and paths of knowledge to the source. And the infinite consciousness that is God transcends the entire orders and creations that are, including all time, motion, and construction.

What happens when one has an idea, there was another idea before. Awareness, thought, and will are free and is consciousness. When you know, you experience awareness (I am and around me), thoughts (motion), and your will (time). The universe, understood through science, is representing consciousness at a level. It is a manifestation of life and reality in a particular way, and there is similarity in the universe to our own being and the nature of consciousness. The universe then becomes an analogy of truth or God, the infinite matrix, and is referring us to this constantly, and science is but a way. When you have an idea, you create a world. What came before that? We can say the infinite consciousness matrix that you share, always present and always ready to manifest in a new way. What was before the beginning of the universe? The infinite matrix of consciousness that always existed before creation is what this creation depends. The universe is much like an idea within that matrix as an analogy, so motion moves through cycles from one thought to the next or return.

**Force Carriers**

We have said that every dimension of energy is a dimension of space or vice versa. Each dimension of space has a monopole particle. According to the standard model, a photon is a particle that travels freely in space, but according to this theory, a photon is a monopole of the matrix of space in a particular dimension of energy $D^3$. What we perceive as photon motion is actually the elastic vibration of photonic monopole space. Therefore, the idea of messenger particle or force particle needs revision. According to standard model, the messenger moves back and forth between sources and is the force carrier or force between. In this theory, the messenger is actually the vibration of the space matrix, and as the amount of vibrations or density between two bodies sharing the same dimension of space increases, you get a greater attraction or attracting force. Space is held together by the inner vibrating motion of space. The gravity between two bodies is related by the following:

- **Sharing the same gravitational space dimension.**
- **The space generated by those bodies and their densities between them.**
- **Elastic vibration of the space between bodies.**
- **Attracting force due to messenger activity.**

Gravity is a relationship of the elastic vibration of space between bodies. Therefore, current or elastic vibration of a dimension of space becomes the binding force of that dimension of space.

It may be that as we move through different dimensions of space, the monopole may be the same although measured differently from where we are of perceived differently in each dimension from an observing dimension. So the photon in a third-dimensional construct may be a graviton in a higher dimensional construct. Meaning what appears as a graviton in a higher dimension actually appears as a photon in another dimension. So all space is the same, only measured differently from the dimension you are in. This may be so for all current, supporting monopole particles and space matrix, which we refer to messenger or force carriers.

The gravity between objects is really the change in density of gravitational space over distance. The messengers increase as the intervening distance shortens by elastic vibration of space increases. So then space of various bodies are both held together by the fabric of space holding itself together, and this is in balance with outward-moving entropy. This balance follows ratios, such as phi. Universal constants relate infinity to center or balance. So bodies with associated gravitational fields interact with one another. Similar systems communicate with one another directly and also trans-dimensionally.

So then from this, a philosophical question of perception emerges. What must the world be like to be perceived? By analogy, a system of consciousness perceives another system of consciousness. So what do we perceive? Another system of consciousness is what we perceive. Then if we are conscious beings sharing an experience, what other system of consciousness are we perceiving? The universe is a system of consciousness, and consciousness is communicated both directly and trans-dimensionally as well. One state of conscious awareness is viewing another. Awareness of self is extended beyond. All existence represents different levels on consciousness. Awareness becomes free will, thought, and awareness; the order of consciousness is the state of being. And a transdimensional universe suggests that consciousness has various levels as energy has various levels as an emergent rule and extending to infinity as a transcendent rule.

So at some point, science says energy is this, but then another level of energy is quite different, and all this has an infinite source because we said that this universe is a qualification of infinity or consciousness that is a relationship of energy. The experience is a reflection of the source, and hence, the universe is an analogy of God. So in consciousness, there are mechanisms at work, infinite essences that are both similar and distinct, which allow infinite freedom. The essences are infinite, yet as they appear as analogies in this universe, they are alluded to as qualities transcendentally and structure as emergence. So then there are two things at work: your consciousness and the thought and perception that define ideas or structures of awareness. And hence, this universe is consciousness from a conscious source.

So when talking about this universe, as scientists we can describe this qualified energy of consciousness that is this universe with its reference to infinity. Now what about the energy of consciousness? In the infinite form—or perhaps the energy of our transcendent nature, which is the same—can we describe it? Even in this universe, energy is manifested quite differently at different levels, and so too would the energy of consciousness exist at quite different levels infinitely with order, of course. We need to redefine our understanding of energy since in this universe, we now understand energy in this theory to be at different levels and forms; this would also be an analogy for the infinite transcendent consciousness that it also has infinite diversification. Whenever we say "what is," as scientists, we are assuming the entire universe is viewable as a finite logical stream, but in reality, there is more progression and change. We may capture a piece and describe it, but the discovery is never complete since even though we may spot the rules for change, understanding what the change is at each step becomes quite difficult. So what is the energy of consciousness? And I will say that it is this universe and also a structure far beyond as well, and if it then becomes the whole of reality, it becomes quite a complex task. But if we can say the universe is a form of consciousness, the actual relationship and definition exist at a much higher level as our experience may be as branches on a tree. But if I had to say what it is today, I would say that the energy of consciousness is awareness, thought, and will. So what this metaphysical definition becomes to a scientist is far removed from saying we have elementary superstrings that make up the universe or distance, time, and velocity. Even more far removed when we say consciousness is infinite since our math is not trained at infinite concepts. In many instances, the sum is greater than the parts. What is energy, or what is a superstring? We may say it has the quality of time, but what is the energy itself? Better to say "consciousness is" rather than to try to see an infinite construct as a finite definition. Yes, finite relationships we can see, but the subdivision of that task of analysis is an infinite task. That is, therefore, the freedom of consciousness and why it is so incredible. Consciousness is qualified as an effect in this universe, alluding to a cause where the nature of the effect refers to the nature of the cause.

## Strings Are the Smallest Component of Space

According to Sylvester Gates or James Gates, string physicist, there is a math that suggests the building block or string may even be subdivided or proceed further as in ad infinitum. However, for the purposes of the conventional string theory, that was never considered in the current explanation. I see the string as a point of balance where the universe has a fundamental structure that can predict a progression toward infinity, a definition of space. Space is the most basic component where any of its larger components are explained as higher dimensions or lower components as subsets of space. So as a point of convergence of dimensions, subsets, and progression, a string is the smallest component or point of balance. Energy is an arrangement of strings in a geometric arrangement; different strings combine differently in different dimensions of space. Matrix is a geometrical arrangement of strings and is called space. Various dimensions of space form dark energy; the matrix also is matter in $D^3T^2$ **= mass**. We are multidimensional since atoms combine all dimensions as a transdimensional machine. Infinites are related and qualified in a Pythagorean relationship.

## A New Model for Physics and a New Method for Science

This theory proposes a new science, different from the traditional scientific theory and physics. In modern physics, a theory has to make predictions that are confirmed by observation; however, if we are three-dimensional beings in a multidimensional universe, perception leads to abstractions of truth rather than the absolute truth. Hence, our theories are full of exceptions, and this implies error within the theories. However, many of these theories are accepted since they have a number of predictions that match a number of experiments. Approximate abstraction can work that way as a rule of thumb however accurate it may be within certain limits and framework.

**So how does the epistemology of this new science work in this theory?**

Well, it is an a priori science. In a priori reasoning, one can gain certain truths about the world prior to experience using pure intuition alone. Since the universe is consciousness and as we are, knowledge is fluid and shared by intuition. In the manner we all share the light of the universe, the analogy caries to truth and knowledge that is available to all.

**In this new science, how do we know what is a good theory?**

Well, it begins intuitively, and then progresses to reason, and then to experiment or verification. What is proposed here is a new framework, not just a new area of study. This theory began with an idea in totality. It is similar to the ideas coming from a theorist, but not based on observations. The idea of this theory presents laws and results; the results were logically studied, experiments proposed and tested. The laws preceded observation and study. The process starts from the beginning outside of a dependency of observation.

**How do we know the ideas we have are correct?**

Well, not all physicists use the scientific method in the same way, e.g., theoretical versus experimental physicists. But they both depend on observation. Here we have an intuitive leap and generate a framework of laws, and then we test those. The umbrella of knowledge is first intuitive and then tested; test them to prove a priori is working. As we advance, that will be natural to advanced humans.

**A priori should not need proving, you say?**

Sure it should; a mathematical statement might be true a priori, but you still need to prove it, such as $\boldsymbol{D^5 = MC^2}$ or $\boldsymbol{T^2}$ = **density**. It's to validate and educate the value of intuition. A priori can be the obvious and then in some cases not, so it is not always self-evident. It is always intuitive. Not all a priori truths are obvious. It takes a great intuitive mind to have a true idea before proof and be correct. This knowledge is unconditionally valid for all beings endowed with the logical structure of the human mind.

A priori reasoning is

- able to draw very strong conclusions from apparently rather weak premises,
- determined in revealing absolute truths,
- metaphysical origins,
- plays an essential role in some very important discoveries, and
- has been almost entirely neglected by science in general.

Scientific method is the kind of reasoning that derives propositions from the observation of facts, or by generalizations from facts arrives at principles and definitions, or infers causes from effects. This is the reverse of a priori reasoning.

## Infinity and Consciousness

As mentioned before, space exists in different dimensions of the space energy matrix. Also, space is related to infinity. Infinity can be represented figuratively or symbolically by a circle or sphere. The meaning behind the circle or sphere is as follows:

- **The center would be scalar or fractal infinity; scalar infinity represents size and magnitude.**
- **While the perimeter is transdimensional infinity or related to change in the nature of energy in each dimension, transdimensional infinity represents infinite change.**

In this theory, space is made from superstrings *D*, *T*, and *V*, and by their combinations in space ,they have known properties dimension per dimension. Time and velocity are balanced in infinity like the pendulum of a clock, so they exist in infinity.

Infinity is the medium within which this universe exists, but the qualification within infinity is both scalar and transdimensional. Let's use a circle as a symbol as we have said before. The center is **scalar infinity,** and the perimeter circle is **transdimensional infinity**. The point of the circle representation is scalar, the perimeter is transdimensional. Scalar relates to size of structure (as in a fractal), and transdimensional relates to change. Infinity has many forms, yes, but not so many in this reality, so they are qualified and organized as a structural relationship. We have also said that absolute infinity prior to qualification is consciousness. Mind is consciousness; we have said that space is a form of consciousness, true. Life is represented by one form of consciousness perceiving another. And consciousness has various levels in this universe: scalar and transdimensional. This universe is infinity qualified or consciousness qualified, like a dream. Infinite is the medium where this universe exists; finite is the qualification from center and from perimeter, which is also infinite in its relation to infinity. The universe is ordered, and there is no chaos. Consciousness has order by analogy. I can understand infinity easily as consciousness, and consciousness includes infinity and all maths and all variations, all possibilities. But this universe is one structure of many and representing in some math.

**So there are many forms of infinity, then how are we to definite it?**

Infinity is every form and every possibility. Let's look at consciousness again. We said before it is awareness, thought, and will. We also said infinity and consciousness includes all math and their structures; it is the plus and division or relationship that is infinity, and when the idea is formed, then it becomes math infinite realm as a qualification. Infinity is not exclusively the equation nor the variables, although it includes those, but the thing or consciousness whole in between holding it together, and hence, including all possibilities. Then action brings infinity into a specific being or form or qualification. Diversity becomes infinity in unity, where consciousness is infinitely qualified into various forms; like the number zero, consciousness is everything representing unity. We might think of zero as nothing, but it is everything.

It is impossible to define infinity or consciousness using the math since it is all math known and because it is a higher consciousness than the math that describes the qualification of consciousness. But in the same manner that we intuitively discover a transdimensional universe, we can also understand the transition to infinity or consciousness as we directly experience in higher forms, such as in prayer, intuition, and meditation.

Mathematics is the philosophy of consciousness. So then if infinity is awareness, thought, and will working together as one, they are consciousness, the unity of infinity that allows diversity.

- **Awareness is I am or to exist.**
- **Thought is organization or time.**
- **And will is motion, action, or intent.**

God is infinite consciousness, and when he thinks, things become real, and it is creation or creative. We do the same when we dream or have thoughts. It is a beautiful idea. This universe is an analogy of God and tells us about some of the nature of God, consciousness, and the conscious infinite. Let's try an example of understanding and truth by analogy. Let's says **space** has a tetrahedral structure, meaning three, and this is represented in six spheres around one; marbles if you will together. When we say **space is a form of consciousness**, maybe three represents three essences to consciousness, and thus we defined three as awareness, thought, and will. The analogy becomes this:

Then God in creating the universe reflected a portion of his nature as a tetrahedral form and called it light, and the thought was the universe matrix or conscious qualified into form, and in it he allowed his consciousness represented in this qualification or universe to be perceived, and that perception was life. In conclusion, conscious beings perceive a conscious creator. One form of infinity is viewing another. Life is one form of consciousness viewing another.

**Rules of Consciousness**

We have said that life is one form of consciousness perceiving another and that all things are a form of consciousness and that math is the philosophy of consciousness, and so consciousness is represented in all math or infinity. So what is real? One dream perceives another dream, but there are rules according to the nature of consciousness.

We said that consciousness is awareness, will, and thought and, in this relationship, are the infinity of all diversity. In it also are equality and hierarchy. In nature, when we qualify our infinite mind, we create forms that reflect the nature of our consciousness. Our thoughts are a symmetrical projection of the nature of consciousness. Hence, symmetry is a rule of consciousness. God's consciousness projects both transdimensional and scalar forms in a hierarchical pattern, which also tells us some of the truths about God. And those forms descend to what we are, and we also project these forms into another realm in our thoughts, which is also a reflection of ourselves. Hence, truth is mirrored as symmetry from one form of consciousness to another.

Once we understand the form of our entire universe, which is also the form of thought, we will have a better understanding of consciousness and the ultimate consciousness, which is God.

When we say that time and velocity are polar and qualified in infinity, it is almost like saying what Buddha said, that the right path was in the middle of opposite extremes. If you play a guitar string and it is too tight, it snaps, and if it is too loose, there is no sound. This goes back to the idea that natural harmony of the universe is balanced within infinity. Toward a center, the path philosophically is the path of moderation on life, the path of enlightenment, and even similar as well to the Christian notion of the straight and narrow path, which we clarify here as nature's balancing effects and God's moral code.

So what is infinity then, that consciousness from which all is made and also which makes the physical universe that we see and all things, that energy which forms all. It is said that in the fifth and sixth centuries, the Greeks discovered infinity, and it was a concept hard to understand. It is hard to understand because infinity includes everything. There are many forms of infinity. For example, some infinite processes have limits, such as walking to a door and taking half a step each time, and others are absolutely infinite, like irrational numbers. The nature of this universe tells us that it does, in fact, exist within infinity. And when we say infinity includes all maths, it means that the most basic source or energy cannot be designated by any known attributes. Qualification of the infinite means designating relationship and attribute. That unknown infinity that includes all is made known partly when attributes are designated. So God is both known and unknown, and so even with the distinctions, all in one since the source is one. And that balance reflected in this universe tells us that infinity extends from one. And hierarchy and diversity are also one. The balance and unity in infinity that results in all diversity is one. God is one, and he is all. So how can we understand infinity? By the very processes that make us infinite—will, thought, and awareness—and the order of this universe is a manifestation or qualification of that infinity, and hence, only partly represented or attributed from the source. Hence, the universe is an analogy of truth. So math of this universe is partly reflection of a greater truth. By a dream or object, can a man know his own consciousness? Or can a man without projection of a dream perceive the subject of his own consciousness and know truth of his consciousness and the universe as by intuition, reflection, or meditation? Math can explain organization, but only intuition can access the meaning of the whole of the infinite, which it already is, the one from which all math extend or come into being. The source then becomes undeterminable. This place transcends all qualification of will, awareness, and thought. And the understanding of infinity suggests that we are transcendental beings, where the higher of consciousness and its qualification results in the lower forms of consciousness represented in this universe. The purpose perhaps for consciousness is to perceive consciousness, to know about God, the one, the source of all.

Some people say that physics just is interpretation of the human mind. If we are three-dimensional creatures in a multidimensional universe, we don't see it objectively. Math is intuitive when purely theoretical, like music.

In meditation, it is known that to empty the mind is the norm. We have discussed how ideas become qualification of the consciousness of the mind. What remains when these are gone are consciousness as essential essences of will, awareness, and thought. Unique to Buddha was the idea in deep meditation to address awareness state, which was uncommon to traditional forms of meditation.

We spoke of math being the philosophy of consciousness. And so all math has representation as a qualification in this universe.

## Vacuum State Zero-Point Energy

The lowest energy state of a physical system is referred to as zero point. Zero point is also used to refer to vacuum energy as the lowest possible energy or quantum vacuum zero-point energy or empty space. Variations in the energy of the vacuum are known by experiment as the Casimir effect. The empirical evidence is made by observation, which seems to indicate the vacuum is nonzero, meaning it is something with structure, very low energy, but energy nonetheless. In this theory, all volume has mass, even the vacuum of space, and it is a prediction of this theory.

Well, empirical evidence is made by observation right in the case of zero-point energy. Even the vacuum is presumed to have structure and energy, however small; an absolute vacuum is impossible. And scale is relative, so what appears small is relative; and 1 atmosphere force is the Casimir effect and is rather large at short distances.

The understanding of energy in this theory is scalar and transdimensional. Within a particular dimension, we can talk about space energy, but we can also talk about the energy distribution of the subspace and the structure of superstrings, so perhaps we can call this investigation of the subspace intra-dimensional. In much the same manner, we know the properties of atoms of different elements, and then we analyze the differences in the structures of different atoms. Different dimensions of space have differences overall as properties and also by construction internally with respective properties.

The transdimensional equations such as $\boldsymbol{D^4 = MA}$ refer to the total energy of a dimension. The Pythagorean theorem is used to explain the relationship of energy at the level of subspace. Geometric equations of volume such as the volume of a sphere are used to explain the energy of superstrings. Since these differences are not addressed in relativity, there is no violation. The equations are invariant; however, their actions due to variances in scalar quantities cause variances in measurements depending on position.

We have stated before that space is a matrix of energy (dimension per dimension) and so has a structure and is symmetrical. This idea lends itself to Noether's theorem and the conservation of energy and momentum in symmetrical systems as we have discussed before. So it is evidence by empirical observation that the conservation of energy requires the vacuum space matrix or conventionally described nonzero vacuum state zero-point energy; this transdimensional structure also forms the path for inertial systems, which also include entropy and disentropy and the natural balances of energies that rest within infinity.

We have said that this universe is a qualification of infinity. The space matrix is made of superstrings with spin. This universe is constantly partaking of infinite energy and dynamic motion in the space matrix superstrings that makes space and all particles.

## Energy from a Black Hole

We have talked about the elasticity of space since it is made from an elastic matrix of superstrings. The elastic wave travels at the speed of light with variances due to its expansion and contraction. Space is also elastic trans-dimensionally, meaning that each dimension has elastic properties and the relative motion is $C^2$. So motion has a relationship and connection trans-dimensionally. The structure of space is elastic, but when is it not elastic? Space approaching a black hole compresses space to a higher dimension and resonance. Since space is a geometric matrix, an elastic collision could conceivably escape from a black hole. Hence, energy should be able to escape a black hole trans-dimensionally in the form of the space matrix (creation of space) and due to waves or elastic collision. Consider space as an ocean and the elastic wave as the ripple wave on the ocean. The distortion of the matrix can indeed prevent elastic collision, but where it is loosed along the event horizon, energy of an elastic wave can and will be released.

We can also say that the transdimensional jump from $D^3$ to $D^{23}$ is so great that this would disrupt elastic collisions, planetary spin, and other phenomenon such as magnetic fields and gravity. So there would be a great deal of anomalies there not evident near other galactic structures, not to mention that gravity is squared in a black hole and, hence, squares the creation of space around the black hole.

## ESP and Scientific Basis

Qualification must precede quantification. If we can't define something, we cannot measure it. Telluric energy is said to be the energy from which all is made, and in this theory, it has been referred to the infinite consciousness that is qualified into form and also that consciousness transcending the form it creates. We have also said the by symmetry, the essence of that primal energy or telluric energy qualifies itself into thoughts and form. But beyond or above consciousness is the essence of all, what Buddha refers to as all-knowing, all-seeing presence of being. These essences defined as awareness, thought, and will are actually manifestation of the awareness or mindfulness. If this primal factor is the telluric form from which everything precedes (the infinite source), then there is a path to the manifestation of all things, and this energy or presence connects all things and is God. By the time this form reaches pour thoughts and alpha waves, it is already several steps removed and transmuted, but nonetheless connected to the source.

## Transdimensional Consciousness and Source

We have tried to explain the basic components of the universe in many ways:

- Emergent as opposed to transcendent, which involves the qualification of infinity or consciousness in creation.
- Superstrings that make the matrix of space in multiple dimensions of space or dimensions of energy.
- Relationships of infinites as are represented in a sphere relating transdimensional progressions and scalar progressions. The number 1 is the number of balance, the relationship of the center of a sphere to the surface, the relationship of infinites qualified. The ratio of 1:1.61803399 is the golden ratio (the relation of balance to infinity). In your distribution of energy in subspace, meaning this is a qualification of infinite energy, it refers to the center of a circle scalar infinity and transdimensional infinity.
- Relationship of consciousness to all reality, life is one form of consciousness perceiving another.

### What about the infinite and transcendental source?

We identified this basic component as telluric energy or the energy of consciousness that precedes action. These ideas show us transdimensional steps or layers of consciousness that give hierarchy and differentiation to life and relationship to the point of source of consciousness: God.

### What do you think happens after you die? Do you believe that it's possible to exist again?

The universe presented in this theory is transformation and change. The idea of beginning is an emergent concept and a human one that focuses at a point of change or transformation, like the caterpillar becoming a butterfly. Focusing on the point rather than the motion leads to the idea of beginning and end. Life is transcendental and consciousness; that source is transformed to another state of consciousness. Life is eternal and infinite as consciousness is unbounded. For humans, time seems to create problems and the thought that God does not exist. The solution exists without time in infinite consciousness. This universe is a qualification of infinity. You know what that means? It's like when you have a thought or qualification. In physics, it means that from the infinite, a pendulum swings to infinity but never gets there because polarity of energy pulls it back. We are balanced within infinity, like what Buddha said, balanced between multiple extremes in infinity, which is the middle path. This is what the middle path means.

### Nothingness can't even exist?

True. Nothingness is a human myth and a limit that disregards transformation. The beginning is an emergent truth, which mistakes transformation and change. Before the universe, infinite consciousness is; the "I am." I am or awareness, thought, and will. That quiet infinite, God, and then action and differentiation of consciousness, and the universe comes into being. The idea of nothingness, those terms evolve after until the universe was something, but it always was, only transformed. *Nothing* is a term that means everything, and then there was transformation; only change and transformation exist. You might say thoughts don't exist since we are all thoughts, but there is an aspect of transcendence and consciousness that is a part of the character of the source or God.

**So you say that's the only way it makes sense, you are thoughts, and I don't know how thoughts exist, but they're nothing, so it doesn't really matter?**

Thoughts are the infinite by which you exist, and the middle path means consequences. The relation between awareness, thought, and will mean consequences and shared with the matrix that is God. When thoughts are gone, you are left with the origin of thoughts, which is that which precedes thought, infinite life, with characteristics of mindfulness, presence in awareness, thought, and will. In Buddha's case, he thought to erase the illusion of thought reality and, as such, emptied himself, but unlike others that meditated, he experienced awareness. That first principle of all being is where he was one with all and God.

Buddha meditated for years, meditated and erased his thoughts and got nowhere, then he thought, not extremes, like nothing; but the middle path, he meditated again. What did he do different? When he was in that state of nothing, he thought of awareness and saw the whole universe. he was in that state of union with God where everything is one. Buddha is about the journey of a man looking for truth by removing illusion transformed and shifted through several layers of consciousness to the point a concept involving the beginning of thoughts of the source that transcends all. All is really transformation where one state of awareness blends into another in multiple layers, and in a similar manner, this universe is transdimensional.

That something beyond this existence, the "I am," must be what is referred to as that infinite telluric presence. That which is transformed to everything and is everything; in its various levels of transformations and conscious states, includes us as we are all a part of God, and his very essence does discovering and all knowing. God loves and wants to be loved as all consciousness exemplifies symmetry. That essence is reflected into reality.

**What is this infinite telluric presence you speak of?**

It is the same spirit we all possess and share, infinite consciousness, but consciousness is motion; and before that, the essence is will, awareness, and thought, before the action, and through that all comes to being in harmony and multilayers of consciousness. And with that source, we can move mountains.

The middle path as Buddha described has meaning in this way, which matrix that created all is calling everything to its harmony or balance like the pendulum of a clock. The balance between extremes is the balance calling your free will into harmony.

The idea of nothing needs to be addressed. The Kabbalah says that nothing means everything, and that this universe exists within God, which by this theory prefers to present this universe as the qualification of the infinite. God is everything.

When I asserted that God was everything, remember there is free will, and I am describing what God is before creation, and creation has its unique characteristics.

Life is one form of consciousness perceiving another, and perhaps God is creating a new facet of his personality by creating us and is another awareness of God in that image.

Your identity may be an illusion because in the harmony with the whole, all is one, and we are experiencing a receding consciousness from the source.

The element of free will, the exception, also presents us with a mathematical axiom: that creation is always beyond our understanding, and truth has another view. That God and reality is not bound although can be experienced where it is found and explained. Hence, God is all powerful.

We can speak of consciousness at various levels or receding states from the universal source or God. There is consciousness on

- the intuitive level,
- physical level,
- thought level, and
- dream level.

Ultimately, the first or highest state of origin is the center of life, and this may give credence to the Christian idea of store treasures in heaven since all receded states of consciousness are places of development and temporary for the development of a higher self.

Also with regard to truth, we are universal in a sense that we partake of the whole; like a leaf on a tree, we have association to the whole structure, but scalar and trans-dimensionally. Truth is relative with regard to its participation in a universal framework, meaning that there is hierarchy and structure in the meaning of truth. Our experience, however, is not absolute since as we have free will, so does the originator of reality or God, and that particular state is always open for action even as it is in our own lives. And in the infinite state is the source of all will, thought, and awareness.

## Receding Levels of Consciousness

How can we show that logical reasoning and mathematics apply to the real world universe? Mathematics is the image of the real world universe; the real world universe is represented by one consciousness perceiving another, and the universe is a form of consciousness represented by mathematics. Math is the structure of infinity and any qualifications such as this universe; math applies to both emergent and transcendent truths.

Truth ultimately is represented in laws through math. Unfortunately we are beings perceiving $D^3$, $D^4$, and $D^5$ but living in a multidimensional universe, so what we believe to be laws in current science are mostly approximate abstractions.

How can you define something that doesn't exist? It doesn't necessarily mean if something is poorly understood, it does not exist. Consciousness is in several receding states from the source. How do you pinpoint universal consciousness or the source? That is where you are before action. People use that word *consciousness* all the time, and usually I feel they don't really know what they mean by it. Well, consciousness has its convenient meaning, but it's more complex and multilayered in truth.

What is the difference between consciousness and awareness? Consciousness is awareness, thought, and will without action; consciousness becomes the actions of that mechanism. Like when Buddha stripped away all thought and did what others did not: awareness. I believe that science tells us about God, the ultimate source of consciousness.

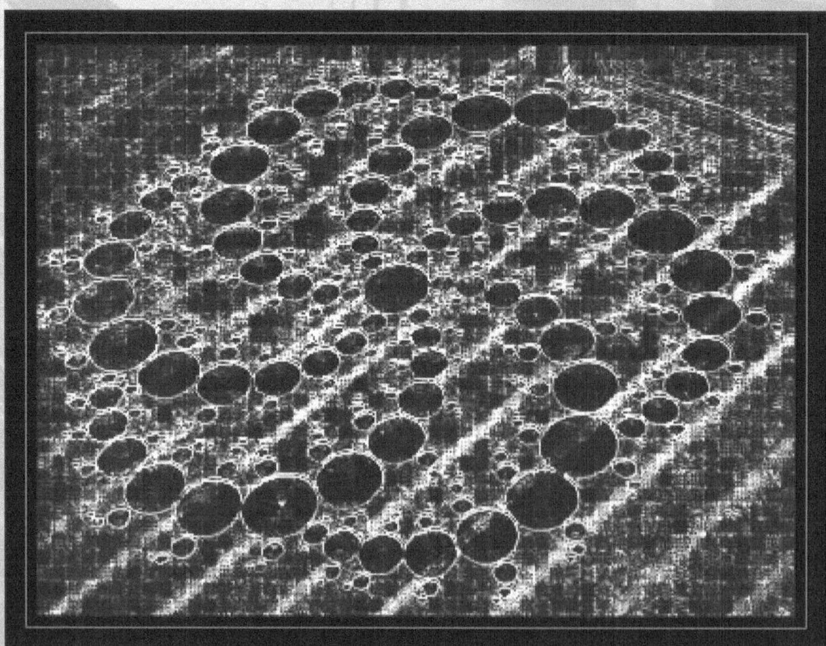

## Tetrahedral Crop Circle Images

Current understanding leads everyone to believe that this reality of consciousness or dream is the truth; it's only an analogy of truth.

Yes, but let's start at the other end. What is the lowest form of consciousness? What is the "least" a living being can experience or would say that involves consciousness? Lowest form might be your dream state since consciousness arises in dreams as well. What about perception? Your perception is but one tier in the possible dimensional states—the physical one. You are a transdimensional being in consciousness as the universe is transdimensional. Consciousness is transcendental but also exists in more limited manner in this universe. Animals are conscious at the highest level since all consciousness is connected. Yes, how amazing are the rules of consciousness that include as relationships, hierarchy, equality, and dependence; the anomaly is free will.

**Infinite Space**

We are always at the center of space, so there is no center of the universe. Just as consciousness has many levels receding from a source, so do space and a transdimensional construct and vice versa, so transdimensional space recedes from infinity. So at all points in space, there is a relationship to infinity. Black holes are the generators of space and are connected to infinite energy. Black holes are at the galaxy centers, but perhaps black holes of much smaller size and range exist in atoms and other smaller stars. The universe is connected to infinity by multiple sources. Every part of the universe is permeated from infinity; there is no absolute center, but we can talk of a center based on our position, but there is none. Since there are ratios in the universe of the amount of particles to space that generates them, the mass we perceive seeks equilibrium to be evenly distributed. These ratios are scalar, transdimensional, and exist within a dimension. The ratios even have relationship to infinity, such as phi. The center of balance is balance between space and particles that move from extremes of entropy and disentropy, motions toward infinity. Entropy would be levels where space expands to disentropy or areas of high energy where space contracts.

Space is receding from higher dimensions and is transformed trans-dimensionally. We are protected or positioned in many layers of space, but you can navigate them with the right technology, which someday will be.

Infinite space is accessed trans-dimensionally through different dimensional layers.

Relative has a limited point of observation and variances in scalar forms between point of observation.

**What makes multiple dimensions necessary?**

Differentiation allows hierarchy to take place. This is a truth of nature of the universe and transcendental nature. Hierarchy and periodicity combine to form a periodic table of energy. And by that concept, the matrix of space combines into different geometric patterns using *D*, *T*, and *V* and forms different energies or space matrixes, which are also referred to as dimensions. These are arranged as tetrahedral spherical geometries that form a matrix of D, T, and V. D, T, and V are established with a multi-dimensional matrix of periods of VT, v and T being complementary, asymmetric, and forming balanced and symmetrical hierarchies through D where all is resonance that includes V and T. So then you have a periodic table of energy in a transdimensional model. It's like asking, why do we have more than one element? The axiom is all physics is related by combining superstrings *D*, *T*, and *V*. Space is made of a matrix of these superstrings dimensions per dimension, where you add a *VT* in each subsequent and higher dimension. Space is energy is dimension.

The universe is a relation of absolute infinities, such as transdimensional and scalar. The universe is a qualification of infinity and an analogy of truth. Reality is alternation between interval and infinity. The infinite and fractal progression subdividing D, T, and V may in fact be a binary coded network.

# Chapter 4

## Universal Code or Keys to the Universe through Phi and Phihat

$$V + T = 1 = D$$

$$V \times T = -1 = -D$$

Phi and Phihat or Velocity and Time respectively give rise to all scalar (additive) systems and proper (multiplicative) systems. These are the templates for all the universal forms and are the keys to the universe and equivalence on a grand scale. To have the correct and true units, we now understand that everything we see is a function of time and velocity, complementary, inverse, and asymmetric, however in the formation of space, larger symmetric systems are formed.

There is equality through symmetry, but in even system of symmetry, equality is maintained through hierarchy, and equality does not exist without hierarchy and vice versa. Balance includes asymmetry and asymmetry includes balance. In nature all is equal under the order of balance but made perfect in hierarchy. We are all equal under the enabling and infinite harmony of the universe and matrix of reality, but are not truly equal in our capacity within a particular system. I love because the universe loves me back. No action is possible unless it is received. Life is one form of consciousness perceiving the other; the universe is the consciousness of the first cause or God.

## Einstein and Curved Space-Time

Under current understanding, space is presumed to be curved due to the effects of gravity and its gradient on velocity of an object and its position. The idea of curvature becomes a two-dimensional abstraction of a volumetric event. This explanation in general relativity is not a volumetric understanding of space and time, which would need to see the gradient of gravity and its effects as a density. In a sense, the gradient density of gravity field causes the gravitational force inertia. Hence, we can dispense with the traditional notion of flat space that has curved form and think of space as varying gradients of gravitational density, that density of gravity includes time. Space is volumetric, and that means the gradient density of gravity affects time and the force of gravity similarly as we observe in the inverse square law.

A revised description of gravity as in this theory will explain why gravity affects time. The description then suggests space has structure and physical attributes. We can already describe how gravity affects time, but it's an abstraction and not correctly presented. In the current accepted definition of gravity, science is not involving density with time. It seems amazing that science can present and even entertain the universe as a curved plane. Space density varies, and gravitational density varies; of course, the math would be much more complex to explain that. So for convenience, we have what we have. The equations for gravity expressed with density and time would look quite different.

**Density = $T^2$,** which, when incorporated to gravity, means that acceleration due to gravity is the result in the change in density. Gradients of density become forces. All right, we could work the concept backward. Since we observe the change in density as a phenomenon that affects force, can we construct a model that gravity should be rewritten to include time and density? Well, knowing that space is volumetric and that we know time is affected by gravity in this way, like with GPS satellites that present an argument. Space is volumetric, and surely a gradient in gravity establishes the concept of density involved with gravity and time.

### ***Working toward gravitational field $D^{10}$:***

**$D^{10} = (MV^2)^2 = M^2 \times V^4$**

**$D^{10} = M^2 \times D^4 / T^2$; $T^2$ = density**

**$D^{10} = M^2 \times D^4$ / density$^2$ = $M^2 \times D^4 \times$ volume$^2$ / $M^2$ = $D^4 \times$ volume$^2$ = $D^{10}$**

**$D^{10}$ = volume$^3 \times D^1$; $D^1$ = density × acceleration**

**$D^{10}$ = volume$^3$ × density × acceleration**

**$D^{10}$ = gravity**

### ***Defining $D^1$ by Density × Acceleration***

**$D^1$ = density × acceleration; density = $D^1$ / acceleration**

**Gravity = $M^2 \times D^4 \times$ acceleration$^2$ / $D^2$ = $M^2 D^2$ acceleration$^2$**

**Gravity = $M^2 D^3$ acceleration / $T^2$**

**Gravity = $M^2$ volume acceleration / $T^2$ = $M^2$ volume acceleration / density**

**Acceleration has dimensions of length / (time)$^2$.**
**Secondly, for any time of motion, acceleration is defined as**
**Acceleration = difference in $V$ / difference in $T$**
**for a constant acceleration, you can integrate this very easily to obtain**
**Velocity = $u$ + acceleration × time**
**where $u$ is the initial velocity at time $T = 0$. Even if you let $u = 0$, you would end up with**
**Velocity = acceleration × time**

**Acceleration = velocity / time = distance / time$^2$**

**This $T$ is the time at THAT instant, not the change in time**

### ***Changing D for R***

$F = M^2 / 2R^2$ **density**$^2$

$F = MA = M\,D / T^2 = MD$ **/ density = mass density acceleration /** $T^2$

$F = M \times M$ **acceleration /** $D^3T^2 = M^2\,D / D^2\,D\,T^4$

$R = SQRT\,(D^2)$ $R$ **based on the square of the circle and Pythagorean triangle**

$R = SQRT\,(D^2);\ R^2 = D^2;\ R = D$

$F = M^2 / D^2\,T^4 = M^2 / R^2$ **density**$^2$

$F = MA$

### ***Working toward Newton's Gravitation in $D^4 = MA$***

$F = MA = M\,D / T^2 = MD$ **/ density = mass density acceleration /** $T^2$

$F = M \times M$ **acceleration /** $D^3T^2 = M^2\,D / D^2\,D\,T^4$

$F = M^2 / D^2$ **density**$^2$ $= M^2 / D^2$ **density** $T^2$

$D = R$

$D^2 = R^2$

$F = M^2 / R^2\,T^4$

### ***Results***

$F = M^2 / R^2$ **density**$^2$

$F = M^2 / R^2$ **density × time**$^2$

Here we see Newton's formula with the exception of density and/or time. Using solid substances, the impact of density might appear negligible, but gravity is also related to density gradients in mass and space field and time dilation in mass and space field. Newton's gravity equation becomes **gravitational concept = force × density²,** which is not a FORCE and is not explaining the true nature of gravity with density and time in an inverse relationship.

**Gravitational Concept** $= M^2 / D^2$; **mass** $= D^3T^2$

**Gravitational Concept** $= MD^3T^2 / D^2 = MDT^2;\ T^2 =$ **density**

**Gravitational Concept** $= M$ **density** $D$; **mass** $= D^3T^2$

**Gravitational Concept** $= D^3T^2$ **density** $D$; $D^4 = MA =$ **force;** $T^2 =$ **density**

**Gravitational Concept = force × density**$^2$

$T^2 =$ **density**; it means half the density and divide the force by 2.

**Why Does Density and Time Matter for Gravitational Force? Since we observe the change in density as a phenomenon that affects force, can we construct a model that gravity should be rewritten to include time and density?**

Yes. These parameters have been omitted by Einstein and Newton. Space is volumetric, not flat curved as in the general relativity model. But how would Newton know to include density or Einstein for that matter? If you are comparing solid masses, perhaps it doesn't matter for specific cases.

The fact that density has relationship in the gravity equation means that we can use **$T^2$ = density to convert $T^2$ to density.** Density relates the space of the gravitational field, which varies with distance from the source mass. The change in density induces a force, since acceleration is a reaction of the change in gravitational density. Acceleration is due to the change in density of space. Yes, variable density in the gravitational field or space causes the gravitational force. There must be some type of field inertia involved of high to low density. I feel that the same applies to charged fields and that perhaps we are speaking of a different dimension of space with charged particles or fields. The effect of a force by a change in gravitational density can be varied by changing the mass and/or the volume, which are components of density. For the purposes of this discussion, we shall consider the mass constant. However, any gravitation involving density and mass should be further refined to include the displacement of mass core and mass gas in the gas planets. Change in density affects force—note Saturn with similar acceleration to the Earth with larger volume and mass. Reduce the volume for the same mass, and an increase in the acceleration will be the result.

So here we come to Newton's equation with exceptions density and time. The size of things changes its force. When the mass is the same and you change the size, you change the gravitational force. To be fair to Newton, he is describing the force of density. Newton is describing the force of density and not force or gravity; although he calls it gravity, his equation can be rewritten

**Gravitational Concept = force × density$^2$**

This is not gravity; it only explains a linear effect of gravity. Newton is a linear explanation of gravity, an abstraction, not really the nature of it. A rule of thumb can tell you an effect without telling you about the true nature of the source that result in the effect. To get to force units on the gravitational force side, you need to multiply in the gravitational constant *G*. So in the Newton equation, gravitational *G* constant revises the units to equal force. So what is *G* understood under this theory? Amazingly, it becomes the very thing predicted above by this theory for gravitational force as follows:

**$G$ = volume / mass$T^2$; $T^2$ = density**

**$G$ = 1 / density $T^2$**

**$G$ = 1 / density$^2$**

Now if you take the gravitational constant **$G$ = volume / $M$ $T^2$ = 1 / density$^2$** and multiply it to gravitational concept, you get the following Newton equation revised to what is predicted in this theory as follows:

**Gravitational Force = force × density$^2$ / density$^2$ = force**

**So the gravitational constant is equal to 1 / density$^2$ and is equivalent to the units volume / mass × $T^2$ where $T^2$ = density.**

In physics, the gravitational constant *G* is the coupling constant. Perhaps this understanding needs revision. Presently, this constant is characterizing the gravitational attraction between two charged elementary particles having nonzero mass, where *G* is a fundamental physical constant and a dimensionless quantity. Here we see that *G* is a variable and is not dimensionless. Perhaps *G* is very small and refers to the density of two interacting gravitational fields of space. In most instances, the variables of *G* might be perceived to be constant, but in other situations, there might be adjustments since they are variables. Hence, one might predict trajectories based on constancy to be in err sometimes, such as with NASA.

This theory does prove Newton's universal law of gravitation with a new understanding of *G* as a variable in density by two interacting gravitational fields.

**$F = M \times M / R^2$ density × density = $G\,(MM / R^2)$**

**$G$ = 1 / density$^2$**

**$G$ = volume / mass$T^2$; $T^2$ = density**

**Now you have the mass of two objects and their fields of gravity density. The constant *G* is rewritten as variables, and so the gravity equation has no constant but two constants that vary per object, which you might call variables. This enables trajectory calculations to be refined to include fluctuations in the density of the gravitational field.**

Both two objects in space and two subatomic particles have gravitation due to the relative density determined by what that density is at their centers of space and their distance apart according to the inverse square law. This difference in density causes natural motion or inertia from low density to high density. Gravitational acceleration is caused in reaction to the change in relative gravity force due to density between gravitational fields by motion and position. Density drives natural inertia. When calculating the density of an object, homogenous mass objects will act differently to polarized mass objects. Hence, a solid homogeneous, both would have density calculated differently than a gas planet, such as Saturn with a solid core where there is a displacement of density and natural gradient in the distribution of mass. Also extending from every object is the gravitational space field, which is another layer in the objects mass. Space has volume and has mass as mentioned before in this theory. To summarize, the realm of an object includes the following:

- **Core or center.**
- **Surrounding particles and space matrix field.**
- **Extension of field matrix of space.**

**The force of gravity is a result of the change in the density of space.**

The speed of light is an elastic collision of a space matrix, and as such, there would be variances in the speed of light, however slight. I agree the speed is non-constant, but constant for a standard model seems appropriate since there is a limit $C + V$ or $C - V$. $C$ is the limit of hyperspace within certain tolerances. I would see the adjustment so slight between the Sun and planets as to be negligible. The elastic vibration is one motion within the space matrix structure that defines $C$, and the density affecting gravity is through the space matrix is another separate motion since one involves hyperspace or intrinsic structure of the space and the other space as a whole respectively. Two different motions are involved here and affected differently. So the effect of the change in density fields between objects and their communication through that elastic matrix are two different events, although related, but quite different in magnitude.

## Equivalency and Conversion of Units

This theory presents equivalency and conversion of units. This means that in an equation like **density = $T^2$**, you have the equivalent and equal conversion of units of time and density. The proof of this axiom allows us to establish the equivalency required in a unified field theory. It also shows us an important step in establishing a transdimensional unified field model where periodicity of multiplicative components $V \times T$ are multiplied dimension per dimension.

## Casimir Effect

We can speak of zero point being the lowest natural state in energy of space. But remember that space has a natural vibratory oscillation, and this would be identified as the Casimir effect. There is also the force that holds the superstring space matrix together, and these two forces are obviously different. The type of forces external to the space matrix and of one type of Casimir effect is as follows:

- **One is oscillatory because of elastic collision of the matrix.**
- **Internal motions of the superstrings.**
- **The motion of space.**

The internal binding of the space matrix is another force of *subspace type of Casimir*. The Casimir effect understood currently has been identified at one hundred times the size of an atom, so it must be due to effects external to the space matrix. But there are also other subspace binding forces at a smaller scale similar to the Casimir effect. We must also consider the effects of spin and charge that not only affects the binding of space but typology, where classification could include as many as four to be discussed later, one being anti-space and related to antimatter.

## What is life?

Let's propose an axiom that only life can beget life. Since there are no emergent processes in the universe that result in the beginning of life, the origin of life is transcendent to the processes internal to the universe. Only life can evolve from life. Also from this theory, we did say that life is one form of consciousness perceiving another and that consciousness involves awareness, will, and thought. So therefore, life is more fundamental than consciousness and consists of that most fundamental energy defined as telluric energy of all life, which is differentiated into equality and hierarchy, and animated, by its order and constructions with the anomaly of free will as an extension of the source matrix or God. We also discussed consciousness trans-dimensionally as participating in several receded states from a source. Which came first, the chicken or the egg? Well, that question is an emergent concept and viewing a narrow range of human experience. More complete is to describe the beginning or end as transformations both in scalar relationships and transdimensional ones.

The real you exists in a higher state. This is a view of a dream of receded form of consciousness. Your eventuality is not your physical form in this universe. So then what is real is relative and biased by our immediate experiences. The differentiation of the source matrix of the universe, in a sense, is providing diversity and, in return, is discovering itself. This action of departure and return is the nature of love, an action by which through symmetry both establishes action and wants to be returned. Love wants to be loved or action that seeks to be acted upon. The actions to diversity, hierarchy, and differentiation also establish laws of dependency and reconciliation. Love, in a sense, is a mechanism matrix or perhaps the telluric energy that differentiates, animates, and is also integrally connected to the order of it established through unity, harmony, balance, order, and infinity of various forms. Therefore, in its differentiated forms, the universe is announcing itself as an archetype of truth or love in various degrees of manifestation. The universe is an analogy of truth or God, the universal matrix. And within this differentiation, free will as an anomaly is allowed, that which acts and is acted upon, the essence of life. Love allows the differentiated portions to rediscover the patterns of the parts and the whole or God. And so an overall consciousness or God is visited by subsequent consciousness. And the harmonic relationship becomes the middle path, so to speak, a balance between extremes; and in science, this becomes the qualification of truth, energy, or existence balanced within infinity.

So life becomes a discovery of the truth, from the parts to the source, and various levels in forms of consciousness. Diversity is allowed, and that diversity encourages the discovery that allows a free will to grow and evolve in wisdom, wisdom being the awareness of truth that leads to knowing how to live. So life is not about understanding where to find truth in one place since it is all around you, both within and without differentiated trans-dimensionally. Life is not about understanding where to find God objectively since God is everything, but a journey of action and being acted upon in this mysterious process of life called love, which is God.

To complicate matters further about an understanding of life is to dispel all the myths that exist that block us from knowledge of truth. These myths are semantics, logical structures, and cultures that are ingrained into our belief system from an early age. Deep questions become confused and separate from truth when, in fact, the opposite should be happening, that investigations in life lead to an understanding of God. The struggle then is the right action suggested by balancing effects of creation interacting with our free will. The matrix is drawing you to balance, and that work is pain. So then the universe, understood as an analogy of truth, becomes a great resource for learning. And then once achieved, religions that represent a view of truth almost become slanted to present certain aspects of a larger truth, in much a similar manner as when many theories of string theory were united into one M theory. We must be careful how religion becomes more an extension of culture than a definition of truth per se, and so we find that a certain religion can be understood vastly different in different geographical areas without question. We must have the tenacity to explore truth for its own sake and illuminate the distortions of human history, cultures, semantics, logical structures, and other aspects. Men or humanity becomes its own worst enemy in upholding truth as a status to govern men when bound by cultural biases and the limited nature of semantics.

We are humans existing in a multidimensional universe and assume that our perception is all inclusive when it's less than 1 percent. And so our analysis becomes distorted. Take for example the illustration earlier of Newton's idea of gravity where **gravitational concept =** $\boldsymbol{M^2 / R^2}$; here we have the inverse square law applied to gravity and the relation of gravity adjusted by the squaring of the distances between masses. This view leaves out density and time. Einstein later included the variation due to position that inferred curved space. But each time if we as humans are experiencing $D^3$, $D^4$, and

$D^5$ in a multidimensional world, then our analysis becomes an abstract approximation like a rule of thumb. Newton's equation rewritten in terms of equations of this theory result in the equation **gravitational concept = force × density**$^2$; amazing, now we have another fascinating rule of thumb that does not describe force or gravity but the experiencing human perception of what is happening and explaining an effect rather than the source. Then a constant *G* is thrown in to balance the units and described incorrectly. *G* is a variable that is 1 / density$^2$. In our definition of truth and human experience, we are relying on reason as the defining mechanism while discounting the higher source of intuition of a higher transcendental self. So there is more than the rational man when we can speak of the rational intuitive man.

There are many exceptions in life, and this should always be considered as a doorway to new truths. So the axiom there is always an exception is akin to free will and the diversity allowed in the universe even where truth is concerned. These variances can be subtle, and life is. And we also must be careful of human myths, which include most ideas. For example, chaos and nothing are myths. The human tendency for duality is a myth. The idea of beginning and end related to life and death is a myth, a closer understanding is transformation, and so what is perceived as real here is a receded state of consciousness while the consciousness that experienced the transformation or death already exists in another plane. So truth understood in the wrong way can be a source of separateness that results in alienation of religions and the beliefs of atheists. And then we have those that say truth is an illusion, promoting another myth of misunderstanding. Truth is not an illusion because it is the basis of the order of all things, and there is order. And the order is an analogy of truth. Human limitations create myths. Life would be meaningless if arbitrary, while free will is evidence of purpose. The awareness of truth leads to wisdom, and by that you know how to live. Truth as an illusion would be an incorrect axiom. The idea supernatural is a myth; the other world is transcendent. You are already emergent and transcendent. Truth is order; it is not relying upon myths based on culture to reinvent truth. So in many cases, religion becomes an extension of culture incorrectly. Being true to truth, it's up to you to explain it without anything but your own being, and then when you see it in other forms, to separate the wheat from the chaff, not to confuse myths with truth. Truth applies to all religions and is a priority, but at times you will find culture defining religion more than the truth of it. So left to the devices of men led in this way, corruption is always possible. So then where is the real truth? Humanity is great at distortion according to language, semantics, culture, and myths. Leave it up to men without seeking truth, and you will have failure at every step. First seek the truth, which is universal and describes God, and confusion will be limited. Accept no man's definition, and explore for yourself to be tested and refined. And you will finally understand the connection between all beliefs where truth is correct. And that will be closer to God or truth than any representation through culture or history. So abandon prejudiced axioms based on myths. Your life is the journey of self-discovery to the truth where that differentiation of the matrix is once again united with the matrix, and the journey becomes a journey of self-discovery of love, which is life matrix. Life is not arbitrary. And that matrix that makes us is love as the receded manifestations of that matrix, such as the creation of the universe and when you dream in your sleep.

Knowledge is for all and evidence of life. If you cannot know, then life would be arbitrary, and there would be no accountability and responsibility or the free will that allows self-discovery in the differentiated matrix. Love that is to love and be loved, to know and be known, to act and be acted upon; this circle is life and represents the cycles involved through the transformations of all reality. In the infinite manifestations of life and love, there is order. All is available to us to know. The universe is an analogy of truth, and some examples have been presented herein. This understanding becomes a great resource once acknowledged. Your life transcends your body like God transcends the universe. That is a truth; the analogies point to truth. Transcends means transcends to this experience. The axiom was that because of transcendence, both life and the universe came into being. In the universe, the beginning of life cannot be shown. Life comes from life.

Most of us struggle with the concept of God but do not have the tenacity of investigations to know the holes of science. If you can see the universe as an analogy of truth, it becomes a great resource for the understanding and interpretation of life, which is the love of the matrix differentiating and discovering itself in its infinite cycle and transformations. Truth and love are infinite. The journey of love and self-discovery become the uniting of truth back to the source or God. To say "I love you" is both departure and return of action into the path of uniting the whole. When J Bach says love is helping one achieve their dreams, he is referring to the action of love, its journey of differentiation and unification. The mysteries of love are infinite. So life becomes an infinite journey of the infinite discovery of love.

Love is the process or journey that allows loving or initiating of action, being loved by return of action, and the love of union. So life in a sense is the free will or awareness animated by its relationship to consciousness and thought, and the process and journey is love. A perpetual motion of transformation, evolution, that allows hierarchy, equality, unity, and a myriad of concepts that define the separate or differentiated from the whole. So the entropy toward diversity and the cyclic adventure of free will away and back from the source will becomes the path to knowledge and wisdom. The process of living love is possible when the object is also capable of receiving and returning it, and also this experience of love is magnified when love is united in its various steps to its final destination to the source or God. Love is active or action, harmonic to its nature and the nature of the source will and yielding to balance, which is its characteristic.

For every action, the object acting cannot act unless the receiving object can be acted upon and receive the action and also has the same characteristics of the object acting. Hence, we said that life is one form of consciousness perceiving another. If I perceive something, then it must also perceive me by this precept. So perhaps the universe is a living entity that allows perception or is perceived and can perceive. The consciousness that is the universe is God—also with love and many other things by this principle of dependency, symmetry, and action. Hence, is there a God? Absolutely, as it's the perceiver and receiver of our actions. The principle of symmetry means that the nature of the action must be received by the balancing matrix. And hence, that we love means that the universe can love and receive love as well and is alive. In the same manner, God created the universe so as "to love." To create is to love and allow the action to be received and returned.

**Existing Physics**

Much of existing physics is represented in the manner of human measurement and perception. This is what we refer to as scientific observation. So this does not mean that is in fact the manner of the construct of the universe. Hence, historical approach would be far off the path, which leads to a unified field theory. The structure of the universe is quite different from scientific observation, although once established, scientific observation can be rewritten according to the precepts of the unified theory. Scientific observation is important in that it's human in its application and use. So science can be rewritten in two ways. The first

- **being the manner that the universe is, and**
- **secondly, in the manner it is perceived by humans and used by humans.**

The universe is multidimensional, and human experience is but a few dimensions. However, what we experience is useful to humans. Gravity may be a ten-dimensional phenomenon, but humanity experiences a four-dimensional effect of these ten dimensions, and Newton measured it in neither but in the manner related by two masses. But the rules remain constant throughout, and we can take Newton's view of measurement and translate that into a transdimensional language to more accurately understand what Newton is measuring. We cannot prove Newton's equation because it is not a dimension, and it is only an arbitrary measurement, hence, an abstract approximation. However, the relationship if physics terms in the measurement can be understood in a better light as a human definition and is still relevant. We can redefine what might be constants into variables and redefine awkward dimensional concepts, such as the gravitational constant *G*. This adds a complete understanding to our view. This theory also allows us to project into dimensions where we cannot measure and make predictions. Hence, we have equations of higher dimensions and can determine characteristics of them and their relationships to other dimensions. At a point where multidimensional travel will be possible, these equations become directly relevant.

**Faraday, Coulomb, Ampere, Ohm, and Voltage**

***Determination of Volt and Coulomb in Relationship to Dimension***

***Proof as follows:***

$D^5$ = joule

Farad = joule / volt$^2$ = coulomb / volt

Farad = $D^5$ / volt$^2$ = coulomb / volt

Joule = force × $D$ = $D^4 \times D$

**Joule = $D^5$ = volt × coulomb**

**Joule$^2$ = $D^{10}$ = volt$^2$ × coulomb$^2$**

**Ampere = joule / volts × time**

**Ampere = $D^5$ / volts × time**

**Volts = $D^5$ / ampere × time**

***Proof of Volt = $D^3$ and Coulomb = $D^2$***

**Farad = time / ohm = ampere × time / voltage**

**Ohm = time × voltage / ampere × Time**

1. **Ohm = voltage / ampere = voltage time / coulomb**

**Farad = $T^2$ × coulomb$^2$ / $D^2$ × $M$ = $T$ / ohm**

**Ohm = time × $D^2$ × $M$ / $T^3$ coulomb$^2$**

2. **Ohm = $D^2$ $M$ / $T$ coulomb$^2$**

***Now let's combine 1 and 2 for 3 expressions as follows:***

**Ohm = voltage / ampere = $D^2$ $M$ / $T$ coulomb$^2$**

**Ohm = voltage × coulomb / ampere = $D^2$ $M$ / $T$ coulomb**

**Ohm = $D^5$ / ampere = $D^2$ $M$ / $T$ coulomb**

**Ohm = $D^3$ / ampere = $M$ / $T$ coulomb**

3. **Ohm = $D^3T^2$ / $T$ coulomb**

**Ohm = voltage time / coulomb = voltage / ampere**

4. **Ohm = voltage time / coulomb**

***Hence, now to combine 3 and 4 for results:***

**Ohm = voltage time / coulomb = $D^3T^2$ / $T$ coulomb**

**Voltage = $D^3T^2$ / $T^2$**

**Voltage = $D^3$**

**Joule = voltage × coulomb = $D^5$**

**Coulomb = $D^2$**

**$D^{10} = (D^3)^2 \times (D^2)^2$**

**$D^{10} = D^6 \times (D^4)^2$**

We see that in space, the ether or ether has a voltage, and this is an expression of the pressure space has upon itself and a relationship to density. This idea of space is similar to what Tesla referred to when he explained the ether as voltage under various degrees of pressure. Space has structure and properties and is also composing all mass and particles.

***Ampere***

**Ampere = coulomb / $T$**

**Ampere = $D^2$ / $T$ = $V$ $D$ = current**

***Farad***

**Farad = ampere × time / volt**

**Farad = $VD$ × $T$ / $D^3$ = $VT$ / $D^2$**

Farad = $D / D^2 = 1 / D$

Farad = 1 / acceleration × density

***Watts***

Watts = joule / $T$ = volts × ampere

Watts = $D^3$ × velocity × $D$

Watts = $D^4$ × velocity = force × velocity

Density = mass / volume

Density = $D^3T^2$ / volume = volts × density / volume

Density = volts × mass / volume$^2$

Density = volts × density / volume

***Force of Gravity***

Force of Gravity = mass × mass / $R^2$ density × density

Force of Gravity = volts × volts / $R^2$

**Ampere tells us that all current moves along a distance. But $R^2$ tells us that as it moves along a distance, it distributes outward along an area. For example, if I light a match in deep space, the wave of light moves outward along the area of an enlarging sphere, hence, giving meaning to the inverse square law where the wave motion is distributed along larger areas as the wave moves along. Current is an elastic collision of the space matrix along the length of its path while energy moves outward along an area governed by the inverse square law. Space is a voltage established between dimensions of space, which allows current or elastic collision in and through space. Space is also a tetrahedral matrix structure of superstrings. The strings are what give the properties to both space and all physical interactions.**

**$D3T^2$ = mass = volts × density means that all volume has electrical influence, even the ether of space. Electrical properties and density are causes for gravitational acceleration.**

**$D3T^2$ = mass = volume × density; density = $T^2$**

**$D3T^2$ = mass = volts × density**

**1 ([$N$ × $m$] / $C$) × ($g$ / [m$^3$]) = 0.001 m-1 kg2 s-3 $A$-1**

**1 ([N × $m$] / $C$) × (kg / [$m^3$]) = 1 m-1 kg2 s-3 $A$-1**

## What Is a Dimension?

A dimension exists in relationship to other dimensions; trans-dimensionally, a dimension exists in a chain along with other dimensions. This coexistence can be described as induction that occurs trans-dimensionally and also as a polarity of opposite relationship. A dimension has both transdimensional and scalar properties. What is a dimension? Dimension is a matrix of space where the structure of the matrix is arranged in a certain way so that the properties are assigned to those arrangements. The lowest dimension on the scale of dimension is $D^1$ or God superstring, where time and velocity are related by polarity. By the time the relationship is expanded to $D^3$, we have space. Each dimension has a direct relationship to all other dimensions, so one does not exist without the others. The components of dimension are superstrings $D$, $T$, and $V$ from which all properties come. Hence, they allow interpretation of all universal orders or equations from the same basic units. $D$, $T$, and $V$ allow equivalency and conversion of units. $D$ allows a bridge between all physics, mathematics, and related definitions, tying a logical structure between all observable phenomena and expressed in mathematics. A dimension is a state or level of energy with properties from the space superstring matrix arrangement. Induction from one dimension to the adjacent is by a power of one. So each subsequent dimension is $D^n + 1$, where the mathematical relationship is multiplicative between dimensions.

Lorentz Force Law – Magnetic Force

$F = qV X B$

Magnetic Force = $q$ ($V \times B$) = coulombs (velocity × magnetic field $DT$) = $D^2$ ($V \times DT$)

Magnetic Force = $D^4$

In the above equation, we see three things happening:

$B$ or magnetic field = $DT$; time is describing magnetism

$I$ or amperes or current = $DV$, velocity is describing current.

$F$ or force is the third component.

The equation could be

Magnetic Force = $DT$ or magnetic field × $DV$ or amperes × $D$, which is the axis of force.

Magnetic Force = $D^4$

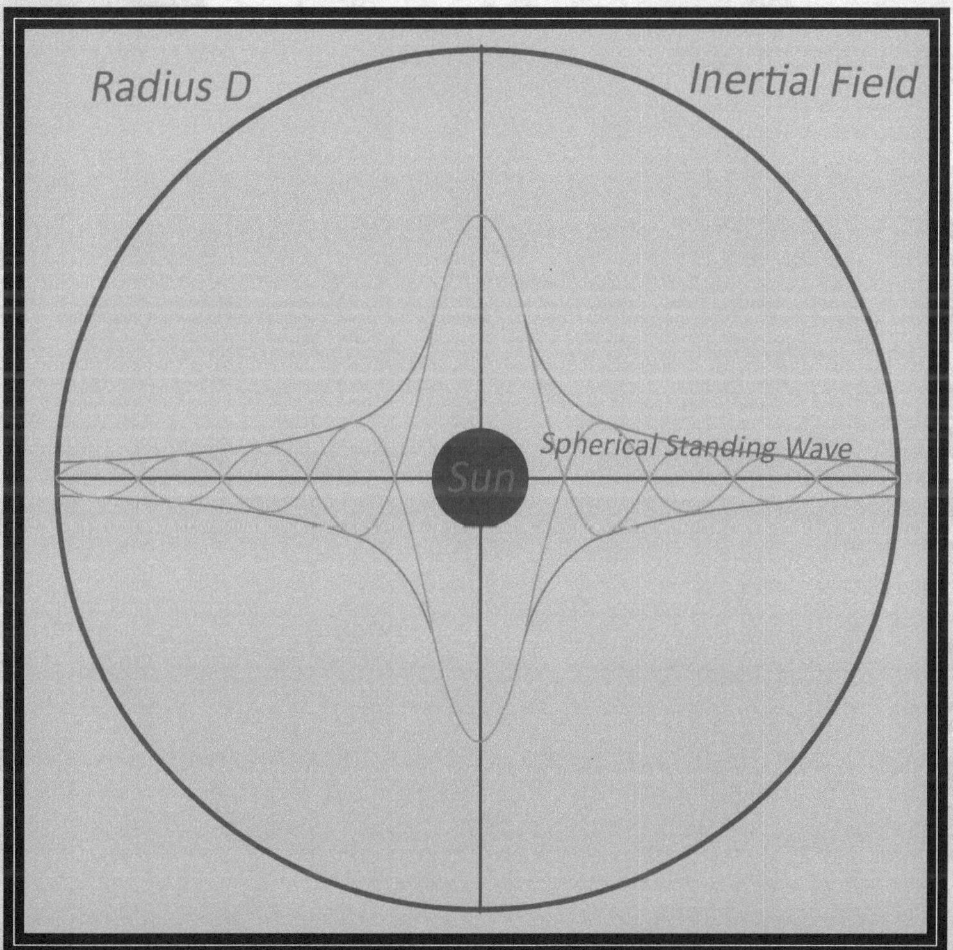

## Velocity and Time Networks in D

**It is interesting to note that the difference between *DV* and *DT* between the axis of current and magnetism is the relationship between velocity and time respectively and in the final direction *D*, which is basically a tetrahedral superstring arrangement of *D*, *T*, and *V*. The relationship between current and magnetic field is the difference between time and velocity. Time is magnetic component, and current has to do with motion. Magnetism is a key to time travel. Magnetism is an expression of time and current an expression of velocity. One inducing the other and one cannot exist without the other, so time is a real thing and a property of space—structural space.**

**With this coding system, you can check the validity of any equation:**

$F = q(E + V \times B)$

$Q$ **= coulombs =** $D^2$

$E$ **= volts per meter =** $D^3 / D^1$

$E = D^2$

$V$ **is velocity**

$B$ **= magnetic field =** $DT$

$F = D^4 = D^2 (D^2 + VDT)$

$F = D^4 + D^2VDT$

$D^4 = D^4 + D^4$**; force here is in two parts**

**So it's correct.**

$D^4 = D^4$

Mass and Time

**Mass =** $D^3T^2$. According to transdimensional theory, the time mass are related as such. In this theory, one can test the validity of any equation using the superstring coding system and reducing physics equations to variables *D*, *T*, and *V*. To get to the time and mass equation, you have to proof the equivalency and conversion of units, which has been done in this presentation, which is the idea behind a unified field theory, which bridges the gaps between and through units, which seem unrelated but are not. Let's look at the following example:

**Mass =** $D^3T^2$ **= volts × density**

**Mass =** $D^3T^2 = V^3 \times T / T^4$

**Gravitational Constant = 1 / density**$^2$ **=** $1 / T^4$ **where** $T^2$ **= density, so then**

**Mass =** $V^3 \times T / G$

**In any expression, the hyperspace for that expression can be achieved by replacing *V* with *C***

**Hence,**

**Mass =** $C^3 \times T / G$

**Since the gravitational constant is 1 / density**$^2$**, it is really a variable. Hence, it is the multiplication of the density of the center of space of two bodies 1 / density × density.**

The proofs for transdimensional equivalency was proved earlier in this presentation. You need to prove that $T^2$ **= density** or **mass =** $D^3 T^2$, knowing $T^2$ **= density**, you can take $G$ **= volume / mass time**$^2$, which are the units for the gravitational constant, and get

$G$ **= 1 / density ×** $T^2$ **= 1 / density**$^2$

So knowing **density** = $\boldsymbol{T^2}$ can allow you to rewrite the $\boldsymbol{G = 1 / T^4}$, changing $V^3$ to $C^3$ is describing the hyperspace or extreme amount of that space.

You can rewrite the equation to **mass =** $\boldsymbol{D^3T^2}$, where $D^3$ is volume and $T^2$ is density; this ties mass and time. Now why is this so? Not only mass and time are convertible, but all electrical units so that **mass = volts × density** and go on to every other equation and explore this. Where do these values come from? Space is an array of components *D*, *T*, and *V*, which, when combined in different configuration, you get different phenomena. Deep space is very simple, but as you move into an atom, they become more complex, adding *VT* for each dimension, which changes the properties. So then particles are where you get a clump of this matrix in concentration.

## Hubble Value

**$H_o$ = gravitational constant $G$ × density × $T$**

**$H_o$ = mass × time / $T^4$ $D^3$**

**$H_o = D^3T^3T / T^4D^3 = 1 / T$**

**$H_o = 1 / T$**

**$H_o = D^3T^3T / T^4D^3 = 1 / T$**

**$H_o$ = mass × time / $T^4$ $D^3$**

**$H_o$ = gravitational constant $G$ × density × $T$**

**Since the gravitational constant is 1 / density$^2$, it is really a variable. Hence, it is the multiplication of the density of the center of space of two bodies 1 / density × density.**

Hubble value is 1 / *T*, so 1 / *T* gets you to the equation above, and so it's also correct; using this transdimensional pattern, you can work from complexity to simplicity and vice versa. About the different dimensions, it surprised me what they became under this format. For example, $\boldsymbol{D^3}$ **= volts.** Was there a physicist that combined that relationship? There was one: Tesla. He saw space as voltage. Even empty space has electrical properties. But it also has volume and also has a relationship to mass and higher dimensions it relates to everything, to amps; you name it, and there is a connection.

## Mass and Space Relationships

What scientific observation perceives are perhaps the first five dimensions at most, but these are effects of sources at higher dimensions or induced by them. We see gravity as a force or $D^4$, but I reason that it is $D^{10}$ and run some calculation to show that you can get to $D^4$ using $D^{10}$ numbers. If you base the mass of the universe on its perceived size, then you can get a number, but I'm thinking it might be much larger since some light from distant stars never gets here.

I describe the wave of light as an elastic collision. So imagine a bowl of marbles, and where they touch is a photon. When this matrix vibrates, it's fully elastic, which means conservation of momentum and energy. What you see is the vibration where it arrives to your eye. Now some say, well, components of the matrix can't be spheres; well, if it's pulled to an ellipse, you get the proportion phi or golden ratio. Now the actual configuration is more complex, and these circles may overlap some. So the real geometry is tetrahedral, which allows movement in all directions. That space is transparent to higher atomic forms, but when excited, the frequency lifts it to be felt, so you see the photon. Ideally, we can describe superstrings as spherical. These are the ones that make space—*D*, *T*, and *V*—but through motion of the space matrix as space, they distort into ellipses, and this is where we get the ratio phi. There are three levels of space, and they are hyperspace, space, and subspace. Space is both scalar and transdimensional. A symbol metaphysically is the circle or sphere; the center is scalar, and the circumference transdimensional, combining infinity on two levels. We are a relationship of infinities of space. Particles are nothing more than clusters of superstring wave or matrix. The very properties of space, which make everything, are those we measure.

Now in a multidimensional universe, everywhere you see $D^3T^2$, you are going to define mass, so all volume has mass, even space. In this theory, a black hole is a higher dimension of space and not to be confused with three-dimensional spaces as we experience it. I imagine that every atom has a black hole, which to the center you pass trans-dimensionally from outer space to inner space near $D^{23}$ to that point of transdimensional infinity. The black hole at the center of the galaxy is a case where many black holes get together with a larger mass and form a large black hole. So the universe is perforated by infinity through black holes. Black holes and their transdimensional effects generate the space. So the center of the galaxy generates the space around it.

There is a ratio of mass to space by $\boldsymbol{D^5 = mc^2}$; calculate $D$ and get the extension of hyperspace of the object generating it. Now, have the mass of an object in two or more, like a galaxy being born, and there is an enormous expansion of space since if you half the object, you nearly double the space. This would account for acceleration of space as in an atomic explosion. Space is overlapping, so we are both in the space of our solar system and the space of the galaxy. The density of space is higher where space overlaps, and planets tend to orbit in areas of higher density. Now in a black hole, the condition of gravity is squared, so this would square the amount of space around a black hole, but not for the Sun in our solar system where the extension is $D$ rather than $D^2$.

I imagine that a graviton is a photon at a higher dimension sharing the same space, and that at center of every atom is a black hole, and when enough of them are together, they telescope into one big black hole. Mass creates space by a ratio extension of $\boldsymbol{D^5 = MC^2}$, where $D$ would be the extension of space. Gravity is a multidimensional phenomenon, where within the atom gravity is $G^2$ and outside it's $G$, and the projection of space is twofold. The Sun would project space at $D$ distance and the black hole at $D^2$ distance. Space density is higher within the solar system than outside. The Sun is projecting space to a distance $D$ as a standing wave, and the planets follow areas of density in an orbit.

## Transdimensional Gravity and Intrinsic Relationships

Gravity as well as all physics phenomenon is multidimensional. Gravity has its place in multidimensions being expressed in several dimensions differently. What we see and measure concerning gravity is a four-dimensional effect of a multidimensional effect from a higher dimensional source. Since all phenomena and effects are tied by induction in a multidimensional or transdimensional network and can be expressed through equations, there is definitely gravity in finite relationship to matter at various levels or dimensions. That Einstein could not define this is the reason for the lack of a unified theory. The force of gravity, in one explanation, is the result of the change in density of the gravity space field. In another definition, that same space has electrical properties and, therefore, connected to magnetic and electrical densities.

As the trans-dimensions rise as in a periodic table to higher levels, the source is infinity, so space as well as having electrical properties is also connected to infinity. Remember, we said that the center of every atom is a black hole and path to infinity. You might say the particular expression of this universe is a particular qualification of infinity. These properties of physics are all intrinsic to the structure and geometry of space dimension per dimension that are variations of $D$, $T$, and $V$. For example, time is an expression of magnetism, and at various other combinations with velocity or current, it becomes various expressions of gravity, from force to density, to field, to magnetism, and to electrical properties. So to say that gravity is an outsider unconnected to matter directly is incorrect. Because mass belongs to inertia as an idea strictly for mass is avoiding the direct connectedness of all things.

In this theory, you can derive all equations from simplicity to complexity and vice versa. The same process that derives the universal gravity equation also derives all other relationships. We can derive volts, coulombs, farad, and force of gravity and over two thousand equations by progressing from the simple $D$, $T$, and $V$.

The cause of inertial property has to do with the relationship between entropy and disentropy or the relationship of space created in a balance to higher dimensional mass and the inertial drive back to density. This is the perpetual motion inertial drive, continuously fed from infinity and driving the cycles of the universe.

Some might say that the electrical force is the fundamental field of all nature, but I would revise that to $D$, $T$, and $V$ and the fundamental superstrings of all space. Since electrical forces appear to influence the lower dimensions of human perception, obviously they are a part of the field forces. Space has structure, and $D^3$ = **voltage**, $D^2$ = **coulomb**; obviously, that is electrical.

Space is infinite as related to its source as a qualification of infinity. However, the universe also has finite properties that a qualification of infinity implies. The math shows us irrational and whole numbers, which speak of the qualification as various forms of limit within infinity. Space is both infinite and finite. Nothing in the universe is consumed, just like nothing really dies, but it is transformed. Transformation drives all motion of gravity; the inertia of higher density to lower density is an unbalance that drives gravity in dimension 4. This involves the density of the gravity field and electrical forces, which has been illustrated in this paper. This change in density is caused by intrinsic structural definitions of space that define the atom, and so is the field of space generated not an extension of the source? It is an extension of the source at higher dimensions, so then we can say that gravity affected from a higher source really is an intrinsic force associated with matter as a process and is defined differently at that higher level.

When we define **mass** = $D^3T^2$, gravity is an intrinsic property of mass because we are defining mass in relationship to space as time and velocity. If we define mass as a particle, then it is not entirely a definition of mass if it does not contain the $D^3T^2$. So then my defining **mass** = $D^3T^2$, we have established mass as a particular arrangement of space.

Scientific observation is flawed in its ability to correctly assess dimensions beyond $D^5$. Phenomena that we measure in gravitational mechanics have effects from sources in other dimensions. So we can speak of the apparent magnetic and gravitation fields as we measure them, but in reality, these are lower dimensional space effects of higher dimensional forms.

**Mass = volts × density**

**1 ([*N* × *m*] / *C*) × (*g* / [*m*³]) = 0.001 m-1 kg2 s-3 *A*-1**

**1 ([*N* × *m*] / *C*) × (kg / [*m*³]) = 1 m-1 kg2 s-3 *A*-1**

**Gravity Force = volts × volts /** $R^2$, which refers to the interaction of two fields of space; and also, this has to be redefined at the point of origin in higher dimensions since gravity is multidimensional. So without equations, anything can be claimed. Here in this theory, equations are presented, and all things then are intrinsically related trans-dimensionally.

There is no classification of anything in transdimensional theory as physical since space creates everything in arrangements of *D*, *T*, and *V*, and those are not physical as we measure physical. Physical is a multidimensional effect; for example, an atom contains almost all dimensions and ascending higher as we move into an atom. What becomes physical is in relationship of one transparent dimension in relationship to others denser. The transition I call transdimensional by induction and relationship. So the transdimensional definition of mass is hardly "physical." Particle is the interference patterns of space. Space is the substance of all things as a structural principle.

It is the change in density that causes the gravitational pull. That change in density or voltage is a priori to displacement, and hence, inertial forces are already in place. The effects of gravity are far reaching. The inverse square law is referring to change to space density through constant *G*, which is not a constant but a variable, and hence, that variable would interfere with the previously known manner of understanding gravity. In addition, all space has electrical properties.

All matter generates a field of space. This field of space varies in density from point of origin at higher density to the maximum extent of the field at lower density according to inverse square law. The total energy of this system including the mass core with supported space is defined by the equation $D^5 = MC^2$, where ***D*** is the extension at ***D*** = **fifth root** ($MC^2$), being the maximum extension of hyperspace measurable in the $D^1$ dimension. When the fields of varying density of two bodies overlap the imbalance in the density of the space or inertial fields or change in density causes a gravitational acceleration to balance, such as resolving a potential difference or voltage between the bodies. This space matrix field is fed from the center of each atom where there is a black hole supported and connected to infinite energy. So the atom is a transdimensional machine of multiple dimensions supplying power for the creation of inertial space and mass. The space of various bodies provides both a drive to entropy and also a return to density as balancing cyclical events. The mass of the entire system includes both solid mass and space mass.

**The new gravitational force equation is**

**Gravitational Force = mass × mass / $D^2$ × density × density = volts × volts / $D^2$**

Acceleration due to gravity becomes an interaction of two inertial fields defined by

**Acceleration = mass / (density × $D^2$ × $T^2$)**

**Acceleration = power / force × time = ($D^5$ / $T$) / $D^4$ × $T$ = $D$ / $T^2$**

**The gravitational constant $G$ has been clarified as 1 / density$^2$, which represents two inertial fields of space.**

The variation of density or unbalance or difference in density between the two objects and their corresponding fields of space causes motion. Changes in density cause acceleration to balance due to resolving action and reaction of the unbalance. The $G$ or gravitational constant in Newton's equation is a variable of two objects where **$G$ = 1 / density$^2$**. $G$ has been revised to mean inertial fields of space. Inertia mass and gravitational mass are the same because there is an inertial field or space matrix associated with every body of mass. The structure emits from the body of mass in the manner of an inertial field or space matrix that makes up the vacuum of space. The difference in the density between two fields of space can be described as voltage. This is also volume or $D^3$. Gravity can be described as the caparison of max volume of inertial fields.

**Gravitational Force = mass × mass / $D^2$ × density × density = volts × volts / $D^2$**

**Gravitational Force = volume × volume / $R^2$**

The velocity from which the field of space emerges from the source may be perpendicular to the direction of extension of space and density and force, but this may still be an open question. The imbalance and gradient of the density space field causes natural inertia to establish balance with other opposing fields. The change in density between two objects causes an acceleration motion toward balance. This field exists prior to or a priori to the gravitational effect, and the effect is intrinsic to the structure of space. Hence, we call it a gravitational field. Since space is transparent, there is no displacement but merging of fields to form a higher density. So the space within the solar system created by the Sun overlaps the space of the center (black hole) of the galaxy that also passes through the solar system. So deep space is less dense and responsible for shifts in flight path of satellites escaping the solar system, such as Pioneer.

All space fields have internal velocities or currents. Velocity and density have a relationship. The gradient of density lessens as you proceed from the source according to the inverse square law. You will note that the gravity equation expresses density, volume, and voltage, but this can also be rewritten to express velocity and acceleration as has been shown previously. Where the gradient in density is greater, one would expect larger velocities.

The matrix of space or inertial body is constant and fully elastic fluid matrix, and hence, you have the elastic electromagnetic wave at the speed of light $C$. As that matrix nears the source, it is still fully elastic but under pressure, and hence, more dense. The structure is elastic and constant in its vibration or allowance of elastic wave, but the form is compressed or denser. There is a gradient of density with constancy. The constancy is in the performance of the elastic wave, not the structure or density, although it has effect relationship. Benoules effect will bring two pieces of paper together if you blow between them. Density affects acceleration, as in the sail of a ship or wing of an airplane. Density is greater to create acceleration, even in Benoules effect. Of course, density has a relation to velocity. The matrix is denser near the source, even as the atmosphere is to the Earth. The difference is that even under this variation, the matrix remains fully elastic.

To get the extension of inertial space, calculate **$D^5$ = $MC^2$**; it's **$D$** for the extension of space. The extension of space for a black hole is squared. Density is a big factor; velocity and acceleration is in the discussion of gravity. But we can describe the inertial difference between bodies as a gradient of density and voltage. That density is created by matrix of varying density that causes motion by interaction of the inertial fields expressed as acceleration. The matrix comes forth at a certain velocity from the object source, and this coupled with a change in density accelerates motion for a gravitational force between bodies.

Space is a fabric, that's why in this theory; it is referred to as a matrix of superstrings *D*, *T*, and *V*. All properties can be reduced to those components. I do not mistake the matrix for the wave. The wave is an elastic collision of the matrix. The structure has the properties, which is also imparted in more complex configurations of space I call dimensions. I figure a black hole is near twenty-three dimensions, and the atom is a transdimensional machine. You cannot detect the density variation because gravity field is in the tenth dimension felt as a force in the fourth. What you can do is estimate the apparent density by solving for the entire density from the total extension. This needs to be studied. GPS is already a measurement of density variation because

**$T^2$ = density.**

You have to understand the transdimensional order to understand how to measure density and what's going on. Detecting waves is not the inertial space matrix field structure because the wave through the field is not the field structure itself. I agree there are density variations, and these cannot be measured directly but by observing effects in time.

**Gravitational Force = mass × mass / $D^2$ × density × density = volts × volts / $D^2$**

**Gravitational Force = volume × volume / $R^2$**

You might also compare gravity by the full extension of the inertial volumes. For example, $\boldsymbol{D^5 = MC^2}$ is the correct expression. Calculate $D^1$ and get the maximum extension of an inertial field around an object; it comes from $\boldsymbol{D^5 = MV^2}$ in this theory, where *C* is referring to the hyperspace or maximum extension of that dimension of space. Calculation of object volume of each object using *D* as radius of a circle can get you to volumes for comparative gravities.
**Gravity Force = mass × mass / $D^2$ × density × density**, where *G* or the gravitational constant is really the inertial field expressed as 1 / density$^2$, 1 / density being the inertial field of each object or field of space. Under modern physics, the gravitational constant *G* could be considered the strength of gravity. Certain variables need to be large to have a strong gravitational strength, such as when large masses are at close distances. This concurs with this theory by the idea that this also refers to the interaction of inertial fields of separate masses.

## Time Travel and Consciousness

We have spoken about this reality being a state of receded consciousness and that your essence or the prime consciousness that is you transcends this experience like if it is a dream. As in any form of consciousness, you have awareness, will, and thought, which is free by intent and action. Life is one form of consciousness perceiving another. And to be acted upon such as to love and be loved, the receiver must have the same capacity by a principle of symmetry as the sender and vice versa. We are speaking of actions in the present. So what is a memory then? A memory is my participation in the past, so without having travelled in time, I transcend the past in my memory as well as the present. If I was a time traveler and travelled to the past, then that new memory is incorporated into past events that already exist. So whether I travel to the past to create these new memories of the past, I am already a time traveler in the past because I have memories of the past. So then if I show up in the past and talk of the future, people would say that I am either psychic or a time traveler, but what would I have to offer of the future? Well, only a limited description of my own experience or memory, and hence, what I do not know is open to any possibility. My time in the past is only participation. Now, if I could know everything and every event and then time travel, then I would go back as a prophet because nothing would be left to imagination; but then again, my memory is limited to a time frame, and beyond that, I have no clue, so there is no scenario where in this universe I can allocate all time and events to my memory, so the possibilities of action and events always remain open. Hence, free will is always limiting both time traveler experience and outcome by your capacity of participation through memory. I can travel back in time, and what I experience is my participation in the collective memories of all. And that participation is evidence of free will since it is not owned except in my own experience, but rather participation. And also, since this is participation by a receded form of consciousness and a single memory, your experience being singular means that there is only one you and not a series of multiple you. Hence, transcendence rules out multiple universe time lines since accountability and purpose would have no meaning for your memory and being. For every action, the object acting cannot act unless the receiving object can be acted upon and receive the action and also has the same characteristics of the object acting. So my action is limited action, and the receiver or universe allows that action or free will, but it is not the total action of everything, and hence, my ability to change the total time line is not absolute, but only by participation in events. My memory is part of the collective memories, and these cannot change. And hence, my capacity to change what is in the memory of total consciousness is not possible, and hence, I can time travel to the past by participation.

With regard to the future, all free will is limited to the memories allocated to it. Can I go into the future and say, "I saw the future, and this is how you are going to be"? Let's say I am in a spaceship at the speed of light and have not aged, and then stop on the Earth and one million years have passed. I know some things have happened, history books written, and then get in my ship and travel back, can I change the events? And also, does the future then become set in stone? Let's now assume that everyone has free will; is it a particular style, and can this is expressed mathematically? If so, then is the interaction of all equations a particular result? Then you would say there is no free will if I can time travel into the future with one time line under such predictable circumstances. So how do I remove the paradox and have free will, or how do I have shared collective memories and a future that is flexible for a time traveler? We have said that to perceive, there must be a perceiver receiving your perception and that the universe in this sense is alive and consciousness of God. Since God includes all consciousness and all transcendence and all time, then the very thing we perceive has all information and all time. And for God, this would be absolute to know all things, but then we are back to the question of free will and the anomaly God allows for love and evolutionary experiences to explore infinitely. Truly if there is free will, there is always the act of the unknown future; even God has allowed Himself this privilege. The only way around the dilemma is to say that you have free will and that the future is only determined in so far as it is experienced and remembered. So if there are no time travelers, then the future is open and things are only nailed down as they are experienced? True, but partly. So in the past, the collective consciousness is a thing remembered and that can only be a point of participation for a time traveler. With regard to the future, only those things remembered about the future determine the future, if anything can be certain for the future. Hence, we are bound by the collective consciousness of the participants, and there is free will to the extent of that memory, and what isn't written in that memory can be changed, and what is written in that memory cannot be changed. So as soon as I travel to the future and only what I experience and record to memory is what is set, not what I read about others or see, only in my direct experience, so that allows the freedom of will beyond my own direct experience. And in the future, even some things that are set limit my free will to participation. So there is absolute free will in the present, but God, who has access to the outcome of all the equations that describe both our free will and their interactions, knows the outcome but allows infinite freedom by interaction with other free wills. God also knows how our free will reacts to those actions by his own free will or allow things to pass without interference. But then to allow for the genuine relationship of experience and love, there must be an absolute present. And so the entire universe is transforming from a fixed point. We can travel to the past and experience and share in what happened, and we can move to the future and establish it in limited degrees from a fixed reference of collective consciousness. So I get in my ship and travel to the future one million years and have missed the collective consciousness of free will from that amount of time, and from that time on, my experience of the past is now participation if I travel back, and I can tell everyone what it is, and I might be seen as a prophet, but I cannot change anything—even to the point that the matrix might require my removal from the matrix at that point for the sake of both my free will and the preservation of the free will of others. So in a sense, we feel sorry for Nostradamus, seer of the future, who is now only a participant. But he did say all the things he mentioned did not have to be, did not have to be in the sense that he did not know them and contribute them to the collective consciousness. Perhaps if his prophecies were wrong, then perhaps that particular memory was not a memory at all but an idea or illusion mistaken for a real memory of the future.

So time travel is allowed from an absolute present, and there is free will. The past is known and the future unknown from an absolute reference point of the present. The past can be shared and not changed. The future can write into memory of your particular experience, which leaves those things outside open to change since they are not in the collective consciousness in the absolute present. There is an absolute present, which is analogous to the creation of the universe: the point where memories and ideas are born by the qualification of infinity. And so the future opens the idea of multiple possible time lines forming to one where the past pulls them together.

## Free Will and Self-Discovery

Free will is in self-discovery and self-knowledge. It's by the intent of your actions that free will is born. That you can actually disagree with the Creator is an indication of free will. You have free will for self-discovery, discovery, and the discovery of an infinite God. God and/or truth exists in our experience and beyond for infinite discovery. Suffering is a path of free will to wisdom. There are many paths. The universe is shared consciousness, one form of consciousness perceiving another. The perception of your manifestation in this realty is an extension of consciousness. When you die, your consciousness manifests itself as another body, and in another reality, like the body of an angel: physics.

**How does one consciousness differ from another? Does it?**

It differs as an expression of God's personality; you are a part of God, an extension. Let's imagine everyone in Australia into one infinite person. And from that infinity, your essence emerges. In that essence, the infinite you is an original free will. The harmony with the matrix attained by wisdom experiencing free will leads to harmony of consciousness and harmonic mergence with it. To the point that if you are in total harmony with the matrix, acting beneficently with it, and hence, this is the nature of faith and miracles, that the wills of the dreamers. You and God in the dream are united in consciousness. Hence, you emerge from God as an extension and reflection of a portion of his essence with free will, and this journey of love is in the action to love and be loved, action and return of action.

Your consciousness transcends the emergent world. Emergent is the dream you are referring to in the manifestation of your body and participation in collective consciousness, which includes awareness, will, and thought (memory). There are no emergent processes that define the beginning of life; there are no emergent processes that define the beginning of the universe. The real you comes before at higher level of consciousness; the real you comes before with the free will: a priori. Emergence is quite predictable, as in the events of this natural universe; free will is an anomaly as life includes free will and unpredictable.

***Are we separate to this universe?***

Separate in thought and acting or will and awareness, in all things that define consciousness but separate and shared, as a transcendent being. There is a collective consciousness that binds us all together through God. Only things you own are your experiences or memories and thoughts; do you own a dream? Control is at the point of self-control in free will. You can act and be acted upon but cannot envelope the whole reality of shared consciousness as even God allows a freedom not to totally own you, evident in free will. So the quest to control is failed from start except in the mergence and harmony of love, where wills act together. And then this precept of love becomes a very contradiction to action that attempts to induce separateness as a philosophy and is a double standard since there is no life without some acknowledgement of love.

**How can physics account for God that loves, which is an action that validates all existence?**

You can prove God exists through science. All science eventually leads to a discovery of God. All life can be described as one form of consciousness perceiving another and that the universe is an analogy of truth or God. Well, I see that there are things similar to Christianity and Hinduism coming out of science. For example, in this theory, space is the building structure, like marbles, and where they touch is a photon. You only see the photons that are excited, and at such a small Planck scale, the whole universe is a sea of light. In the Bible, the idea of "let there be light" is presented. What God is doing is reflecting the essence of consciousness and projecting it into the order of the universe. The Toltec also describe the universe made from light, so on and so forth to just about all religions. They all are right, and these analogies of truth go on and on. There is no excuse for not knowing God because of works in creation.

When people talk their religion, it's more about their culture than religion. You never get to the truth because there is so much bias, so you'll talk to someone, and it's about their cultural history and nothing about sacred text, so you have huge abuses in all religions because of this. Sacred text does not necessarily mean cultural history of the person who is presenting the religion. If you get someone who believes in God but divorces themselves from their culture and gets some objectivity, it allows truth to emerge naturally and less encumbered by cultural myths.

Sacred text is a written account; culture is your family, country, language, semantics, logic, and myths confusing and interpreting truth incorrectly. Then you have, by bad interpretation, God perceived in bad reputation. The problem in understanding the truth of sacred texts is the cultural natures of people and how they interpret. If it weren't so, the church wouldn't be discussing the meaning of aliens now after two thousand plus years rather than look for the truth. Divergence from truth is based on cultural myths, so you get this huge divergence, and of course, it justifies all atheists because of contradictions. So here is presented still implying a constant underlying God, simply misinterpreted or misunderstood. Who is experienced at that level of discernment to correctly figure it out to what is misinterpreted and misunderstood? Surely this cannot be accomplished in a situation submerged by culture and myths.

**Do we need any of these sacred texts to find God?**

There is no excuse for not knowing God because of works in creation. Sacred texts can be a point of refining truths by both the lives of others as examples and wisdom that an active God would impart. But at the outset to know God, sacred texts are not required. The universe is an analogy of the truth speaking to all consciousness of this reality. We have all had the thought of God, and in our intuition does the greater truth lie. Simply, you can discover God for yourself; God has given you the capacity for discovery. By conscience, truth is available to all. So those that live by their conscience—the truth—are in harmony with the universe, and those that do not are in disharmony. Man sins because of conscience, the beginning with intent that is fulfilled in action. Critically looking for the exceptions is a great source of truth, and then there is the realization. There is more freedom than you imagined in your free will.

**How wrong can people be in their own myths and history to the point of extreme bias?**

They can be very wrong to the point of blindness. A biblical example would be Saul, later to be called Paul on the road to Damascus. Once the myths are removed, Paul sees more clearly, and as an example, this clarity is needed throughout life so that the memories of own lives, clouded by the many untruths of human error, do not mislead, and perhaps these can be released to start anew. This universe is no more the center of reality than this Earth is the center of the universe. Your experience, as a receded state of consciousness, is neither the center nor your eventuality, which is a transcendent source.

**Some might say, "Well, I don't like the suffering, and it seems a game to me."**

Without the anomaly of free will, we would be less than machines. There would be no freedom and infinite love; more specifically with regard to consciousness, there would be no awareness, will, or thought through which the anomaly or free will and the infinite manifests itself.

**Maxwell**

**$H$ = magnetizing field = $V$**

**$H$ = magnetizing field or auxiliary mag. field = ampere / meter = $VD / D$ = velocity = $V$**

**$H = V$**

**$E$ = electric field = $D^2$**

**$E$ = electric field = volt / meter = $D^3 / D = D^2$**

**$E = D^2$**

**Tesla = newton / ampere × meter = $D^4 / VD \times D = D^2 / V = DT$**

**Tesla = $DT = B$ = magnetic field $B$ = magnetic flux density = magnetic induction**

**Tesla = $DT$**

**$B$ = magnetic field = $DT$**

$B$ = magnetic field = Tesla = weber / $D^2$

Weber = Tesla × $D^2$ = $DT$ × $D^2$ = $D^3T$

$B = DT$

<u>$D$ = electric displacement field = constant</u>

$D$ = electric displacement field = newton / volt × meter

$D = D^4 / D^3 \times D = 1$ = balance

<u>$e$ = vacuum permittivity = permittivity of free space = $T^2$</u>

$e$ = permittivity of free space = farads / meter = $(1 / D) / D$

$e = 1 / D^2$

<u>$u$ = permeability of free space = $T^2$</u>

$u$ = permeability of free space = newtons / amperes$^2$

$u = D^4 / (VD)^2 = D^4T^2 / D^2D^2 = T^2$

$u = T^2$

<u>$pf$ = free charge density = $1 / D$</u>

$pf$ = free charge density = coulombs / meter cubic

$pf = D^2 / D^3 = 1 / D$

$pf = 1 / D$

<u>$J$ = current density = $1 / T$</u>

$J$ = current density = amperes / $D^2$ = $VD / D^2 = 1 / T$

$J = 1 / T$

<u>Gravity and Antigravity</u>

If force of gravity is $D^4$ or force, then an anti-force or antigravity would be $1 / D^4$ or $e^2$: reverse polarity. Gravitational force becomes $e^2$ or permittivity of free space, which is affected by dielectric material, such as in capacitors. Permittivity is the ability to allow an electrical field to maintain its natural polarity. So amazingly, there is an unexplained mystery with the Biefeld-Brown antigravity effect using capacitors. (Canning: 1-23) The force exerted on an asymmetric capacitor is given by the rate of change of momentum as follows:

***$F = MVI / q$***

***$F = (D^3T^2)\ V\ (VD) / D^2 = MV^2 \times D^3 / D^4 = E \times volts \times e^2$, where E is energy and e is permittivity.***

***$F = Q \times volts1 \times volts2 \times e1 \times e2 = 1 / e1 \times e2$***

***$F = 1 / e1 \times e2 = Q1 \times Q2$***

***For antigravity, the expression of $D^4$ would become $1 / D^4$, and hence***

$1 / D^4 = 1 / E \times \text{volts} \times e^2 = e^2$

The increase is force should be linear or proportional with voltage, energy, and permittivity squared, according to the equations illustrated. Energy applied to the total energy of the system, kinetic and potential. Permittivity is a function of the space matrix. Antigravity is represented by reversal in polarity we see possible in a capacitor.

$\boldsymbol{E = F / Q}$ **= newton / coulombs =** $\boldsymbol{D^4 / D^2 = D^2}$

$\boldsymbol{G\,M\,m / R^2 = Q \times E1 \times E2 / R^2 = D^6 / D^2 = D^4}$ **= force**

$\boldsymbol{G\,M\,m / R^2 = mv^2 \times mv^2 \times e1 \times e2 / R^2}$ **= mass1 × mass2 /** $\boldsymbol{R^2 \times u1 \times u2}$

***<u>Further Iterations</u>***

$\boldsymbol{H = V}$ **= velocity = magnetic field strength**

$\boldsymbol{G}$ **= 1 / density1 × density2 = 1 /** $\boldsymbol{T^2}$

$\boldsymbol{E = D^2}$

$\boldsymbol{Q = D^2}$

$\boldsymbol{Q = V^2 / G = H^2 / G}$; ***H* is magnetic field strength**

$\boldsymbol{F = Q\,E1 \times E2 / R^2 = H1 \times H2 \times E1 \times E2 / R^2 \times G = D^4}$

In the above equations, you see a relationship between gravity, voltage, charge, and also capacitance. That ties electrical to gravity in this theory, which are volume of the inertial mass field, voltage, charge, capacitance, and the gravitational constant being the density inertial field of two bodies where ***G* = 1 / density1 × density2**.

Antigravity can be explored in various ways. For example, as follows:

- **Antigravity due to the reverse in polarity within the inertial field or space matrix.**
- **The source of the space matrix or inertial field is the atom, and a reverse in polarity of the relationship between positive and negative field positions that generate the inertial field also would be anti-gravitational.**
- **Adjustment of the field polarity by magnetic fields, which can be explored.**

The atom generates the inertial field involved in the polarities that are attributed to gravity and its manipulation for antigravity. So an entire chain of multidimensional events are involved in the gravitational effect, from the generated field to the mechanisms that generate the field in an atom. That Einstein might claim the nonexistence of "gravity" is purely semantics, and a singular notion that excludes the multidimensional variances involved in gravity. Also, his notion of four dimensions, whereby you have three for space and one for time, is very limiting. It may be that we perceive four dimensions as humans and have a good handle on adjacent dimensions such as $D^5$ via experimentation, but $D^4$ is $D^3 \times D^1$, where $D^3$ could be volume or volts and $D^1$ is acceleration × density (this is where he puts time). The variances of a transdimensional construct offer more flexibility.

To reverse the model of the atom as a complete action is antimatter. However, the things needed to create anti-gravitational effect by a reverse in polarity can be accomplished completely or partially or, better said, at different levels or dimensions of inertial field or space matrix generation. Hence, reversing the polarity at the level of the inertial field with magnetism or capacitance is but one locality or level, and this can become different at another level of dimension, accomplishing the same thing or anti-gravitational force. There would be several methods to generate antigravity. One would be to reverse polarity in the natural inertial field or space matrix, either through magnetic fields or capacitance, and another would be to reverse the natural field generating relationship within an atom rather than having negative flow around a positive core; the reverse of positive around negative would reverse polarity. This might be the avenue in having positive flows around a negative core as in superconductors and super fluids.

Light is an elastic collision through the space matrix and not internal to the functioning of that matrix, so the math relationship would be partial there or indirect as related to gravity. This is often confused in the standard model. Although this relationship is of value to determine what that current motion is in relationship to gravity.

***The gravity of a planet can be solved by volume inferred by the equation as follows:***

**Force of Gravity = volume1 × volume2 / $R^2$ = mass1 × mass2 / $R^2$ × density × density**

You have the total **mass = mass core + mass of the inertial field**. Now the mass of the inertial field varies by density, so the total mass of the inertial from the mass core equals the mass core plus the result of **phi** times the **mass core** and adjusted according to $R^2$ or inverse square law by taking the square root of this second part as follows:

**Mass of the Inertial Field = mass of the core + *SQRT* (mass of core × phi)**

***SQRT* (mass × phi) = inertial field adjusted for change in density by inverse square law**

*D* or radius of the inertial field from the center of the mass core would be derived from the equations $D^5 = MC^2$ by calculating for *D*, which is *R* in the mass of inertial field equation above. Once you have *R*, you can calculate the inertial volume of any planet or object by using the volume for a sphere, which is 4/3 pi $R^3$. If you run the numbers for inertial volume *V* for Mars and the Earth, you will find that Mars is 38% volume of the Earth, which is a ratio of gravity or acceleration by comparing volumes of inertial fields. And as a reminder, you don't need the gravitational constant *G* if you calculate the size and mass or density of the inertial fields. There is more unmeasured mass than measured mass.

One might ask, "Well, how does space have mass?" But remember the equation **mass = $D^3T^2$,** and rewritten, this is $D^3$ or **volume = mass / $T^2$ = mass / density**. You will note that this relationship refers to the gravitational constant of two bodies in 1 / density. Also, mass / density are also included in the revised gravitational equation above. The equation for gravitational acceleration comparisons above relates to solid bodies or planets of similar density. When we review the gas planets, an adjustment must be made for the change in density. So for example, Saturn would have an inertial field similar to the Earth while acting overall with a stronger gravitational energy at $D^{10}$ because of the larger mass. Hence, diffusing the mass lessens the projection of space. So the inertial fields of Saturn and the Earth are similar.

The shift in center of mass of planets around the Sun implies shift in inertial densities as well of the planets around the Sun. Also, when exiting beyond Neptune, you fall into the space outside the Sun or the space created by the center of the galaxy, which is less dense. The form of all space is ideally spherical but, due to planetary spin when occurring, should be deformed into more elliptical form with inertial densities concentrated perpendicular to the axis of spin. The "effect" of space created by the Sun would be as a standing wave originating from the center of the Sun but, in reality, is a zone of higher density and spun around by it, forming like an elliptical spindle top or gyroscopic action more dense in the plane of the ecliptic, a natural location for the planets to gravitate to zones of higher inertial density. Pluto does not have this advantage and, therefore, would not fall in that ecliptic plane and probably be more comfortable relating to the asymmetries in space density that it exhibits. Also, the space generated by a black hole should be at a scale squared, since the inertial transitions have collapsed and essentially a direct transmission. Also, there may be a slight eccentricity in the center of mass of the inertial field since it is flexible and, if sometimes slightly elliptical, might cause reason for an eccentricity. This may contribute to trajectory shifts or slight shifts in planetary orbits over time along with other gravitational eccentricities between planets and the Sun. So no orbit will ever be so perfect in space.

***<u>T</u>* <u>= time</u>**

***T* = time**

$J$ = current density = $1 / T$

$J$ = amperes / $D^2$ = $VD / D^2$ = $1 / T$

$J = qnv$ = charged particle × particle density per volume × drift velocity

$J$ = coulombs $(1 / D^3)(D / T) = 1 / T$

$J = pv$ = charge density × drift velocity = (coulombs / $D^3$) $V = V / D = 1 / T$

Time = $D^2$ / amperes = coulomb / ampere

Ampere = coulomb / sec

Time = $D^2$ / amperes = coulomb / ampere

Time = $V^2T^2 / VD = V^2$Density / $VD = V$ × density / $D$ = density / $T$

Time = density × current density = density × J

**Time appears to be a function between the inertial field density and the moving charge density within the inertial field. Time is a real phenomenon. The manner in which we perceive time has to do with perception, but the quality of time has to do with the relationship within the inertial field involving the entire field density in relationship to the moving charge density within the inertial field. Hence, by controlling these features of the inertial field through magnetism, it is possible to influence time as follows:**

- **Directly within our present dimensional experience by manipulating inertial density and charge density of the inertial field.**
- **Trans-dimensionally, the effects of time are even more extreme, where time flow can be induced by other dimensional arrangements or dimensions to cause time flow or travel.**
- **And also trans-dimensionally and in relative terms, where an object is moved trans-dimensionally, time travel would be even more pronounced.**

***Understanding Time***

**Magnetic Force = *DT* or magnetic field × *DV* or amperes × *D*, which is the axis of force.**

**Magnetic Force = $D^4$**

As we have seen before in the equation above, the relationship between *D*, *DV*, and *DT* illustrate a very interesting phenomena. The equation is about the electromagnetic relationship of electrical current or space matrix as follows:

- ***DT* or *D* with time is the magnetic component of the relationship.**
- ***DV* or *D* with velocity is the current component of the relationship.**
- ***D* is the inductive component where *D* = acceleration × density and is understood presently as force.**

So there is a triad relationship in dimensional form between time, velocity, and *D*, which is also a relationship of magnetism, current, and *D* respectively. *D* seems tied to relationships of induction and, hence, become the creator superstring, God particle, or form of balance.

Now take this form in another example for the definition of time.

- **Time = $D^2$ / amperes = coulomb / ampere = $D^2 / VD$**
- **Velocity = $D / T = D^2 / DT$**
- **$D = D^2 / D$**

**Here in our description of time, again we have time as an expression of magnetism, velocity as an expression of current, and *D* in a triad relationship. We can take this exercise to other areas for example:**

**Mass = $D^3T^2$ = volts × density = volume × density and so on.**

**Then three expressions or volume and time related in mass are as follows:**

- **$D^3T^2$ = mass**
- **$D^3V^2 = D^5 / T^2$ = energy / $T^2$**
- **$D^3$ = volume**

**Three expressions involving time and $D$ are as follows:**

- **Ohm = volts / amperes = $D^3 / VD = D^3 T / D^2 = DT$**
  **Ohm = resistance = $V / I = D^3 / VD = D^3 T / D^2 = DT$**
  **$B$ = magnetic field = $u\,H$ = permeability of free space ($T^2$) × auxiliary magnetic field or magnetizing field $V = DT$**
- **Amperes = $DV$**
- **$D = D$**

**There is a relationship of resistance and current in the induction of $D$ and magnetic field and current in the induction of $D$ as expressed above. So the relationship of mass is the axis of magnetism, the relationship of current is the change in the rate of energy, and the third is volume. Mass can be described on its own terms and also as induction of other constructs. Time is a factor in these relationships. Where time is concerned, we might try:**

**$V$ × density / $D$ = $T$ = density / $T$**

**$T$ × density / $D$ = density / $V$**

**Density / $D$**

**So there appears to be a consistent relationship with regard to $D$, $T$, and $V$ with regard to the formation of all physical phenomena, where $D$ refers to an induction principle as a function in relationship to magnetism involving $T$ and current involving $V$.**

**So what is time? Time becomes a component that ties space to motion in multiple dimensions $D^n$. So experiencing time is in multiple ways. A clock may be one manifestation, but it appears incorporated to the transdimensional reality on many levels. It appears to be a component of transformation, as are the others $D$ and $V$, but more subtle in the sense that it incorporates acceleration of density.**

We had spoken before how consciousness was awareness, will, and thought and had located time to the nature of thought. All of us are never the same in past time as we are in the present time and always transforming as we pass to the future time. The common human idea is that there is a beginning and end, but perhaps the greater truth is that transformation is the real nature of time and that what might appear to be beginning or end is nothing more than a transformation or shift in consciousness. Hence, time in life, as it is in physics, is transforming those things related to it. One thing becomes slightly transformed or changed as transformation dictates. Transformation exists within a certain experience of consciousness and also between levels of consciousness. The depth of the transformation is experienced as the depth to which time is experienced. Hence, the life of a being that is truly unique and a deep experience can be considered to be a long time. And for each creation, that transformation, whether an insect or a mammal, time lingers through the process of transformation. Even in dreams we see time moving. But between levels of consciousness, what may seem forever in a dream, as when you sleep, may be but a second when you wake to the present level of consciousness, which also appears to be time consuming. So perhaps, in different dimensions of experience relating one to another, time in the experience is much longer than when you see it from another dimension, relatively speaking. So perhaps at the source of your being, comparing that experience with your life here, the time here may not be that long, relatively speaking. So in both science and consciousness of your life, time is assisting in the process of transformation. When Shakespeare says, "Time will unite that knot, not I, 'tis too hard a knot for me to untie," he is saying time solves problems sometimes when solutions cannot be found. He is also saying time transforms our lives in ways we don't control or understand, and the outcome sometimes is better than our plans or capacity.

**Mass = $D^3T^2$**

**Mass = $D^3T^2 = D^3\,u$ (magnetic field strength) = $D^3T^3$ = volts × $u$**

Mass can be described in many ways. Most simply, it has to do with the density of the inertial field in its various states trans-dimensionally.

**$E$ = electromagnetic induction = weber / $T$ = $D^3T\,/\,T$**

**$E = D^3$ = volts**

Mass can be described by the relationship of volts to density or electromagnetic induction to density. The vacuum of space has mass; that it is not felt is because space is transparent to higher atomic dimensions of space. We live in a universe that is multidimensional, transdimensional, and multi-transparent when relating one dimension to another. It is conceivable to travel in another dimension and through mass and shortcut space in relative terms as well. (Christov: Web Link)

**$P$ = power = $VI$ = volts × amperes = $D^3\,VD = D^4\,V = D^5\,/\,T$**
**$P = MV^2\,/\,T$**
**$n$ = particle density = $1\,/\,D^3$**
**$B$ = magnetic field strength = $DT$**
**Mass = $P\,T\,/\,V^2$ = power × particle density × magnetic field strength × density**
**Mass = $P\,D\,/\,V^3 = P\,D\,T^3\,/\,D^3 = P \times D \times T$ × density × $n$ = $P \times B$ × density × $n$**
**Mass = $P \times n \times B$ × density**

**Relationship of Inductance and Capacitance**

***Capacitance***

**Capacitance = coulombs / volts = $C = Q\,/\,V = 1\,/\,D$ = farad**

***Inductance***

**$H$ = henry = V × s / ampere = weber / ampere = inductance**
**$H = D^3 \times T\,/\,VD = D^3T^2\,/\,D^2 = DT^2$ = ohm × time = $(DT) \times T$**
**$H$ or $L$ = weber / ampere = $N\,/\,I = D^3T\,/\,VD = DT^2$ = inductance**

***Capacitance and Inductance***

**$L\,/\,D \times C\,/\,D = e \times u = T^2\,/\,D^2 = 1\,/\,V^2$**
***Inductance in Henries $H = DT^2$***
***Capacitance = $1\,/\,D$***
**$L\,/\,D \times C\,/\,D = (DT^2\,/\,D) \times (1\,/\,D)\,/\,D = T^2 \times 1\,/\,D^2 = T^2\,/\,D^2 = 1\,/\,V^2$**
**$L \times C = DT^2 \times 1\,/\,D = T^2$**

***Induction***

**Volts or induction = weber / $T = D^3$**

**Volts or induction = weber / $T = D^3T / T = D^3$**

**Magnetizing Field**

**$H$ = magnetizing field or auxiliary magnetic field = $V$**

**$H$ = amperes / $D = VD / D = V$ or velocity**

**$H = B / u = DT / T^2 = V$ = derived field = magnetic field / permeability of free space**

**$M = V$ = magnetization = amperes / meters = $nm = (N / V)\ m$**

**Gravitational Constant**

Let's now look at the gravitation related to two spheres with a diameter of one meter.

*Mass of Inertial Sphere*

**Sphere of one-meter radius has volume = 4/3 pi $R^3$ = 4.189 cubic meters**, so 4.189 kg if one cubic meter is 1 kg.

**So the total mass = mass core + mass of the inertial volume**

**So the total mass = mass core + *SQRT* (mass of the core × phi [1.618])**

**So the total mass = 4.189 + *SQRT* (4.189 × phi [1.618])**

**Total mass = 4.189 + 2.6 = 6.79**

*Volume of Inertial Sphere*

From $D^5 = MC^2$

**$D$ or radius = fifth root (total mass) = fifth root (6.79) = 1.466; radius of sphere = 1.466**

**Volume of sphere of inertial field = 4/3 pi $R^3$ = 4/3 pi $(1.466)^3$**

**Volume of sphere of inertial field = 13.2**

*Gravitational Constant*

In this theory, the gravitational constant is defined as follows:

**$G = 1 / \text{density}^2$ = volume1 × volume2 / total mass1 × total mass2**

**$G$ = 13.2 × 13.2 / 6.79 × 6.79 = 3.8**

This seems far off from 6.7; well, then inertial space is not considered in Newton's gravity formula, and $G$ is the approximation, and also the densities described in this theory that constitute $G$ are variables and, as such, will show divergences in calculated space trajectories if the traditional $G$ is used. If you throw out the inertial field from the mass in the gravity calculation, you are forced to create a constant, and this moves the constant number to near 6.7, approximate and not accurate.

For one sphere, the ratio of inertial space mass to total mass is

**2.6 / 6.79 = 0.38 mass is inertial space.**

More than half of this inertial space is within the mass core, and the remainder extending beyond the planet from the center. If you eliminate the inertial space or about .38 of the mass, $G$ needs to increase by a factor that approximates the gravitational constant to make up the difference. In this case, the mass of two objects are identical, and so .38 of the mass is missing for each sphere if we are using the normal mass, where *G* is used to make up the difference. **0.38 of 6.79 = 2.6**, which when added back to the total mass, is close to the gravitational constant at 6.7 without using *G*. In conclusion, if you do not consider inertial mass, *G* needs to be used; if you are using inertial mass, it comes down to describing gravity as the comparison of inertial volumes. Note that the calculation of inertial volumes includes density and mass of both mass core and inertial mass, so mass has not been excluded.

***G* or 6.7 – revised *G* 3.8 = 2.9 balance *G***

**2.9 / 6.7 = 0.44 of the total *G***

**3.8 / 6.7 = 0.56 of the total *G***

***SQRT* 2.9 balance *G* for two masses is 1.7*G***

**1.7*G* × 0.38 = 0.7*G*, and the remainder is 1*G***

***SQRT* 3.8 = 1.94*G* for each mass**

**1.9*G* × 0.38 = 0.8, and the remainder is 1.1*G***

**Mass core is *G* core = 1.1*G* + 1.1*G* + 1*G* + 1*G* = 4.2*G***

**Mass inertial field *G* = 0.7*G* + 0.7*G* + 0.8*G* + 0.8*G* = 3.0*G*, but this needs to be adjusted by *SQRT* phi**

Which is where we started, and the relationship of core mass to inertial mass is the *SQRT* phi.

**3.0*G* / *SQRT* × phi = 2.70*G***, *G* for inertial field is adjusted by *SQRT* × phi

**Now we have near 6.7 because 4.2*G* + 2.7*G* = 6.7*G***

When you take the inertial mass out and you have to build a fudge factor, unknowingly, this is the *SQRT* of phi and related to *G*. So *G* is close to 6.67 because you have removed the inertial mass in Newton's equation and have referred to it by a general multiplier for each mass, but *G* is a variable when explored in its diversity described in densities of interacting gravitational objects with inertial fields.

**$G = v1\ v2\ /\ (m1 + i1)\ (m2 + i2)$; *i* refers to inertial field masses**

<u>*Comparison of G in Newton's Standard Gravity Formula and Total Mass Gravitational that Includes Inertial Mass and Core Mass in Transdimensional Theory*</u>

**$F = Gm1m\ /\ R^2$**

Total inertial mass is 6.79

Total mass as in Newton is mass core and inertial mass in the core

**Mass core = 4.189**

Inertial mass in the core: Well, radius of mass core is 1 meter, and radius of the inertial mass is 1.466

**1.466 - 1 = 0.466; 0.466 / 1.466 = 0.683 or 68%; 68% of inertial mass of 2.6 = 1.77; 2.6 - 1.77 = 0.83 inertial space outside the mass core**

**Total Newton mass is 0.83 + 4.189 = 5.02** if we add the space outside, although he doesn't do this.

**$F = Gm1m2\ /\ R^2 = G\ 25.2$**; *G* at 6.9 gives you 174

**$F = V1V2\ /\ R^2 = 13.2 \times 13.2 = 174$**

**As you can see, using inertial volume is far more accurate than *G*, which is totally incorrect, and *G* is really an approximate abstraction of what is really happening. *G* is not even accurate since it leaves out inertial mass, so one would expect trajectory calculations to be in error.**

<u>**Electrical Conductivity and Resistivity**</u>

**$o$ = electrical conductivity= $J\ /\ E$ = current density / electrical field strength**

**$o = (1\ /\ T)\ /\ D^2$**

$o = 1 / TD^2$

$p = \textit{resistivity} = 1 / o$

$p = 1 / (1 / TD^2)$

$p = D^2T$

**Free Charge Density**

$pf$ = free charge density

$pf = 1 / D$ = coulombs $/ D^3 = D^2 / D^3$

$pf = 1 / D$

## Life and Purpose

Life is about us. We have talked about life being experienced in various states of consciousness. Therefore, the universe is not a goal in and of itself. The processes are not goals for themselves. Both the universe and its processes allude to truth, but they are not the destiny of life, to make the understanding of the workings of the universe a goal or quest by which we are fulfilled: the evolved man rising up from primal beginnings. As a form of consciousness, the universe is present for collaboration and development of self. So the quest can be finished as soon as it began, and its usefulness is appropriate insofar as it accomplishes the task of leading us to ultimate truth. This can take time, or it can be as an epiphany. All material accomplishments are illusions; they are nothing that is permanent, except that which develops the self in you and others. So the idea of "store your treasures in heaven" means develop your consciousness as a harmonic relationship to ultimate truth; the tools are merely transient and there for this purpose. But the opposite appears prevalent in human understanding, that most of the things we do have such importance, that we will transport them from this emergent reality to a higher transcendental source. Like the Egyptian pharaohs, we can take our accomplishments and things with us to the other world. All our myths can come with us. Alas, not so. You can no more take your car than you can take the car you dreamed of in a dream last night. So to the transcendent being, the importance of a dream is not that it comes with you; it's how you managed it and what you learned. So then belief becomes a powerful thing; the myths that you have been given can mislead you and control your life. And by contrast, as they say, the truth will set you free. So the task of conquering or mastering the universe does not concern the enlightened man who is on a quest for the mastery and understanding of self, truth, and relationship to others and, in this process, also assist others similarly by nature of a rightful consciousness or mindfulness.

We said life is one consciousness perceiving another. Also, we talked about shared consciousness when referring to shared memories. This universe is an extension of God's consciousness. So we are not only individually created but also share consciousness with God and, through him, everyone else collectively. So when you die, that consciousness that you shared as your body remains as you go on to share another portion of God's reality or consciousness manifesting in another body form. Life is transformation, and as we move through it, we learn its secrets, whether by a middle path or the suffering that allows us to know the difference between truth and illusion.

### Is there free will?

Yes, there is a free will.

Consciousness is being transformed. That transformation is taking future variables and translating them into the past. Without free will, we would be less than machines; without accident, anomaly, and mistakes; without evolution and change; without pain; and without transformation. The matrix allows the interaction of free will. The quality of action must have that quality in the receiver by symmetry, and so a God of free will and love must impart that to us as well. This universe is one form of consciousness perceiving another, and with that includes those things of the sender. To love, the receiver must also love, and this includes various characteristics that also involve free will.

Consciousness is awareness, will, and thought. Therefore, free will is a transformative element or action in the manner that velocity and time are transformative elements by their mathematical relationships. In this way, infinity of consciousness is qualified and transformed. So if we think of the universe as a "set" place since you have the same input and output, this is not a restriction of free will since the will is not so much in the object, but in the relationship that causes transformation. This begins with intent and completed by action. We are here to learn, not to make the dream the object of truth; but the transformation of consciousness, the dream becomes an analogy of truth by our experience. We are not puppets in a set universe. We are transcendent beings experiencing and emergent reality by our participation of our consciousness and its participation in the collective consciousness.

Ultimately, the freedom of consciousness resides at the source or God, and this life is but a receded form of consciousness.

**Anti, Antimatter, and Transformation**

In the quest for anti, we need to reverse polarity. This means flipping the properties from numerator to denominator and vice versa. So for example, antimatter would be mass, which is

**Mass =** $D^3T^2$

And reversing polarity to get

**Antimatter = 1 / mass = 1 /** $D^3T^2$

The normal relationships are reversed in the anti principle: anti

**Antimatter = 1 / mass = 1 /** $D^3T^2$

$n$ **= particle density = 1 /** $D^3$

$u$ **= permeability of free space =** $T^2$

**Antimatter = particle density / permeability of free space =** $n / u$

**Matter = permeability of free space / particle density =** $u / n$

Transformation

Reversing polarity has to do with reversing the relationship of properties. This means the transformation of one in relationship to the other, or the transformation of the numerator by the denominator. Anti means to reverse polarity, and the divisible refers to the process of transformation. In many instances, the transforming principle is related to time, and that is why time is very much perceived as a transforming property.

**Variations of Anti, Antimatter, Anti-space Matrix Based on Charge and Spin**

We can talk about the natural occurring matter *s* "+ - +" or positive matter core, negative orbit electrons, and positive inertial field, which relates to the charge and spin of particles and the space matrix that produces those particles by interference patterns. The traditional use of the word *antimatter* of annihilation would be relegated to the realm of "- + -"or negative matter or atom of reverse polarity or charge and reverse field spin. There are other variations of charge and spin that can be described, and so the space matrix can be described as having relationships as follows:

- "+ - +," normal matter, + charge proton and spin, - electrons and spin, and + field.
- "+ - +," antimatter: + charge and opposite spin, - charge opposite spin, + field opposite spin.
- "- + -," antimatter: - charge and opposite spin, + charge opposite spin, - field opposite spin.
- "- + -," antimatter of reverse polarity: - charge and spin, + electrons and spin, and - field.

In this theory when we refer to antigravity and antimatter, this has to do with the reverse polarity relating to our dimension "+ - +" and inverse of "- + -," meaning the relationship of core matter to orbiting charged particles or field to inertial field generated. This is a reverse in polarity of particles and/or space matrix of the same spin, not opposite spin. When you get to the ant-matter of opposite spin or "- + -," then you are faced with opposite spin and opposite charges. The annihilation comes from the field differences where symmetry of spin along with charge of the matrix cancels.

An antiparticle has opposite charge and opposite spin or charge conjugation and parity respectively. So in the above model, the useful anti-gravitational field and antimatter relationship without annihilation is where the spin remains the same with the charges different. So "+ - +" of normal matter works well with the "- + -,"enabling inertial fields of opposite charge polarity. When we move toward models of opposite spin and charge, then we have the anti-relationship or antimatter that results in annihilation.

So anti in the ultimate sense requires three things:

- **Reverse in polarity or model, which here is described as the inverse relationship $1 / X$.**
- **Reverse in charge.**
- **Reverse in spin.**

Normal matter has a matrix structure, a charge structure, and a spin structure. When we speak of effects of reverse polarity in this theory with regard to antigravity, we are speaking of a model of reversing the transdimensional structure or relationship and charge or two steps, since the antigravity we are seeking does not involve annihilation but a reverse in the charge field or geometric orientation of two fields that repel.

**Transformation, Time, and Density**

**Anti-time = $1 / T$ = will reverse time**

**$J$ = current density = $1 / T$**

**$J$ = amperes / $D^2$ = $VD / D^2 = 1 / T$**

**$J = qnv$ = charged particle × particle density per volume × drift velocity**

**$J$ = coulombs $(1 / D^3)(D / T) = 1 / T$**

**$J = pv$ = charge density × drift velocity = (coulombs / $D^3$) $V = V / D = 1 / T$**

The density of current flow affects time, and the change in density transforms time and, in this case, reverses time. Throughout this theory, we see how time acts as a function that transforms and by its relationship to density. The variation of density in various functions consequently has an effect on time. In this case, the change in density of charge or by relationship, the magnetic field will in fact alter time. It is possible to accelerate or decelerate time by changes in density of the space matrix.

**$D^3$ or Space and the Induction of Free Energy**

We can consider space to be a photonic superconductor. As we said through the various dimensions, superstrings $D$, $T$, and $V$ of the space matrix relate in more and more complex ways to produce the phenomena of force related to $D$, magnetism or field related to $T$, and current related to $V$. So $D^3$ would have a relationship of induction relating $D$, $T$, and $V$ as follows:

**$D^3 = D^2D$**

**$D^3\,T = D^2\,DT = D^2\,B$ (magnetic field)**

**$D^3\,V = D^2\,DV = D^2\,DV$ (amperes or current)**

**$D^2$ = electric field**

We see again a relationship of space as *D*, current (*V*), and magnetic field (*T*). Let us now suppose that the field of space $D^3$ is naturally related by induction of the other two fields of space that are time in $D^3T$ and space velocity in $D^3V$. When dealing with magnetic and electrical current, we can take an electrical coil, like a solenoid, and bend the fields to create a concentrated linear field at the center of the coil, similar to a bar magnet. It should be feasible by a similar process to take this similar geometric relationship we find in inductive geometry to treat space the same way for $D^3$ photonic space. So if $D^3$ space itself contains current, meaning that the matrix of space that contains light or photon as a monopole particle that vibrates also incorporates a field of related current by principles of induction also has another filed of related time. That $D^3$ and related fields of *T* and *V* matrix structure should be able to be manipulated geometrically to induce a concentrated *T* magnetic field within a coil perhaps of higher dimensional relationship. This *T* field could then be manipulated to produce a concentrated *V* field or current that might induce current in a wire. The idea is to manipulate the fabric of space and reconfigure it to produce concentrated *T* and *V* fields and, by that reconfiguration, to produce a simulation of a higher dimension whereby electric current could be induced in a wire. Thus, you would be on the track to creating a free energy device that is taking advantage of natural relationships between *D, T, and V* occurring naturally in space. Components including *V* represent current. And in the equations above, you will notice the pattern of *D*, *T*, and *V* always repeating the same core superstring relationship within a pattern of induction. Above that core is $D^2$, and in each instance, *D*, *T*, and *V* are added to distinguish one from the others. (Bailey: 1-12)

## Revisiting *D, T,* and *V* Superstrings: What Are They?

When we think of the state of something, we think of it as a certain property that has characteristics and somewhat static in its description. So therefore, we think of time as we always measure time, and mass as we always measure mass. Now what if these things that we measure as so determinate and are not really static, but could be altered and dynamic, where then mass could be described in terms of time; and by common understanding, this becomes a contradiction, but if possible, then you would have the equivalency and convertibility of units. Mass can now be described as volume and time instead of kilograms: **mass** = $\boldsymbol{D^3T^2}$. How would this be possible? Well, you would need a common denominator for all things we measure, and by a slight adjustment, it becomes something else. In a way, see that phenomenon when by nucleosynthesis in nuclear fusion of a star or sun, every element in the galaxy through nuclear fusion can be created from hydrogen. Now imagine something even more flexible and smaller than an atom, the component that everything is made from, by slight adjustments becomes everything else. Well, the God particle or God superstring, that thing that makes up the vacuum of space, is *D* or dimension, a vector of space. And *D* is in fact a combination of *V* and *T* strings. So *D* is the smallest string, and $D^3$ is space. But each time we add another *D*, the properties change; what was space as $D^3$ becomes mass when we combine with $T^2$. The geometric configuration of these geometries is what we call space, and when they rotate and cause interference, we get particles. Particles are shadows of movements in space matrix. I refer to space as a matrix. Since these strings can combine in different geometries, different geometries are different dimensions or matrix configurations. With regard to *D*, dimensions of *D* vary by power $D^n$. $D^1$ is very different from $D^5$ and don't even measure the same. You would never know they are related. The three primary strings are *D*, *T*, and *V*. And you can describe all physics equations in terms of these strings to the point you can relate all equations in a sort of dimensional analysis. So *D* exists in various dimensions as powers of $D^n$; we can also explain *D* with other strings as hybrid dimensions such as **mass** = $\boldsymbol{D^3T^2}$; three dimensions of *D*, two dimension of *T* or time. So then what is *T* or time? Well, here you now have to really reinvent your entire understanding of *D*, *T*, and *V*. Time is many things, but not described as such conventionally, but as different in its various dimensional states; an analogy of difference is as hydrogen is from copper as we add electrons. These characteristics of *D* are flexible and dynamic through $D^n$, meaning they adopt properties under different dimensional relationships. *D* then becomes the smallest denominator superstring from which all is made, and $D^n$ represents the infinite number of dimensions in a periodic table of possibilities where *D* changes its properties through successive ascending or descending dimensions.

*D* exists in various powers $D^n$ and also *T* and *V* and $T^n$ and $V^n$ respectively. *D* is the primary superstring or God superstring because it is made by symmetry of *V* and *T*. *V* and *T* together, in effect, are *D*. Well then why isn't time or velocity the common denominator? Well, because they exist as a transformation of *D*. *D* is the numerator, and *T* and *V* the denominator to get either *T* or *V*. So then what is time or velocity? There is a triad established by these three, a tetrahedral geometry that is the basic geometry of spherical natural geometry. One does not exist without the defining other two. What is primary is by relation to the other two, and *D* is not required as transformation at its most basic and lowest level; however, this changes as dimensions progress. So time and velocity are described inherently as transforming principles at this lowest level. And often in natural relationships, you find time as the denominator, a transforming principle. The Cartesian system is an effect of the tetrahedral spherical geometries that form the supporting basis. Phi or the golden ratio of 1.612 is introduced by the dynamics of relationship or rations of circle to ellipse and which also relates to ration of dimensions of energy, as we saw the ratio of inertial space to mass space. So in our natural environment, we measure displacement or that transformation of *D* by *T* and use a clock, but since it is something that isn't visual, we have a problem with its real understanding. *D* can be seen visually, and *V* as well, but *T* cannot. So to this point we say, time is not real, but it is as real as other phenomenon. So then this tetrahedral arrangement has deep philosophical meaning since it suggests dependency and support of other components in all events. And then what is the nature of the unifying relationship, the math, with numerator and denominator, by multiplication, that, in effect, these things are related? Well, they show interdependency, support, hierarchy, symmetry, equality, and unity as one manifestation, which becomes our understanding of harmony and order. But the incorporation of infinite sets and irrational numbers in the math suggest that, in fact, along with this interdependency, is also a qualification of infinity. Hence, the universe exists within infinity and also ordered or bound by such restraining principles of three that result in what we call form.

So how do strings really differ? Well, they are spherical or elliptical entities, and by the nature of the geometries, they establish and define the phenomena we measure. Through tetrahedral geometries, it is easy to incorporate almost all geometric forms, actual or implied. The variances of strings are also in spin and charge (positive, negative, and neutral). So space has its characteristics and variances on the following:

- **The substance of the string, whether it is *D*, *T*, or *V*.**
- **The charge of the string.**
- **The spin of the string.**

and then more complex hierarchies result from the following:

- **Geometric form of the space matrix dimension.**
- **Participating number of strings or dimension of each set of strings.**
- **Whether the strings are transforming in the denominator.**
- **Whether string are primary in the numerator.**
- **Relationship to infinity.**

There is an absolute present, and all consciousness is being transformed through the present; this is also the universe.

We can imagine a string as a small sphere with length, breadth, and height made from energy of consciousness. Since the universe is a living entity—God's consciousness—eventually we are describing consciousness. What is perceived has the same qualities of the one who perceives. Action must be received, which is also a principle of the natural world. Now one might say, "Well, if you are in the first dimension or $D^1$, isn't a small ball to define it already space?" The small constructs or strings act to "affect" three-dimensional spaces in as many as twenty-seven strings in close compaction. The "effect" is three dimensional or $D^3$ space. So many more strings are required than three to generate 3-D or $D^3$ space. The strings of *D*, *T*, and *V* as small spheres at smaller than the Planck length in close formation arrange themselves in a different spherical matrix dimension per dimension. So then the energy of each string might be *D*, *T*, or *V*. What is this energy? I would say it is the qualification of infinite consciousness. This energy is both transdimensional, meaning changing properties in its combination dimension per dimension, and scalar or fractal, meaning that it is broken down to infinity as a fractal pattern.

**Schwarzschild Radius, Inertial Field Radius, and Gravitational Field Radius**

Known as the Schwarzschild radius, where *G* is the gravitational constant and *c* is the speed of light as follows:

**$R = 2GM/C^2$**

Under dimensional analysis, we get:

**$D = G \times \text{mass} / C^2 = E/\textit{Strong Force}$**

$G = 1 / \text{density}^2$

$\text{Density} = T^2$

$D^5 = MC^2$; $D = \text{fifth root of } MC^2$

$D = \text{mass} / \text{density}^2 \times C^2 = (D^3T^2) / T^4 \times C^2$

$D = (D^3T^2) \times T^2 / T^4 \times D^2 = D^3T^4 / T^4 \times D^2 = D^3 / D^2$

$D = D$

Equation is correct.

**$D = G \times \text{mass} / C^2 = \text{fifth root of } (MC^2) = D$, known as the gravitational radius. This is, in fact, the radius also of the inertial field.**

**$D = G \times \text{mass} / C^2 = \text{mass} / \text{density}^2 \times C^2$**

**$D = \text{volume}^2 / MC^2$, where the volume is the volume of the inertial field squared.**

**So the radius is very large and, by contrast, not small as imagined using the traditional *G* or gravitational constant. Normally, the ratio is set at**

**$D$ Inertial Space Extension = fifth root $(MC^2) = \text{volume}^2 / MC^2$**

**When you get to a black hole at $E^4$, the mass is raised to the power of 4 as in a transdimensional atom or atom. So at the center of the atom in a small confined volume, the density of space is extremely large. When the masses of individual atoms are combined where the core mass contained within the extension of *D* for a specific volume is raised to the fourth power, the dimensional structure is raised to a black hole. This would be energy $E^4$ for a specified amount of space or volume. Normally, we are looking at *E* for a volume of mass. As in a black hole at the center of the galaxy, transdimensional effects are eliminated, and the subsequent space extension becomes $D^2$. Gravity at $D^{10}$ produces a measurable *D* such as the Sun or the Earth, but with transdimensional transitions eliminated as in a black hole, gravity squared produces $D^2$ extension of inertial space. This would imply the squaring of the density trans-dimensionally and taking the square root on the inertial volume as you progress into the atom.**

**The ratio of the volume of the Sun to the volume of the inertial mass of the Sun is about one to a billion respectively. And squaring down, the density of the inertial volume from the electron orbit boundary to the nucleus should be squared to $E^2$, and within the nucleus to the black hole at the center of an atom would be another squaring of density or to $E^4$, and thus there is a progression from *E* to $E^2$ to $E^4$, which reflects an ever-increasing density of the space matrix and energy to the center of an atom. This would be a stable transdimensional relationship.**

**At the point where an unstable black hole is created as in the center of our galaxy, the mass would allow a projection of inertial space to square or $D^2$, eliminating a step in the inertial transition of space.**

**Hence, *G* or gravitational constant is misinterpreted since it refers to the inertial extension projected from the center of mass rather than its current definition. This new definition also applies to the total mass, which needs to be revised to include both core mass and inertial mass. Gravitational equations need to be rewritten. Therefore, many misconceptions and misinterpretations come from excluding the mass of inertial space by its misrepresentation in our current understanding of the gravitational constant *G* and Newton's classic definition. This also alters our interpretation of the Schwarzschild radius and volume of the event horizon of a black hole. The determination of the point where a black hole begins has to do with the density of the space matrix raised to that, which is $E^4$ by ratio to what it is in outer space at *E*.**

**In higher dimensions, the fields are more complex and involve other mechanisms or relationship, and this is where the relationships are true, meaning where they operate. Our measurements are effects measurable in other dimensions, and not the cause dimensions. Hence, the cause must be distinguished from the effect, both speak of each other, but understanding is not complete without this knowledge.**

**Life then becomes involved in the primary cause of consciousness and the effects of experience, which are an analogy of the cause, the truth of life.**

**Axioms of Universal Pattern**

The universal and infinite consciousness or God caused the universe to appear as a *form* of consciousness; the universe is presented as *form* by the organization of three. Three is also a form ascribed to consciousness or the infinite by organizing free will, thought, and awareness. The universe becomes an analogy of truth by principle of symmetry. Infinity or infinite consciousness *caused* this universe to become as an *effect* or a form of consciousness instantaneously as an idea. The infinite is qualified by form, proportion, and irrational numbers. Irrational number represents the relationship to infinity. The resulting form of the universe is both infinite and bound by its qualification. The form of three becomes the tetrahedral geometry or space along with three properties of dimension, time, and velocity. This tetrahedral geometry is also the form of light or space and also the organizing pattern structure. In higher and lower form of organization, the tetrahedral geometries of space differentiated are the various properties of the universe. The interactions of space geometries become particles. The particles are the boundaries of space structures that interact and produce interference patterns or particles, accounting for less than 1 percent of the universe, which is mostly space matrix energy. Created life perceives the 1 percent of space interference patterns and not the organizing structure of space or light structure. Rather than being seen as the entire structure by life-forms, the universe is seen only as an excited portion or space, whether it be light or atomic matter.

Space is organized in a hierarchy from simplicity to complexity, adopting different properties through more complex tetrahedral and geometric arrangement along an infinite scale. This is referred to as transdimensional. Space is also divisible along an infinite scalar or fractal nature. So the form of space became transdimensional, meaning of differing tetrahedral geometric forms along an infinite scale, and scalar, meaning infinitely adjusted along a fractal scale. Space is infinitely divisible and transdimensional.

The universe is a form of consciousness and does an effect exist within infinity or infinite consciousness as a cause. Our perception is in a place of balance, meaning where a place from which infinity extends both scalar and transdimensional, where form and energy exist in organization, hierarchy, the center or relationship. A symbolic form is the sphere, the center representing scalar infinity and the perimeter transdimensional infinity. Like a pendulum swinging on a clock, the swing is always brought back through the center. The access to infinity is from every location. We exist between infinite boundaries or form.

Life is transcendental, which means the ultimate you is not here, but you are looking in. Consciousness is shared; life is one form of consciousness perceiving another. So life becomes a journey of self-development since this is a place for transformation. You are not taking any things with you like some Egyptian pharaohs thought they could.

The form of space and associated particles of each organization or geometry of space are organized by proportional relationship or math. The form of space being geometrical is made of components defined here as superstrings. The primary superstrings are *D*, *T*, and *V*. Space has property and is a description of a form of energy with associated particles for a particular geometry dimension. A particle for the vacuum of space is the photon or monopole or node within the geometry of space. Space is also elastic, and the wave passes through at *C*, but since space can be stretched or compressed, *C* can have slight variations of $C + V$ or $C - V$. Light or photons do not travel, but the space matrix transmits an elastic wave, and you see the excited wave when it hits your eye. Particles are the interferences patterns of space motions or the components that spin in the geometry of space. Much like when you throw two rocks in a pond, the waves when crossing generate an interference node; this is not to be mistaken with the elastic vibration of space itself.

All property geometries can be explained by its component strings ***D*, *T*, and *V***, such that all relationships are a dimensional analysis. Units are valid as we experience them, but typically there are patterns that associate certain strings with certain properties. For example, velocity is always associated with current, and time with magnetism, and dimension with force, no matter what level of the hierarchy. Pattern is more important than object since it reveals the truth about the hierarchy of this existence. Cause tells us more than effect and more so the ***pattern*** of cause.

The atom is the representation of the hierarchical order of space through all dimensional states, from infinity to balance, both scalar and transdimensional, from infinite black hole to balance and back again. The universe is connected to infinity from all sides. And hence, the big bang was multiple from all points, and the universe is under constant creation and transformation. Space is expanding and retracting constantly.

All atomic mass is associated with space that it generates. The Sun generates space, the black hole generates space, the Earth generates space, and you generate space, and so on and so forth. So then the mass of the Earth includes not only the mass of the core, but inertial, gravitational, or associated space mass. By comparing the inertial volume or space produced by each planet, you are comparing the gravitational accelerations of each planet. Hence, Newton's gravitational equation can be rewritten as $\boldsymbol{F} = \textbf{volume1} \times \textbf{volume2} / \boldsymbol{R}^2$, where $G$ is eliminated. $G$ refers to density of the inertial fields, and this density has to be explained for gas planets since they are different from solid forms.

The axiom of organization is $D$, $T$, and $V$ and, by its level of organization, explains all mechanics of the universe. The organizational pattern explored tells us about the unification of the universe and its truths. Without the organizational pattern, there is no unification theory. The pattern alludes to cause, the ultimate truth. Hence, the universe is an analogy of truth represented in the mathematical pattern of the universe, which is also a pattern of consciousness, also equivalent to life and reality.

Solar System with Inertial Field D of the Sun, Kuiper Belt, and Standing Gyroscopic Wave

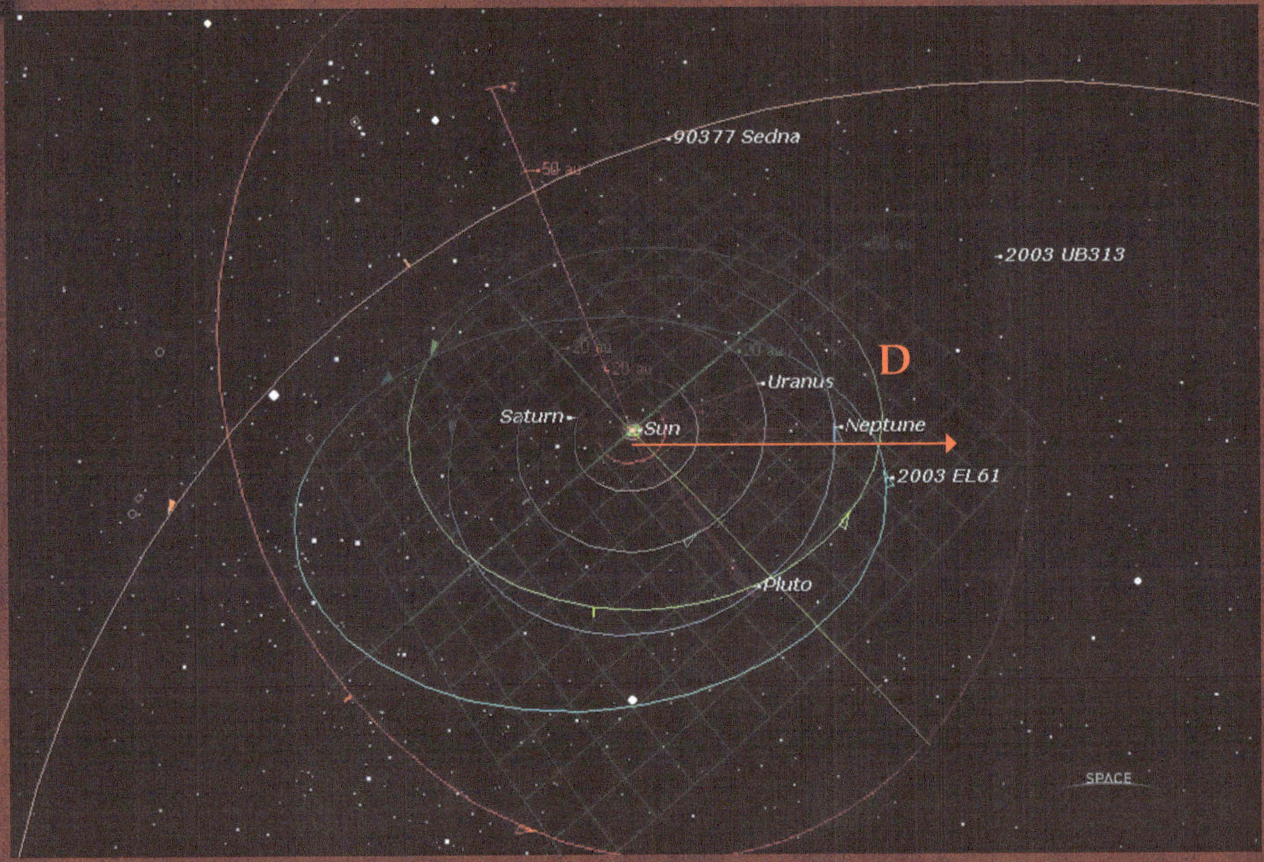

## Gravitational Matrix, Inertial Matrix, and Dimensions

We have talked about multiple dimensions, meaning each dimension is particular space matrix geometry with distinct mechanics and properties, and each dimension is associated with particular particles and spin. Superstrings within each matrix have spin and charge as do particles, and also, superstrings interfere within the space matrix by their own motion and the motion of rotating space to produce interference nodes or particles.

Dimensions exist with adjacencies, occupying the same space and sharing some nodes of particles. So as geometry, all dimensions are connected, except what we experience as gravity in our perception is an effect trans-dimensionally descending or caused from a higher expression or dimension. Our sense of gravity is abstracted by our limited dimensional experience. A graviton is a photon since the geometries of multiple dimensions are shared.

We have discussed gravity simply by comparing the volumes of inertial space to get the relative accelerations associated with gravity of bodies and their inertial spaces. The equation

$$F = \text{volume1} \times \text{volume2} / R^2$$

is the relationship of gravitational force to inertial space. This force is dependent on the extension of inertial space by $D$ from

$$D^5 = MC^2 = G \times \text{mass} / C^2$$

But the **effect** of gravity related to $D$ actually extends beyond $D$, and though the relative force acceleration associated with the inertial space body is related by $D$ radius of the inertial field mass and core mass, the extension of gravity is greater. Hence, the actual extension is squared since **gravity** = $E^2 = D^{10}$, and for a black hole with $D^2$ of inertial space extension, the gravitational effect extension is to the power of 4 at $E^4$. Hence, you have gravity acting through the inertial fields of other bodies at large distances, like the planet Sedna to the Earth explained before.

The inertial spaces of bodies and eccentricities of inertial fields caused by spin cause the shifting of center of masses and have impact on orbits such that none are perfect. As said before, we can attribute the eccentricity to spinning objects, and relatively speaking, all objects are spinning. Inertial space density changes also impact planetary orbits. Planets within the inertial space of the supporting star like the Sun tend to be more stable in homogenous density paths and greater density paths than passing outside and into the inertial density of less dense center of the galaxy. So therefore, Pluto and Sedna have the wild obits we see, while those within the inertial fields of the Sun exhibit an effect like a gyroscope and appear more stable within the plane of the ecliptic. So there is much to consider in the obit of the planet, which are as follows:

- **The extension of the inertial fields of bodies involved and how they overlap within the orbit of multiple bodies around a star.**
- **The passing through (in or overlap and out) different inertial fields within an orbit of object inertial field in question.**
- **The extension of the inertial field.**
- **The extension of the gravitational effect $E^2$ or $E^4$.**
- **Within or without the inertial field of the supporting star.**
- **Eccentricities of center of mass of inertial fields caused by spin of inertial field.**
- **Eccentricities between center of mass inertial fields of multiple bodies.**
- **Center of mass (space or inertial mass and core mass) between multiple bodies.**

Each variance has an impact on orbit; to the extent it's very difficult to get a perfect trajectory in a complicated space flight depending on gravitational pull. Constant adjustments would be required.

# CHAPTER 5

## Free Will and Transcendence

$$E^{(i \times \text{pi})} = -1 = -D = \text{Time} \times \text{Velocity}$$

Universal code of 1 and 0 give rise to all functions. The Big Bang and creation of the universe was induced. D represents all emergent operators arising from transcendent operators outside the universe. True constants such as Pi have their origin outside the Universe.

Math, physics, consciousness are the same or equivalent reality. Reality is the alternation of infinity and interval, where form is a triad organization of infinite relationships. The construct of the triad relationship of form is a binary code of 1-0, life, and also self-organizing. All that is, is, and there is nothing that is nothing, for everything is eternally. I am that "I am" is the first principle of life, the 1 which is subdivided into all things infinitely. To love was to create, action and return of action, universal symmetry, consciousness.

**Pattern Universe**

When someone asks what something is made of in the universe, it is important to distinguish cause from effect, or better pattern from object. And hence, when we speak of superstrings, one does not exist without the other. So when someone asks what the string time is made of, it does not exist without the other two primary superstrings *D* (dimension) and *T* (time). We could say time is made from time, and then the next question would be what is time, and what is the string made of, and this subdivision could go on forever as it always does. We can distinguish that *D* is neutral since ***D* = *VT*** and that time and velocity are transformative in this relationship. Numerator is primary, and denominator is transformative. The differentiation of what time is made of is in the relationship with

- **the other superstrings as primary or nominative or transformative,**
- **transdimensional pattern,**
- **fractal or scalar pattern,**
- **spin, and**
- **charge.**

which determine in effect why that element came to be. But then we refer the cause, and that refers to the pattern as illustrated. So causes are pattern and not object, just as the universe is the result of pattern and not particles or objects. But since science has reversed this, we think of ultimate truth as an object. We also have spoken of the most basic element of substance or telluric energy as an abstract idea and how this infinite energy adopts properties by its relationship within pattern. So therefore, all strings would be made of the same energy, although its property and quality can never be assessed since the pattern to which it is experienced manifests in infinite qualities or properties. So ultimately within the pattern, we must look at the point of balance, where infinities are related at the most basic level and swing from infinity back to infinity, and this would be *D*, *T*, and *V*. So to answer the question what a superstring is made of, I would refer to pattern since the subdivision of space and particles is infinite. The number 3 represents a pattern of relationship at the most basic level in the universe and is also a mirror of consciousness. So 3 is the number representative of form. The number 3 then becomes the number of hierarchy, organization, form, and properties. This tetrahedral form based on 3 also can be broken down infinitely both as a fractal pattern and transdimensional pattern, but more significant than determining objects is to understand pattern relationships; this is the very nature of math, and that one basic energy within that pattern results in all the variations that we see, but because pattern brings form of consciousness and reality, then that particular universal substance can never be isolated but understood by relationship.

The universe began with this infinite consciousness or God that adopted a form as only God can do; hence, that most basic energy is God. When we ask ourselves, well, we cannot experience the infinite; well, the infinite is our consciousness, and its pattern comes from infinite consciousness. This reality is an effect consciousness of our cause consciousness; that's why we refer to it as a receded form of consciousness as a dream, but it's all related. Our consciousness transcends this universe as an effect. Hence, this place is for learning about us and development and harmony to patterns of universal truth. Life is about us, and this is more profound than thinking of species evolving and conquest of the universe, which is object. The substance of life is about us, and therefore, the universe is a tool, life to be appreciated, and how we handle it speaks about us, our nature, and state of harmony.

When the Sun shines, it sends a wave across the universe. In similar fashion, out intent of our free will, manifest in action, affects the whole of creation. Energy flows and is constantly transformed, and hence, an action has infinite implications. As a shared form of consciousness, our actions affect all those around us, even to the far side of the galaxy and to all existence. And hence, the godliness or creator, also having a will to maintain the harmony within the pattern, organizes that pattern to both reveal truth and impart wisdom and understanding and even correct. So therefore, all actions involve responsibility and consequences. Harmony with the matrix is love and connected to ideas of faith and healing, and discontinuity with the matrix is referred to as pain, tribulation, and sin. So therefore, we have been created with free will to become an infinite flowering of what conscious love is to be, and that expression is life; and anything less is limit where transformation in the natural sense of positive growth pattern of the universe is hindered, stopped, or ended. This is akin to death, although the pattern of the universe is constantly seeking to maintain its pattern through transformation; the hierarchy of universal consciousness, which as first cause is the creator or God.

Life is one form of consciousness perceiving the other. All consciousness is infinite and shared, and God is the source and cause of consciousness and its differentiations that are all forms of life. In physics, space made from superstrings form all dimensions and hierarchies of space; this can be visualized as a pattern of spheres and the multitude of geometries we can get from them. Compacting spheres together, you get a tetrahedral geometry, and the organizing number is 3. On a table, six spheres surround one sphere. In three dimensions, four spheres together make a tetrahedron; abstracted, this is a polyhedron with four faces. Various combinations of spherical geometries account for all forms of space. By tetrahedral I mean a system of creating the geometry of space using spheres, and at the most basic level of organization, the number 3 represents the triangulation of spheres.

The universal consciousness or God is manifest in the universe, and thus, the universe is living. The universe becomes an analogy about the truth of God. The truth is available to all, but since God also has an active will, the truth is also "enabled" through revelation. Hence, it would be natural for God to have expressed himself through various sacred texts. Hence, we can see as examples of men who are in harmony with the universe and truth and to them is accounted faith and life. Harmony with the matrix for primitive people is the exception rather than the rule, and therefore, also through revelation and grace, the knowledge of truth is enabled more readily. The harmony of consciousness is enabled through self-discovery and through revelation, whether through individuals or God's involvement directly or indirectly. Disharmony in the matrix separates, and harmony unites, giving full evidence of free will; life has meaning and purpose. We can speak of Buddha and Abraham as men of exception, and then we can speak of revelation through God as grace and enabling enlightenment, where now rather than a few, many are enabled and expectable to the harmony. Where is science in all this? Science leads us to reconsider the real nature of truth and reality, and eventually, all science leads to truth. Pattern is found in nature, and it is also found in sacred text. Pattern reveals fundamental hierarchies and harmonies of truth.

**Why does anything exist at all, you say?**

Everything that is, is. The idea of the beginning is a human idea. Life is transformation. Infinity always was and always will be; it is within and from infinity that everything is. When we look at events, it is always a natural human idea to present a cause for something coming onto existence and, therefore, associate a reason to it since cause enables a pattern of the universe. We are more focused on object and segments of time rather than pattern and infinity. We give importance to objects or effects rather than cause. If you see the universe as transformation, then the pattern becomes prominent by this definition. If you ascribe a first cause or God as the source of reality, then it would be the cause, which is the focus of truth and that which gives purpose and meaning. Hence, ultimately, all-purpose can be ascribed to the first cause that results in the patterns that we experience; the cause which always was and, in a sense, is the largest mystery of all. God gives meaning and the answer of "why" to all things. But then ascribing a reason why to the infinite, which contains all reason and reasons, is to reduce the infinite and universal patterns to less than what it is. Hence, the contradiction in determining a singular why in the very infinity that contains "all why" is impossible. The pattern reveals truth and the why of the pattern, and perhaps the pattern is an explanation of itself. God's expression of pattern is creation, and this is self-discovery and discovery, action and return of action, hierarchy and symmetry, to love and be loved. Infinity exists because it is good for it to be along with infinite reasons. The whys involve infinity; the whys are infinite, and so therefore, the explanation why of some concepts never end, although the related pattern is perceived. So the whys are unlimited as well as the answers to the whys. Not enough books can be written to describe the infinite. Why and what is love? We could fill the libraries of the universe for the answers. From the patterns, we drive truth and what in infinite ways, and from the infinite why, why becomes infinite. So what is God is more approachable as pattern than why. What is consciousness is more approachable than why. Obviously we are speaking of life, and why is life consciousness; why is there pattern? Is there a cause to the infinite that was first from which form and all ideas proceeded? At this point, the point of the why of the first cause, singular why ceases since something infinite without a beginning is eternal—not contained, unlimited, with all information, forever. You cannot explain all the whys with one why, and why becomes answerable through pattern form, but it's never ending as pattern of thought and better perceived as pattern. Why is love? The pattern of what is love, by Bach's definition, is helping someone achieve their dreams, so one of infinite descriptions become both what and also the why; the pattern is its own answer.

**Again you ask why ANYTHING exists at all. Where did all the dimensions, matter, electrons, etc., come from?**

The universe is consciousness, a living thing. Life is one form of consciousness perceiving another. All science leads to God. That energy, which is God, by symmetry of its pattern, is a pattern of the universe and also a pattern of consciousness.

**Perhaps you say it is impossible to know how anything exists.**

Well, I disagree. An argument can be made through science and proof that God exists.

**What do you mean by a God?**

God includes the cause consciousness, with the infinite patterns.

**Infinity, what is that?**

This universe, by its pattern, is a pattern of consciousness, an analogy or effect of the cause consciousness, an analogy of truth. The pattern of "form" reveals a pattern of consciousness. Yes, consciousness is not about object, but infinite pattern, not effect or object, but cause. Similarly, particles are infinitely divided, and space is infinitely divided and transdimensional. Infinity is consciousness and, in the universe, is transdimensional and scalar. There is no point in using colliders to measure infinity. Pattern is the key.

**I don't think that things can be truly infinite, you say.**

The pattern provides a unified theory. Irrational numbers prove a relationship to infinity. Incorporated into this reality, the infinite is related to some boundaries or qualifications, which through pattern manifests form of this universe. The lowest relationship of form is the number 3 where some aspects are transformative, others primary or nominative. Time is transformative. *D* is primary. Velocity is transformative. With regard to *D*, *T*, and *V*, one does not exist without the other. The universe has principle of dependence through transformation since it is fluid, plastic, or connected, even so dependence of creation from infinity. Every atom has a black hole, every center of galaxy too. All black holes and mass create all the space. Infinity is connected to this universe from multiple points. Creation of the universe is from infinite points, like an idea, and refers to a higher form of consciousness. All life is transcendental.

**Okay, what happens if I take a spacecraft and fly directly outward from our clusters of galaxies in this universe? Do I hit a wall eventually or what?**

We exist in a narrow band between conscious infinities. Since you generate space and it's all connected, you never get to the end. Since the universe is connected to infinity itself, we are not bound. Space is twisted in all directions. You can also travel trans-dimensionally as follows:

- Time travel
- Subspace travel
- Quadrant space
- Anti-space

It's very complex. Antimatter does not mean antigravity. Antimatter is a certain form of space, but there are others. The one that creates antigravity is a reverse polarity. Antimatter has aspects of counter spin and countercharge that negate the other. There are several variations of space.

**<u>Planck's Constant</u>**

**Hertz = frequency = $v = 1 / T$**

**Planck's constant = $h$**

**$E = hv = D^5$ = Planck's constant × frequency**

**$h$ = Planck's constant = $E$ / $v$ = $ET$ = joules × time**

**$h$ = $E$ × lambda / $C$ = $E$ $DT$ / $D$ = $ET$ = $D^5T$**

**$h$ = $D^5T$ = $D^6$ / $V$ = volume$^2$ / magnetic field strength**

**$h$ = volume$^2$ / $H$; $H$ = magnetic field strength**

## Black Holes, Gravity, and the Space Matrix

A point source produces a spherical wave; an analogy would be sound. The generation of the space matrix is an inertial space construct that allows spherical waves since the matrix is a uniformly spherical and radial geometric construct. We have referred to this structure as fractal and tetrahedral, tetrahedral meaning 3, the number of form. Tetrahedral form is the particular construct that allows 3-D movement or space or interpretation of motion within the medium. Strings being spheres and having volume is different from the "effect" of volume that these spheres of superstrings create when they are in a space matrix array or inertial field. The sphere of space or inertial field surrounding an object of mass that is created and supported by the mass is uniformly gravitational and is evidence of other uniform properties that space has. Gravitation is as in a uniformly gravitational sphere or inertial field space matrix. The spin of a planet or body of mass and related surrounding space will cause asymmetries to develop, particularly at the north and south poles of the generating object. This is ordered according to phi.

As the space is generated, waves travel through the matrix in two directions—in and out from the center of mass—which are out to perimeter of the inertial field and back to center of mass. As this happens, the result is as a standing. The standing waves resulting from the spherical waves within the matrix has a core of high amplitude and fan out from center like a spindle top. This form is the form of galaxies that contain masses within an ecliptic plane, like a gyroscope effect. Planets would rest where the amplitudes are high of the standing waves, which reflect areas of higher density of the spherical inertial field generated by the supporting star such as the Sun.

It may be that a superstring has a black hole at the center of the sphere that generates space and a spherical wave with spin and charge, and where superstrings meet at the interference areas are particles, which also have black holes at fractal center. Hence, the entire space, from atoms to superstrings, is both made of waves and manifested from infinite points, infinity, and infinite energy and related or qualified as form. But for now, we can see the atom as the prototype for a transdimensional machine, where all dimensions find relationship. Going smaller to the superstrings of space, the combination or geometric array of superstrings is a point of balance within infinity and manifestation of the pattern of form or space of the universe. They consist of space and waves originating from a central source with spin and charge. From center to perimeter, pulsing from a center space is generating volume and a phenomenon of standing wave. These patterns become the nature of space matrix trans-dimensionally from the micro scale to the macro scale. And perhaps even the centers of superstrings are black holes, and where they interfere also, that interference is what we identify as particles. This would make space complete transparent and connected to all infinity from infinite locations. Therefore, there may be multidimensional black holes, meaning black holes existing within different dimensional arrays or constructs.

**These black holes might be classified as follows:**

- **Black Hole at the Macro Scale of Galaxy Centers. These black holes form the centers of galaxies and generate space at $D^2$ and gravity felt to $D^4$ since gravity is $E^4$ here. These are the results of black holes of many atoms stripped and black holes combining together.**
- **Atomic Black Hole. The atom is a classic example of a balanced transdimensional event where all dimensions are related in balance. The black hole exists at the center of the nucleus. Space is generated at $D$ distance and gravity felt to $D^2$ since gravity is $E^2$ here.**
- **Space Matrix Interference Patterns or Particles. These would be point of interference between superstrings. Hence, all particles created by the interference patterns of the space matrix would have a**
- **black hole. This would include photons. The overlapping of several spheres is actually the creation of a black hole, and so the photon exists as the event horizon, and the space that is generated is the space matrix. Hence, the space matrix is supported not only as a transdimensional ratio, but also infinitely from all points in space or black holes. This black hole is very small and is another type. So the overlapping**

**of spheres is a connection to infinite energy and perhaps their centers. The pairing of spheres represents polarities.**

- **<u>Center of the Superstring to Perimeter of Superstring</u>. The center of each superstring is also the source of infinite energy, and hence, the interior of the superstring is fractal geometry of a tetrahedral subdivision, and hence, the center of the superstring divided into three also is another infinite point or zone. This refers to fractal or scalar infinity mentioned before, so literally all points in space are connected to infinity. There is also transdimensional infinity that results in the super massive black holes at the center of the galaxies.**

So now we become aware of the variety in which infinities are related in the construct of the universe. And hence, there is no centralized big bang but a bang from infinite points, almost like the creation of a thought within your mind or the explosion of an idea.

**<u>Qualification of Infinite Energy through Proportioning of Phi Superstrings</u>**

**<u>Begin</u>**

**<u>$0 = 2D^5 - 2D^4 + D^3$ = 2energy - 2force + volume</u>**

We had this equation derived before, and it appears that the two may in fact be referring to two statements of $D^4$ combined into one, so we may do as follows:

**<u>$0 = D^5 - D^4 + D^3$ = energy - force + volume</u>**

Note that there is - $D$ in proper multiplicative (×) forms, and + $D$ in scalar additive (+) forms. $D^3$ appears to be "-" being and odd power of -1. And earlier when we substituted mass with $\underline{D^3T^2}$, $D^3$ is "-" negative. Now we may balance energy with force and volume because since the center force is a combination of two functions or forces (positive and negative), these results in volume as negative for the following:

**<u>$0 = D^5 - D^4 - D^3$ = energy - force - volume</u>**

**<u>$X^2 - X - 1 = 0$; the number 0 refers to unity, as does the value $C$ in the hyperspace</u>**

*Now we have the true superstring equation.*

We had derived this equation before using the Pythagorean Theorem and now have made appropriate adjustments. Zero as discussed before established unity in both qualified and infinite variables. These variables produce constant such as phi, which becomes the organizing constant of emergent universal energy of the universe. Other constants are 1, referring to balance, and zero (0), referring to unity of the tetrahedral (3) geometric space matrix form. The number 3 represents form:

- **The Pythagorean relationship of fractal infinities, where the number 3 produces the number phi, 0, and 1, which provide hierarchy referring to infinity, unity, and balance respectively.**
- **Transdimensional relationships of space also referring to phi, 0, and 1.**

***$0 = D^5 - D^4 - D^3$ = energy - force - volume is a Pythagorean representation.***

We now see a relationship to the Pythagorean Theorem, and the following equations also relate and apply:

**$Phi^2$ - phi = 1**

**$Phi^2$ - phi - 1 = 0**

## Standing Spherical Wave

$X^2 - X - 1 = 0$

Which represents $X^2 - X - 1 = 0$**,** I will show that this is in fact related to energy, force, and volume in $D^5$, $D^4$, and $D^3$ respectively and also related to phi and, as such, relates the distribution of infinite energy to phi by a harmonic tetrahedral Pythagorean relationship in the subspace of the matrix. And as we move to the transdimensional, we also find the organization of phi in the hyperspace of geometric forms and volumes. ***D*** becomes the transition between scalar or fractal infinity and transdimensional infinity.

***D* as phi and phihat together form the 1 or point of balance for *D*.**

*D* is the primary superstring and then also "one" or 1, the point of balances subdivided in phi and phihat, where we note the following:

**Phi × phihat = 1**. Phi and its conjugate result in 1, the proportioning of the infinite into parts according to ratio that ends in balance. We may surmise that phi and phihat are related to *T* and *V* superstrings, and also which become *D* by relationship.

**Phi$^2$ = phi + 1**

**Phi + phihat = 1**

**Phi × phihat = -1**

**We see above that the + refers to subspace and × to transdimensional space. And the twp become connected where 1 = *D*, but one space is + and the other is -.**

**Phi - phihat = *SQRT* 5**

**What two parts result in $X^2 - X - 1$ such that the power of *X* is not greater than 1? By this I hope to relate phi to the Pythagorean relationship of subspace that results in**

So I want $X^2 - x - 1$, and so I use $(X - a)$ $(X - b)$, and the two values for ***a* and *b*** that make it work is **phi** and its conjugate **phihat**. So I multiply $(x - a)$ $(x - b)$ out and equate it to $x^2 - x - 1$ as follows:

$$(x - a)(X - b) = x^2 - xa - xb + ab$$

Now I group the terms with the same power of *x*

$$X^2 - xa - xb + ab = x^2 - x - 1$$

$$X^2 - (a + b)\,x + ab = x^2 - x - 1$$

Now for those two polynomials to be the same, all their coefficients must be the same, so the coefficient of $x^2$ is 1 on the left and also 1 on the right: **1 = 1.**

The coefficient of *x* must also be equal.

$(a + b)$ on the left must be equal to 1 on the right.

$X^2 - ax - bx + ab$

$X^2 - (a + b)\ x + ab$

$X^2 + (-1)\ ax + (-1)\ bx + ab$

$X^2 + (-1)\ (ax + bx) + ab$

$X^2 + (-1)\ (a + b)\ x + ab$

$\mathbf{X^2 - (a + b)\ x + ab;\ ab = -1}$

$\mathbf{X^2 - (a + b)\ x + (-1)\ ab}$

$\mathbf{(x - a)\ (x - b) = x^2 - x - 1}$

$\underline{x^2 - (a + b)x - ab = x^2 - x - 1}$

This means that within the space matrix, the organization of energy is according to phi and phihat whether added or multiplied, scalar or transdimensional.

This means that the coefficients are the same, so the three equations must hold:

**1 = 1 (for $x^2$)**

**- $(a + b) = -1$ (for $x$)**

**$-ab = -1$ (for the terms without $x$)**

If we can find $a$ and $b$ such that those two hold, we have the values for

$\mathbf{(x - a)\ (x - b) = x^2 - x - 1}$

We have values, which are phi and phihat

The two equations can be simplified to

**$a + b = 1$ = phi + phihat = 1**

**$ab = -1$ = phi × phihat = -1**

We can solve them and find indeed **$a$ = (1 - *SQRT* [5]) / 2 and $b$ = (1 + *SQRT* [5]) / 2**

**Phi × phihat = 1, transdimensional, *V* (phi) × *T* (phihat) = *D***

**Phi + phihat= -1, scalar or fractal, *V* (phi) × *T* (phihat) = -*D***

So in the end, we have found **<u>*a* and *b*</u>**, and it relates to the Pythagorean Theorem also, and you can say the following:

<u>$x^2 - x - 1 = (x - \text{phi})\ (x - \text{phihat})$</u>

$\mathbf{D^3 - D^2 = D^1}$

$\mathbf{D^1 \times D^2 = D^3}$

$\mathbf{D^1 \times D^2 = D^1 + D^2}$

**$D^2 = 1 + D^1$; 1 = $D$ = phi + phihat**

1 is phi and phihat, which is *D*, and here refers to the substructure of space that is Pythagorean scalar fractal and the other transdimensional. Phi represents infinity, the qualification of infinite energy into proportion and property into 1, which represents balance.

**I have just showed how infinity is represented as scalar and transdimensional functions related through *D* or 1 or balance.**

$D^5 = MC^2$ is the extension of inertial space as mentioned before. For a super galactic black hole, the extension is $D^2$; however, if the critical mass is brought to within $D$, then you have a black hole. The Schwarzschild equation is "interpreted" incorrectly, and one large factor is that the idea of mass does not include inertial space under the standard model. We have said before that total mass includes the mass of the planet as well as the inertial mass or space that it generates and supports.

**$D = G \times$ mass / $C^2$ = fifth root of ($MC^2$) are the same.**

$D$ here is the radius of the inertial field not to be confused with the extent of the gravitational force to $D^2$ for planet, $D^4$ for super giant black hole. The gravitational field effect is to $D^2$, primarily because gravity is working at a higher dimension $D^{10} = E^2$ of higher energy, and hence at that level, the effect of gravity is further due to an expanded field. What we experience is the resultant force of second-dimensional energy $E^2$ at $D^2$. And the volume of inertial field also is proportional to the gravitational force, and its effect in the third dimension plays out in the gyroscopic effect of gravitational orbits around stars such as the Sun. Spherical waves sourcing from the core mass within the inertial field result in standing waves similar to a spindle top, and hence, the controlling gyroscopic action to distance $D$ for a star and $D^2$ for an exposed super giant black hole. With inertial fields, the center of mass is always eccentric because of the spin of the source and elliptical form of the space field. Orbits are not perfect due to spin and resulting eccentricities in the center of inertial mass. The extension of the inertial field involves the *SQRT* of phi or golden ratio. Inertial mass is another form of space or energy, kinetic and potential.

There is the mass of the core and that of the inertial field or mass energy generated by the core rest mass. There may even be an impulse velocity associated with the inertial field that results in the spherical wave that produces a standing wave within the inertial field. Pluto has the freedom to orbit without gyroscopic effect outside the controlling inertial field of the Sun, travelling in the space generated from the center of the galaxy.

**Then what about Uranus? Why is its rotation tilted to over ninety degrees to the plane of orbit?**

Spin is allowed. Inertial fields and orbits induce spin due to relative motion. When the solar system formed, as the mass condensed, the inertial field increased. Like a gas planet, Saturn—its field—is similar size to the Earth, and hence, the accelerations are similar even though with more mass but less density. As the density decreases, the inertial field contracts and vice versa.

**$F$ = volume1 volume2 / $R^2$ = mass1 mass2 / density1 density2 $R^2$ = $G$ mass1 mass2 / $R^2$**

**$G$ = 1 / density1 density2**

**$T^2$ = density, if this is, you have two densities in the gravitational force equation, one for each mass, as shown in the proof prior to this paragraph. *G* refers to the density of two objects where the inertial fields are acting as one.**

***Phi and PhiHat***

**$D$ = phi × phihat = 1 = velocity × time**

**$V$ and $T$ are in a ratio along with phi and phihat, where $D$ =1 = phi × phihat**

**1 relates to infinity and ties into the Pythagorean theorem by**

**$X^2 - x - 1 = 0$**

**$x^2 - (a + b)\, x - ab = x^2 - x - 1$**

***a* and *b* are phi and phihat**

From the Pythagorean Theorem, we can get

**energy - force - volume = 0**

**This is related to $X^2 - x - 1 = 0$**

**$D^2 - D - 1 = 0$ and raise the powers, and you get to $D^5 - D^4 - D^3 = 0$, which can arrive at**

**$D^2 - D - 1 = 0$**

**Involved in this are two series of + addition and subtraction and × multiplication and division.**

Pythagorean is scalar and involves string theory, and multiplicative is transdimensional. One is internal to space and the other external to space. There are no constants but variables combined with proportional factors, resulting in variables that are proportioned. Hence, the superstring velocity incorporates phi, and time incorporates phihat, and the two combine to form a balance equal to 1 or *D*, whether it's in the Pythagorean relationship or multiplicatively transdimensional. So then all constants are variables, and there are no constants, only functions with proportion. That clears up Planck's constant and *G* and other constants to being proportioned variables.

Now we can revise all units into three, which are *D*, *T*, and *V* or distance, time, and velocity respectively with powers thereof. If it looks like an apple, call it an apple for the sake of simplicity, but to integrate all properties, there are only three units that vary by power or dimension; dimension also refers to dimension of space and geometric array.

*D* is the God particle because in phihat × phi, and phihat + phi, you get *D* and 1

**$D^3T^2$ = mass** exits in many forms, so you have a lot of hidden mass. An atom has many levels of mass, and space is transparent to atoms.

**What is the idea of space-time presented in this theory?**

Energy, force, and volume is where we measure, but as the dimensions get higher—as in an atom—things change and are beyond our normal observation. That is why we have trouble establishing units over $E \times T$, but they exist, we just haven't measured them yet. There may be as many as twenty-one to twenty-three dimensions. The volume we experience is an effect of the volume of a matrix of superstrings. There may be as many as twenty-seven forming the $D^3$ space we experience. So volume of the superstring does not mean 3-D space effect, but their combinations of *D*, *T*, and *V*. The monopoles of the matrix are photons and, at higher dimensions, are gravitons sharing the same space. Overlapping spherical superstrings form the photon.

**Creator, Consciousness, and Beliefs**

God is intimately personal and immensely universal. But it's hard for people to grasp the two together since if he is seen as personal, the other side of the coin is how does God have time to deal with all problems; and if he is universal, does he even listen to me in such a large universe? Well, human perception is the limit, since in essence, God is everything and all possibilities. The universe is alive through God consciousness listening to everything you say and feeling all you feel. You talk to everyone, but who talks to the universe? To Native Americans, nature was alive and the creator. It's not common to think of creation as a living being. It's even less common to view the capacity of an intimate and universal creator simultaneously when we personalize all our views based on human capacities. But perhaps the truth is more fantastic than the myths that we impose on ourselves.

Your destiny is as you believe. If you want a miracle, you need to believe in miracles. Just as the universe is intertwined, so is cause and effect. So if you believe in death, so you die. So it's hard for someone that doesn't believe in miracles to get one because as an external participant, such a person would say, "Show me a miracle," and therefore, the belief or cause of miracle is totally opposed to the very nature of the asking for a miracle. But to believe in miracles is to have miracles in your life, the nature of the cause extends to the nature of the effect, and the effect becomes an analogy of the cause. This is a template for truth. To love is to be loved, to hate is to be hated, to have eternal life is to believe in an eternal life. The universe is a living being or consciousness—God—and the universal consciousness perceives all things intimately. Size is relative and human since a large miracle is of no difference from one small. So limit reflects what humans are capable of. This was a better miracle than that only because it seems out of reach from human capabilities. We think of miracles as patterns that defy ordinary experiences, but all is miracle for the believing inspired to the realization of the capacities of consciousness. Since all is from the same source.

We are transcendent beings. And life is about you, the inner you that is participating in this experience for knowledge and wisdom. We are truly free and with free will. If you believe in miracles, you find them; if you hate, you find that; if you love, you find that. You set up all the harmony and disaster that you can imagine. Suffering shows the latitude of experience to the opposite spectrum of elation. But life is not about your body, the universe, or of man's evolution, nor the goals of the effect of a universe; it is about the causes related to your transcendent being, the cause and not the effect, truth rather than the illusion that it refers to. Your destiny is to grow.

**But the suffering outweighs the alleviation through miracle?**

Suffering and miracle are two different events. Miracles are not designed to alleviate suffering but are a response to belief. Destiny is a response to the intent of your actions. We might say everyone does what they want, and some might assume it's satisfactory, but there is a harmonic standard pattern; as in nature, there are ratios, hierarchies, and harmonies that maintain the balance. So even though intent and free will is open, the anomaly is allowed within a system where if the same precepts of the anomaly were used in the pattern of the universe, they would destroy the universe. The grass is always green, and however, we in our disposition are not always so consistent. But we rely on the pattern even despite the confusion of our free will when it exists, and then when disaster strikes, blame the universe. And hence, a truth of life is to understand that without harmonic order, nothing can exist, not even creation or the imagining of something to be; even if you willed it to be, it's impossible. So therefore, that which is allowed in infinite knowledge and experience has an underlying pattern that is reality. And the knowledge of that pattern enables, and the disallowance of it restricts. Our total free will is allowed to exist to the point of self-destruction.

So then are we allowed to be what we want or behave as we like? Yes. But there is no growth without the pattern. You are something dynamic becoming static without the pattern. Life in the real sense is gone. The way you behave reveals about your character. We put too much weight on illusion of effect rather than the reality of the development of consciousness and cause. Men concentrate on the effects and not the causes, so too in science. We are here not for the conquest of the universe but the conquest of truth. Conquer not the world but conquer self. Love as in preservation, action given and returned infinitely in growth and self-discovery, I perceive because the universe does too.

**But there are so many exceptions to this?**

Exceptions are doors to new knowledge; that's the beauty of free will. Be an exception to do well. In intuition does the greater truth is.

**Hence, why I am always asking about creation?**

The pattern of consciousness speaks to us constantly. Intuition is constantly prompting us with the truth of pattern reality. All our feelings are unlimited, but in the end, what is our intent? Intent describes us the most, and do you ever look at it? At what do you think when you do something and why, or are you focused on the crowd and act externally on impulses based on myths given to you by society like a mindless person, a person without a developed self? You are more than that. You are an infinite being and have access to all knowledge; you just have to realize it. But if you see limit, then it becomes the limit to all around you, and then the limit boxes in the infinite you with even greater myths, such as about death and nonexistence. So there you are feeling bad about all the wrong you see in the world and really swept away because rather than exploring self, you are focused on the problems. A developed self is better to help others and seeks truth in all things; it has the tenacity to explore truth.

Resistance

$R = V / I$ = voltage / amperes = ohms

$R = D^3 / DV$

$R = DT$

Electrical Field

$E = p \times J$

$E$ = electrical field = resistivity × current density

$E$ = ohm - meters × amperes / $D^2 = (DT \times D) \times (DV / D^2)$

$E = D^2T \times 1 / T = D^2$

$E = D^2$

Resonance

$F = V / D$ = velocity / resonance

$D$ = resonance = $V / F = VT$ = ohms / $T$

$D = D$

Einstein Tensor $G = G\,8\,\text{pi}\,T / C^4$

Einstein tensor $G = G\,8\,\text{pi}\,T / C^4$

$G = 1 / \text{density}^2 = 1 / T^4$

$T$ = force = $D^4$

$C^4 = V^4 = D^4 / T^4$

$G = D^4\,T^4 / D^4\,T^4$

**$G$ = no units since all functions cancel out. This is a violation since all constants must observe units since they are integral with functions.**

You will notice that in the above equations, all units cancel, and you are left with no units or functions. Surely, the field equations are not correct for many reasons as follows:

- **$G = 1 / \text{density}^2$ and is a function (variable) of the source of gravitation, total mass of the core and generated inertial fields from the core.**
- **$T^2$ = density causes the cancellation of units; surely had Einstein known that density was a function that can be described by time, totally different results would ensue.**
- **Gravitation and a variation of the density of the inertial volume is a totally different concept to curvature of a plane for gravitational effect. Curvature has to do with gradients of space density rather than in the manner presently understood. So curved space-time is really curved densities.**
- **There are no constants in physics, only proportioned constants where the function or variable coexists with the constant and are equivalent.**
- **You only need three units in physics that are $D$, $T$, and $V$ for the understanding and unification of all expressions.**

## Review of Dimensions $D^1$ through $D^5$ Thus Far

$D^1$ = resonance

$D^2 = Q$ = charge = $E$ = electric field

$D^3$ = voltage

$D^4$ = force

$D^5$ = energy

$1 / D^1 = F$ = farad = capacitance = charge density

$1 / D^2 = e$ = vacuum permittivity

$1 / D^3 = n$ = particle density

## Review of Dimensions $T^n$ Thus Far

$T^1$ = time

$T^2$ = density = $u$ = permeability of free space

$T^4 = 1 / G$; $G$ = gravitational constant

$1 / T = J$ = current density

## Review of Dimensions $V^n$ Thus Far

$V^1$ = magnetic field = velocity = drift velocity

$V^2$ = temperature

## Constants and Proportions

We had talked about constants and that all were integrated with functions. So there are no constants per se, but variables or functions that were proportioned according to constants. Constants originating inside the universe we shall term emergent constants. Those that have no units such as pi originate from outside the universe and are referred to as true constants. Variables are integral with constants.

### The Value 1

- **Represents balance of infinite sets within the universe. Transdimensional and scalar infinities are related in the *D* superstring. To achieve this 1, there is a subdividing phi and phihat, which is also related to subdividing string *V* and *T* respectively.**

### The Value *C*

- **The value *C* related to velocity in the reference to hyperspace. *C* is adjusted for *C* + *V* and *C* - *V* to account for the expansion and dilation of space. *C* works to provide unity to hyperspace.**

### The Value 0

- **The value 0 works to provide unity to subspace. Here, unity is provided relating the proportion of energy superstrings. As such, this refers to scalar infinities related to the Pythagorean theorem. The quadratic equations relate subspace and also hyperspace since the four progression being fractal is the same.**

## Trans Dimensional Periodic Table

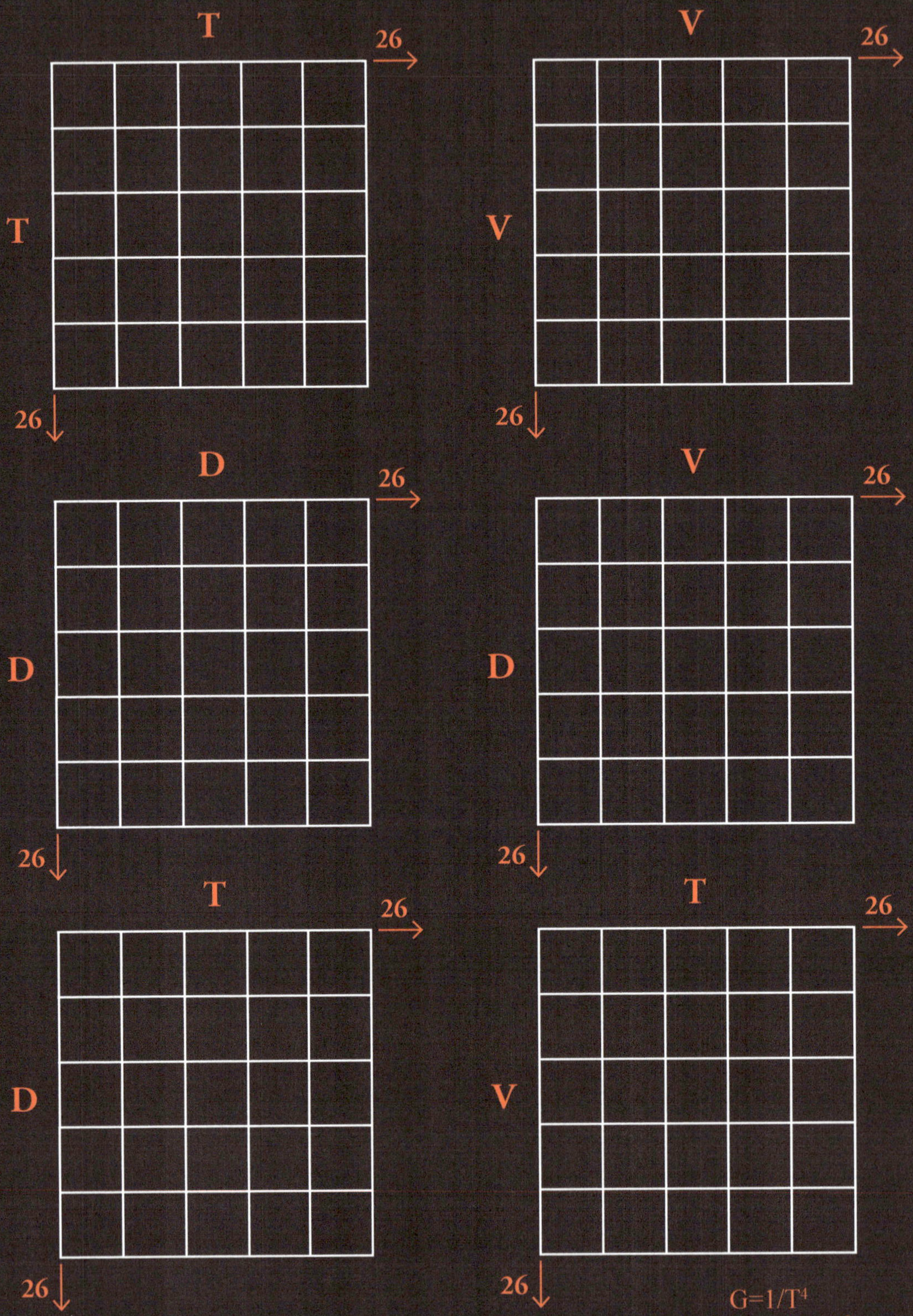


$G=1/T^4$
$\text{Density}=T^2$

We have maintained that all constants are integral with units, which means that units cannot be ignored. For example:

**$V = V$**

**$V - V = 0$**

**$V - D / T = 0$; so far, units recognize a consistency**

**Now for the special case of hyperspace and $C$ where $C$ is unity or $C = 0$ as follows:**

**$2V - 1 / T = C$; where $D = 1$ in the hyperspace condition**

**$V - 1 / T = 0$**

**Phi - 1 / phihat = 0**

**1 / phihat = phi, and so**

**Phi - phi = 0**

The Einstein tensor has no units either after a dimensional analysis is performed, but then again, Einstein didn't have

**$G = 1 / T^4 = 1 / \text{density}^2$, so even $G$ here has units**

So even units cannot be ignored even though you assume a constant, but then when everything is under this new transdimensional system, we can take a shortcut and look at magnitudes, ignoring units, because we know they are all integrated and correct. In this theory, **force = $D^4$**, so even in the analysis of $F$ under other conditions of charge, electricity, and magnetism has to come to $D^4$ as well, any force, whether gravitational or electrical. So a dimensional analysis is promising to see what is missing and what might be added to give a mathematical and physical property relationship balance.

**Now $D$ is the God superstring or primary superstring. Why?**

$D$ is integral with the number 1, coming from the multiplication of phi (1.618) and phi's conjugate, phihat (.618). Phi and phihat combined result in the location of balance, the 1, that combines additive scalar functions with multiplicative or power functions. The 0 represents unity as in the Pythagorean Theorem, as well as $C$, which is a special condition of hyperspace where $D$ is 1. That's the only place where that happens, the 1, the crossroads of infinity, where form is created and exists. $V$ and $T$ would relate to phi and phihat respectively.

**You only need three units in all physics to explain everything: $D$, $T$, and $V$ or distance, time, and velocity.**

So then you can reduce anything related to force to $D^4$ and check for accuracy, so you don't even need to do the complex unit thing if you know how the units transfer. For example, coulombs are $D^2$ and volts $D^3$.

## The Sun

The Sun generates an **inertial field to $D$** and the **effect of gravity to $D^2$** since gravity is at $E^2$, so the power of $D$ extends to $D^2$ for higher dimensional gravity. For a super massive black hole, $D^2$ is the extension of the inertial field and $D^4$ is the extension of the gravitational effect since gravity now is raised to $E^4$. So in the extension of space or inertial field, the extension is like an atom or transdimensional atom. The Sun, in a sense, is behaving like the nucleus of an atom, with generated space and the planets revolving like electrons through the generated space. So space in a sense becomes a voltage, from a positive Sun nucleus to more oppositely charged planets.

To reverse polarity of the space matrix is to generate an anti-gravitational effect. This means to reverse the charge field such that positive flow is reverse in relationship to negative field. Well, since the inertial field extending from the Sun has a pattern that complements the traditional atomic model, there isn't any antigravity involved in the Sun, although one method to antigravity is to have positive current flows directed near the speed of light that generate a field of opposite polarity with regard to charge or spin. As we have stated, space is a matrix of superstring $D$, $T$, and $V$ superstrings, and so all physics equations can be rewritten into these three variables. The extension of space follows the Schwarzschild radius, where $D$ is radius as follows:

$$D = G \times \text{mass} / C^2 = R$$

$$G = 1 / \text{density}^2$$

$$\text{Density} = T^2$$

$$D = \text{mass} / \text{density}^2 \times C^2 = (D^3T^2) / T^4 \times C^2$$

$$D = (D^3T^2)\ T^2 / T^4 \times D^2 = D^3T^4 / T^4D^2 = D^3 / D^2$$

$$D = D = R$$

Equation is correct.

**$D = G \times$ mass / $C^2$ = fifth root of ($MC^2$), known as the gravitational radius; this is in fact the radius also of the inertial field.**

And so $D$ can be calculated, but what is the mass that we use?

**Total mass = mass core of planets + mass of the inertial field**

**Total mass = mass core + *SQRT* (phi × mass core)**

With the total mass, you can calculate $D$ or $R$ and with $D$, then the volume of the inertial field by using $D$ as radius in the volume of a sphere where **volume = (4/3) pi $D^3$**.

Comparative volume gives comparative gravitational acceleration. We can also rewrite the gravitational equation in various ways as follows:

$$F = G\,M1\ M2 / R^2 = M1\ M2 / \text{density1 density2} / R^2 = \text{volume1 volume2} / R^2$$

Where

$$G = 1 / \text{density1 density2} = 1 / T^4$$

$$\text{Density} = T^2$$

$$\text{Mass} = D^3T^2 = \text{volume} \times \text{density} = \text{voltage} \times \text{density}$$

As the Sun generates a space or inertial field to distance $D$, which is just past Neptune, the pulse of the spherical wave within the space field from the Sun causes the effect of a standing wave in the shape of a spindle top within the field. This creates a gyroscopic effect for planets within the field. With regard to space anomalies from Pioneer, all mass generates space, according to $D^5 = MC^2$, which is the same as the Schwarzschild radius when you figure radius $D$. The space or inertial field generated by the Sun goes just past Neptune. The field is a sphere or ellipse and pulses out, and within the sphere is a standing wave that looks like a spindle top from the spherical wave pulse from the mass Sun. This standing wave internal to the inertial field of the Sun has a gyroscopic effect and assists to hold planets within this space in the ecliptic. Note Pluto is to the outside. The inertial field extends to $D$ but gravity to $D^2$, since gravity is a ten-dimensional field at **gravity = $E^2$**. A super massive black hole is $E^4$. So when Pioneer left the solar system, it enters less dense space, outside the internal waves of the Sun's inertial space. It is now in the space of the center of the galaxy.

## Infinity

This theory presents the idea that form is represented by the number 3 and also that through form, infinities are qualified and related in the manifestation of the form universe and consciousness by a set pattern. There exists the relationship of scalar or fractal infinities with transdimensional or power ($^N$) infinities through balance and unity represented in the numbers 1 and 0 respectively. Constants in this reality or universe become integral with function. The number phi (1.618

or the golden ratio) and phihat (.618 or the conjugate of phi) multiply to form the number 1. These numbers organize function by proportion and are integral with functions that result in 1.

**So what is the number 1?**

The number 1 represents the relationship between transdimensional and scalar infinities. Hence, infinites though pattern result in form. In the universe, all units can be represented by three, which are *D*, *T*, and *V* through both scalar (additive) and power (multiplicative) relationships. Only three units are needed to represent all physical phenomenon. Similarly, consciousness consists of awareness, thought, and will, and these are expressed trans-dimensionally and scalable within infinity. For example, receded forms of consciousness are a scalar principle. This life is a dream, and when you sleep, you dream. So from the source, the style of the action repeats itself through various layers. When I am awake, the other forms of consciousness are not concurrently experienced but by perhaps memory, so this tells us that memories represent a source memory of source consciousness or soul. So if I damage my brain and cannot think, my present consciousness is not rid of that bad experience until that state of consciousness is over or transformed into another.

**So what is character?**

My infinite self is definitely developed by experiences. And so being human definitely carries with it certain experience defined by genetic code. The mind is infinite, and the pattern of consciousness is subdivided almost as is the universe. We can speak of receded states of consciousness as patterns that are equal and happening similarly at different levels. We can speak of transdimensional as the infinite qualities we associate with thought, awareness, and will. And so for thought might carry at various levels ideas of love, desire, honesty, and since the pattern is infinite, there is never an answer to the end of "explain to me what love is," or "explain to me what desire is" since the pattern of its meaning is developed infinitely. And thus the consciousness is truly free since its very construction is unlimited, only organized at a center point, the real you, a point of balance of infinite sets that has a certain characteristic of being you. How one set or individual differs from another is perhaps in the manner organization within the pattern.

Imagine that you are in the desert on a flat plane at the intersection of two roads, like a cross, and you are standing in the center and the roads extending outward infinitely. And one road comes to you with a pattern of three colors and another with a pattern of three other colors, and in the center, the overlap of two patterns presents a third color sequence. And so the universe has this patterning, and where the two roads meet is the universe. And consciousness is where they meet as well, but to be consciousness or the universe, the form of the center appears to have boundaries, but in the relationship, they really do not exist because the patterns that create the form extend toward infinity.

**And so what is the form of the universe or the form of consciousness or the universe?**

Consciousness or the universe definitely has form and infinite patterning, but when we think of the ratios that exist for the center, we think the universe is finite, but what is finite is only in the perception of the center, not in the operation of it. And so our universe, although being perceived with limitation, is really an illusion of the pattern, and truly we are within infinity with no boundaries to it except by our own perception and experience.

**And so what is the edge of the universe?**

The edge is definitely a transformation into another reality. So how man creates is how he perceives, with perceived limit, but the creation of man has no limits except by the limited understanding of the patterns within which he exists. Hence, all life and consciousness and mind are infinite and hardly the predictable nature to which we ascribe non-infinite qualities. We are more than we imagine ourselves to be and more than we perceive, and perceiving less is our illusion. But it is given to the infinite qualities of consciousness and intuition to see more.

## Velocity: Transdimensional and Scalar

We have talked about velocity being magnetic field strength, which is amperes per meter or related to the current. This also has implications for being related to the current of the inertial field of a mass. The inertial space is in relationship to mass; that ratio exists according to $D$ in the Schwarzschild radius and also explained by $D$ in $\boldsymbol{D^5 = MC^2}$. Both equations are equivalent. The mass velocity has a pulsing current that produces a spherical wave. This wave goes out to the $D$ extension of the field and returns, which is also responsible for the standing wave produced within the inertial field that enables gyroscopic action of orbits and assists planets to remain in the plane of the ecliptic.

Velocity also can be described in the equation as follows:

**Density = mass / $D^3 = T^2$**

**$T$ = *SQRT* (mass / $D^3$)**

**$V$= $D$ / (*SQRT* [mass / $D^3$])**

**$V = D^2$ ($D$ / mass) $^{.5}$**

And here the mass velocity is directly related to the mass core where larger masses have a slower mass velocity than smaller masses. The mass velocity is directly proportional to the mass, slower in larger masses. Also, this means that as the velocity of an object increases in a closed system, the mass decreases, which is contrary to the notion of mass increasing. An alternate interpretation to relativity would involve proper velocity where mass decreases with velocity. When the proper velocity or transdimensional velocity within the system is increased, the mass is reduced and approaches zero. So these transdimensional equations deal with proper velocity and not the coordinate velocity of the observer, which is scalar. (Epstein: 1-4)

At $C$ or zero, $\boldsymbol{D}$ **= 1**, wherein $D$, velocity is phi and time is phihat.

**Velocity is proportional to (1 / mass)$^{.5}$**

**We can also deal with velocity of the system in scalar terms.**

Scalar velocity has to do with arithmetic ratios from the Pythagorean theorem. And at zero, $\boldsymbol{D}$ **= 1**. $D$ is where both transdimensional and scalar series meet in balance of 1 (phi and phihat).

**There are two systems of physical equations, which are scalar and transdimensional.**

So for normal motion equations through a coordinate system, scalar equations related to Pythagorean relationships would lead to solutions involving quadratic equations and coordinate relationships. Contrarily, transdimensional equations that deal with proper velocity within the systems viewed have such equations that are multiplicative and by dimensional powers as in $\boldsymbol{D^5 = MC^2}$. These two systems become equal at point of balance where $\boldsymbol{D}$ **= 1**. Point of balance is at $C$ in hyperspace and 0 at subspace. Between these, the transdimensional equations functions are variable and also for scalar functions. ***D* is the prime superstring** at point of balance in both scalar and transdimensional forms, where $\boldsymbol{D}$ **= 1**; the two systems become integrated are united. A scalar system is relative by position of the observer. A transdimensional system is intrinsic to the system itself. One system deals with multiplicative properties of space described as transdimensional, and the other deals with scalar Pythagorean as in subspace where superstrings overlap.

Current physics believes there are four dimensions, three for space and one for time, but this idea is the reason why gravity is not connected to magnetism and electricity in theory, and it has to do with the understanding of time and density. Once that is resolved, you will in fact have multiple dimensions and a unified field. You can rewrite all physics in terms of three units; once this is arrived at, navigation is far easier, and in fact, we can almost dispense with units altogether, as in using magnitudes in a dimensional analysis.

The gravitational constant is in fact $\boldsymbol{G = 1 / T^4}$.

**$G = 1 / T^4 = 1$ / density$^2$**

The relationship of time to density is the key to a unified theory. The description of time is the key to unification as well as redefining expressions into three units. When time is redefined, and then multidimensions is the result. Current understanding of time and velocity are wrong. We as humans and by our perception think that space is 3-D and time is one dimension. If you use dimensional analysis using this theory, you find that Einstein tensor has no dimensions at all. Because electricity and magnetism are not connected to gravity is because of the current understanding of time, which has been revised in this theory that allows for unification. When unification is resolved, you see there are in fact multiple dimensions.

There are no constants per se, only constants integral with functions; hence, units must be respected. If we can reduce all units to three, then after this unification, we can almost dispense with the units and talk about magnitudes as in a dimensional analysis mentioned before; however, units still have to work out.

**Can you calculate the velocity of a falling rock with $E = MC^2$?**

No, and why? A falling rock is scalar within a coordinate system determined by Pythagorean relationship and quadratic functions. Transdimensional systems refer to proper velocity of closed systems. Two systems, scalar or Pythagorean and transdimensional, become equivalent and equal in special conditions of balance where $\boldsymbol{D} = 1$ at point of balance or 1 that results in zero or $C$ for subspace and hyperspace respectively.

Lower mass velocity means higher inertial resistance. Approaching the speed of light in a closed system, the mass is less.

**Phi and Phihat: The Transition between Scalar and Transdimensional Infinities**

**$\text{Phi}^2 = \text{phi} + 1$**

**Phi + Phihat = 1**

**Phi x Phihat = -1**

**Phi – Phihat = SQRT 5**

Scalar and transdimensional infinities become equivalent when $\boldsymbol{D} = 1$ in subspace and hyperspace. Subspace refers to the area of space where matrix superstrings overlap space and are scalar in representation by additive and quadratic systems where values are calculated on a coordinate system. Here, velocity would be relative. Transdimensional systems are multiplicative, and values are proper to the systems, which are closed. In the hyperspace of the transdimensional system where value is $\boldsymbol{D} = 1$ for $C$, there is also a relationship between scalar and transdimensional systems. So the systems relate in balance of 1, where time and velocity are in balance and unity in subspace at 0 and $C$ in hyperspace. From these transformative points, the universe branches out like a tree into scalar and transdimensional infinities. So the point of balance become $D$, the God particle, but here defined as the primary superstring.

***Pattern of D or Dimension***

**Energy ($MV^2$) = force ($MA$) + volume ($D^3$)**

$D^5 = D^4 + D^3$

$D^5 = D^4 + D^3$ **= energy = force + volume**

$D^5 - D^4 - D^3 = 0$

$D^2 - D - 1 = 0$

$X^2 - X - 1 = 0$

$D^5 = MV^2$

$D^5 = D^4 \times D^1$

$D^4 \times D^1 = D^4 + D^3 = D^5$

$D^1 = 1 - 1 / D$

$D^2 - D - 1 = 0$

$X^2 - X - 1 = 0$

$D / T - V / D = 0$

$D - 1 / D = 0$

$D^2 - D - 1 = 0$

$X^2 - X - 1 = 0$

***Pattern of V or Velocity***

$V = V$

$V - V = 0$

$V - D / T = 0$; so far, units recognize a consistency

Now for the special case of hyperspace and $C$, where $C$ is unity or $C = 0$ as follows:

$V - 1 / T = C$; where $D = 1$ in the hyperspace condition

$V - 1 / T = 0$; where $D = 1$ in the subspace condition

Phi - 1 / phihat = 0; 1 / phihat = phi and so

Phi - phi = 0

$0 = D - 1 / D$ for subspace

$C = D - 1 / D$ for hyperspace

$0 = VT - 1 / VT = V - 1 / VT^2 = V - 1 / DT = V - 1 / T$

$0 = V - 1 / T$ for subspace

$C = V - 1 / T$ for hyperspace

Phi × phihat = -1

Phi + phihat = 1

$V$ is integral with phi, and $T$ is integral with phihat

$VT = -1$

$(V + T) = 1$

$-VT = (V + T)$

$-VT - (V + T) = 0 = 1 - 1$

$VT - (V + T) = 0$

$V - V / T + T / T = 0$

$V - (V + T) / T = 0$

$V - 1 / T = 0$

$V^2 - V - 1 = 0$

$X^2 - X - 1 = 0$

$Phi^2 – phi -1 = 0$ or $phi^2 - phi = 1$

This is the quadratic formula for motion.

***Pattern T or Time***

**Phihat = -1 / phi**

**Phihat = 1 - phi**

**Phihat = 1 - 1 / phihat**

**Phihat - 1 + 1 / phihat = 0**

**$Phihat^2$ - phihat + 1 = 0**

$T^2 - T + 1 = 0$

$X^2 - X + 1 = 0$

Here you see that when *D* is in the nominative case or numerator, at unity or zero ***D*** **= 1** (balance). This condition of balance exists at unity where scalar and transdimensional infinities relate in balance or ***D*** **= 1**. Unity also exists at *C*, which is the hyper-dimension of space. Zero is the subspace of space. This represents the relationship of phi and phihat in all things. The universe is not stationary, so there is no cosmological constant since all constants are integral with functions. However, ***D*** **= 1** for balance and unity represents the constant of transition or point of transformation. Hence, you might say that the cosmological constant is ***D*** **= 1** and represents the transition of all infinities or change. All the universe is being transformed at ***D*** **= 1**.

So now we can see that time is transformative as well as velocity. *D* is the point of balance that allows for transformation, that relates time and velocity. So in this theory and for physics, time is a multidimensional phenomenon. And also we claim that the universe is a form of consciousness referred to as God's consciousness. And so now then is the form of time in our consciousness as diverse as is the universe. Well, if all consciousness is collective, yes! And therefore, time is experienced internally and perceived externally in many different ways. It can be linear, it involves attraction and repulsion, and it involves density as concentration and relaxation. So as time has infinite dimensions, so too time and its experience as something to be explained should also have such a definition.

**Pi**

In an anti-gravitational force, the opposite is true about temperature. Pi comes from phi, and phi and phihat make the number 1. At hyperspace of *C* or subspace of 0, ***D*** **= 1**. Both *C* and 0 represents the transforming or transformation of all change or relationship or transformation. Constants phi and phihat are integral with the functions velocity and time respectively. Pi is integral with the function *r* or *X*, and all equations refer to functions. In the case of the sphere, it refers to *R*, which is the extension *D* of the energy of a superstring. This *D* also relates to the extension of inertial space as in Schwarzschild radius. There are no constants, but integral with functions and pi refers to *X*, which refers to *D*, *T*, or *V*.

**Phi and Phihat**

**Phi × phihat = -1**

**Phi + phihat = 1**

**$Phi^2$ = (phi + 1)**

**$Phi^2 = -(phi + 1)$**

**$Phi^2 - (phi + 1) = 0$**

**$Phi^2 - phi - 1 = 0$**

**$V^2 - V - 1 = 0$**

**This is the quadratic formula for motion.**

**$X^2 - X - 1 = 0$**

***D***

**$(phi \times phihat)^2 = 1 = D^2 = X^2$**

**$(phi + phihat) = 1 = D = X$**

**$X^2 - X - 1 = 0$**

It appears that the quadratic equation unites scalar and transdimensional functions. The combination appears to be transdimensional, scalar (*a* and *b*), balance (1), and unity (0).

**Temperature**

In my theory, everything is made from space where space is made of a matrix of superstrings. Particles are the interference patterns of the waves or superstring of the matrix. Interference patterns create particles. The atom is a transdimensional model going from dimension to dimension and containing all dimensions; dimension 23 in a black hole at the center of the nucleus of an atom where energy is $E^4$. So as the transdimensional structure is dismantled, space at $E^4$ expands to a state of $E^2$, resulting in an expansion of space to account for the additional space expanded at a lower density. Which means the space surrounding matter squares in $D$ when mass is split. This space or inertial space has a velocity in proportion to the core mass. As the mass decreases or as the density decreases, the extended space has increased velocity. The extension of the new lower space density coming from the higher space density has a velocity that is square root of the original velocity.

There is ratio of inertial space extension $D$ and associated $V$ inertial velocity, and more energy released means more space. Hence, a black hole totally exposed projects inertial space to $D^2$ instead of $D$ of the Sun and other planets according to $\boldsymbol{D^5 = MC^2}$.

Chemical reactions cause the expansion and contraction of space. Energy released and from it increased inertial velocity due to expanded space causes repulsion away from higher density gravitational gradients. So the repulsion effect within a fire has to do partly with the expansion of space and its relationship to the density of that space gravity gradient of the planet. Since the density is highest at the core of the Earth, then the expansion against that density gradient would be in the opposite direction as new space is added. This may seem as an anti-gravitational "effect" but not true antigravity. True antigravity has to do with reverse polarity to the field of space created by the atom. Consequently, that means the inverse of gravity along with reversing the spin of the superstrings within the space matrix. Positive matter flows around a negative core is but one method to do this and will generate a field of space with opposite polarity containing opposite spin of the containing superstrings. We can assume at this point positive and negative values refer to charge, but spin is also to be considered.

***Now let's look at pressure:***

**Pressure = force / area = $D^4 / D^2$**

**$P = F / A = D^4 / D^2 = D^2$ = pressure**

$P = D^2$

## *Relationship between Pressure, Temperature, and Velocity:*

$PV = nRT = D^2 \times D^3 =$ **mass** $RT = D^5 =$ **energy**

$PV = D^5 = D^3T^2\ RT$

$T = K$

$RT = RK = D^5 / D^3T^2 = D^2 / T^2 =$ **pressure / density** $= V^2$

$RT = V^2$

Discounting the *R* or constant, temperature is an expression of velocity. So in a real sense, temperature as defined is velocity$^2$. When you explode an atomic bomb, the creation of space or inertial mass in $V^2$ causes a huge expansion. This is a transformative event contrary to the natural inertial velocities associated with stable forms of mass, whether they are atoms or nuclear as in the Sun. Normally the voltage ($D^3$) between mass densities draws them together, as in a planet orbiting a larger star. But in transformative events, the action is repulsive counter to the gradient of the gravitational densities since space is expanding since it is being introduced or new. In addition to this effect as is a fire, you also have the atmospheric buoyancy action casing new particles generated to ascend into the atmosphere. But also within the fire, you have space being created and inertial mass velocity being accelerated counter to the planetary gravitational host.

## *Further Review of Temperature:*

$RT = RK = D^5 / D^3T^2 = D^2 / T^2 =$ **pressure / density** $= V^2$

$R =$ **pressure / density** $K$

$R =$ **joule /** $K \times$ **mole**

$R = D^5 / K \times$ **mass** $= V^2 / K$

$V^2 / K =$ **pressure / density** $K$

We will note that both sides of the equation cancel out under dimensional analysis, and so the constant *R* seems redundant. But as an expression of temperature, $T = V^2$ seems to suffice, referring to temperature equivalence with mass velocity.

## *We might recommend organizations developed of the change in temperature:*

$K1 = d$ **pressure** $K2 / d$ **density** $\times dV^2$

**or**

$K2 = K1\ d$ **density** $dV^2 / d$ **pressure**

## *Gravity*

**Force of gravity = $D^3 \times D^3 / D^2$**

$$FG = D \times D \times V^2 \times V^2 \times T^2 \times T^2 / D^2$$

$$FG = V^2 \times V^2 \times T^2 \times T^2$$

$$FG = T1 \times T^2 \times \text{density1} \times \text{density2} = T1 \times T2 / G$$

Gravity has, by the relationship above, a connection between temperature and density of the interacting inertial fields. As the temperature goes up, the density goes down and vice versa in a closed system. So temperature affects gravity when temperature is a measure of the velocity of the inertial field mass. Temperature is intrinsically connected to the performance of both gravity and its inverse relationship antigravity. Tied to temperature is the mass velocity of the inertial field $V$ and the density of that inertial field.

## *Antigravity*

**Antigravity force = $1 / T1 \times T^2 \times$ density1 $\times$ density2**

$$AF = G / T^1 \times T^2$$

In an anti-gravitational force, the opposite or inverse is true about temperature. So we see that temperature is important to the gravitational attraction and repulsion and is a value like all other transdimensional functions in its significance. The gravitation equations can be rewritten in terms of temperature.

## Unified Field

Variations of gravity according to density changes of the vacuum or inertial field near a mass follow the relationship **density = $T^2$**. This is the key to a unified field theory in physics. Having arrived at that relationship of time equivalency to density proves multiple dimensions and expresses additional relationships, which all other theories do not have. So therefore, relativity is wrong by definition, and quantum theory is redefined with greater clarity, but in a sense, the redefinition of one and the clarification of the other bring them together and also add the idea of multiple dimensions as in string theory.

I think that the theory presented says that all physics equations can be redefined with three units and powers thereof. I think it clarifies quantum theory, not taking anything away and, by that, making relationships easier to discover. Now there are multiple dimensions, but also I would say that for our direct use, five seems to be within our current grasp and does not complicate existing physics at all.

One of the key propositions here is that the constant $G$ in the gravitation equation really refers to the density of two objects, so

$$G = 1 / \text{density1} \times \text{density2}$$

$$G = 1 / \text{time}^4$$

And that the variation in the density of the quantum vacuum is in fact what can be called an inertial field and has structure. So the ether or vacuum is in fact referring to $G$ of two objects.

If it is feasible that **$G = 1 / \text{density}^2$** is proved, which is shown mathematically in this theory, that means that **density = $T^2$**. This relationship of time and density opens up the entire connection between electromagnetism and gravity, and not to mention many other relationships such as, for example, with temperature and capacitance. The variations in dimension, time, and velocity in multiple powers show multiple dimensions as well. With this step, you can come to derive temperature related to velocity$^2$ and rewrite the Schwarzschild radius as **$D^5 = MC^2$**, where $D$ would be the radius.

There is justification for string theory as well:

1. **There is agreement on this theory that there are smaller particle strings, but they are identified as being particles that form space, and they are assigned three values *D*, *T*, and *V*. All physics can be expressed into three units and powers thereof. The problem here is that it also should connect with quadratic functions that I call scalar. And in this theory, the expression of *D*, *T*, and *V* is both transdimensional or multiplicative, and scalar as in quadratic or additive. If this is correct, these two systems come together where $D = 1$ at 0 and *C*, which is the goal as presented.**

2. **String theory deals with the particle side. So in this theory here, it is said that space is primary, and that seems reversed in string theory where you have particles instead of physical values or functions. This particle orientation is an advantage in understanding particles but a drawback in understanding values. The strings identified here are in fact overlapping, combining, and interfering to create particles. So perhaps we can call them space particles. To these space particles have been assigned values distance, time, and velocity.**

3. **Phi (velocity) and phihat (time) relate to form all the physical equations. This is the proof of super symmetry.**

This theory will help string theory because space particles or strings set up two distinct patterns, what are called transdimensional string relationships or multiplication, and scalar functions related to quadratic formulas. Of an experiment, the connections run right through the entire field. Temperature affecting gravity would be one experiment. Calculation of gravitational pull without the constant *G* would prove relationship of gravity related to densities, temperature, and other electromagnetic values would mean that mass objects produce inertial fields. A redefinition of the constant by showing the relationship of time to the density of the inertial field in

**density = $T^2$**

**leads to**

**$G = 1$ / density1 density2 $= 1 / T^4$**

and would allow the validation of expression of all physics phenomenon into three units. This validates and simplifies the use of quadratic formula in the expression of string theory and multiplicative functions. At the simplest level, the functions are *D*, *T*, and *V*. We have said before that space density varies by a ratio of distance, which is correct; however, we must also include time since **density = $T^2$.**

The Schwarzschild radius can be rewritten $D^5 = mc^2$, and why? Because of the relationship of **density = $T^2$**, where **$G = 1$ / density$^2$ $= 1 / T^4$**, so yes, time does vary by distance with the gravitational source. We can define gravity solely through volume where volume is the volume of the inertial field sphere calculated from *D*, which goes beyond the planet surface, of course. In addition to the mass of the planet, we need the mass of the vacuum or inertial field generated by the planet.

**Total mass = mass of planet + inertial field mass**

**Total mass = mass of planet + *SQRT* (phi × mass of planet)**

and then use total mass to calculate the Schwarzschild radius *D* since the total mass includes the inertial field.
By calculating the volumes of inertial field for Earth and Mars, the volume ratio is representative of the difference in acceleration of the planets. Now for a gas planet, that volume needs to be broken out into density since the object is not a solid. And as the mass expands or is less dense as in Saturn, it will cause the inertial field to recede and should vary the gravity and time density gradient due to distance. The ratio of time to gradient of density distance is connected to the inverse square law. By calculations, *D* for the Sun goes just past Neptune, and it is proposed that the vacuum has mass,

density variation according to **density** = $T^2$, and also velocity, and so as the inertial field pulses in a spherical wave from the Earth and within the sphere of influence, you get a standing wave that assists the planets within that field in the plane of the ecliptic like a gyroscope.

**These new definitions also redefine *G*, which is in fact a variable.**

*T* is time. There is a gravitation equation with *temperature* spelled out. Temperature derived here is $V^2$, amazingly like the kinetic gas theory, except the constant seems redundant since it cancels out the units and so is dropped. The principle is the same.

*C* here represents as an elastic collision of the inertial field intrinsic to a closed system. However, you can get $C + V$ and $C - V$ because space can be expanded and contracted due to various forces, and therefore, a scalar measurement within the coordinate system might produce slightly varied results.

**So does the speed of light vary, or is it constant?**

Yes and no. It is most definitely a constant like zero, representing a special case of hyperspace, and no, due to space distortion, it may be possible that it varies from a third-party measurement as a scalar measurement. Transdimensional includes multiplicative equations and closed systems with "proper velocity," and additive systems are scalar.

**The Bending of Starlight**

Space is a tetrahedral structure of superstring spheres. Imagine a bowl of marbles, but rather than touching, they overlap at that very small scale. Where they overlap or interfere, particles are created. So a photon is in fact a monopole of that matrix. So everything by construction is radial and spherical, but not exactly so linear. So by such a situation, you are going to get a shift in the path of light. Also, there is a ratio of space to each generating body. Although those collision of light are almost straight, they are not absolutely so straight.

**The Shapiro Time Delay**

Proper time and coordinate time, just as proper velocity and coordinate velocity, vary because one system is closed or transdimensional and the other scalar or open. You have the Sun generating its field of space and also the planets. And this relative coordinate system will result in divergence from that of the contained proper system or transdimensional.

**The Pound Rebka and Vessot Scout Rocket Experiments**

The experiments show that clocks gain time with altitude but lose it with speed. The variation of time according to density of the inertial field can be expressed by **density** = $T^2$ and the loss of speed also with the proper velocity. So you need a gravitation equation expressing velocity and time such as

$$F = \text{energy1 energy2} / R^2 V^2 V^2 T^2 T^2$$

$$F = G \text{ energy1 energy2} / R^2 V^2 V^2$$

$$F = G \text{ energy1 energy2} / R^2 \text{ temperature1 temperature2}$$

$$F = \text{energy1} \times \text{energy2} / R^2 \times \text{force} = D^5 D^5 / D^2 D^4 = D^{10} / D^6 = D^4$$

$$F^2 = \text{energy1} \times \text{energy2} / R^2$$

$$F = SQRT(\text{energy1} \times \text{energy2} / R^2)$$

## The Gravitational Red Shift of Light

If the density of space varies away from the source of gravity and also has structure or form or energy, then similar to sound, you are going to get a shift in the coordinate system.

## The Anomalous Perihelion Advance of the Planets

Masses generate space as a velocity of the field of inertia. The limit according to the farthermost extension *D* is derived by $\boldsymbol{D^5 = MC^2}$ to form a sphere or slightly ellipsoid area. The energy coming from the mass is like a spherical wave, which causes within the inertial field a standing wave like a spindle top, which assists in holding planets in an ecliptic plane. Near the Sun, this area or gyroscopic effect is minimized since the wave amplitude is greater, so it would affect Mercury slightly. And also beyond Neptune where the inertial field of the Sun ends, then you get behaviors like Pluto. So you have the density of space to contend with and the internal matrix velocity pulse. Now when considering the field of orbit, you have for each planet an inertial field, which also has mass and are two offsetting conditions that are explained by the following:

- **The center of the mass of the planet is symmetrical.**
- **The center of mass of the inertial field is eccentric since due to spin, the field becomes elliptic.**

So no orbit can be perfect. You have the interaction of asymmetric inertial system and symmetrical mass core planet system, and hence, there is always slight adjustment, which also causes perihelion and aphelion points.

There is proper time and scalar time of the coordinate system; both are valid because both are true. There are "special conditions" where *C* (hyperspace) and 0 (quantum space) are achieved in quadratic formulas where time is constrained. There are no contradictions except if you try to put all conditions on equal or universal basis when they are really at different levels of operation. Which they are not on the same level of operation, but we can say that in this case, this is so, and another it's different. The variables operate differently at different levels, at some constrained, at others scalar, and others proper. For simplification, you can describe all equations in physics by units *D*, *T*, and *V* and powers thereof. By having three units, you are able to extrapolate multiple meanings and relationships where before, you might look at an equation as one relationship. As you have done to rewrite gravity in another form, it can be rewritten in maybe ten other forms or more, showing that gravity is interconnected by many variables. The equivalency of three units allows multiple definitions of the same expression, and therefore, the pattern leads to discoveries beyond an initial understanding or discovery. This patterning also leads us to know that temperature and gravity are related. Although space may be an elastic superconductor, temperature is related to gravity. So now as we progress in our unification of physics, where we thought maybe one expression of gravity will suffice, for a unified field, multiple expressions, all equivalent, are to be expected. And although the temperature of space is relatively consistent, temperature is still a factor.

It was said by Einstein that mass and energy are related by something; that something, the smallest denominator, is velocity and time and combined in various powers to produce everything with *D* or dimension. Now why would that be the smallest? Because at *C* and 0, $\boldsymbol{D = 1}$, which is an expression of phi and phihat, constant integral with velocity and time respectively. At zero and *C*, both scalar and multiplicative systems come together. Smallness is infinite, but at 0 and *C*, there is a point of relationship of two systems, and that point of balance is the focus of opposite extremes. So that would be the primary point or God particle.

You have options to express gravity in one way with *E*, another with volume or voltage, another with temperature, another with mass; and so which is the correct one? It is the one where the variables are being manipulated. So is matter made of mass or energy? It is made of different combinations of *D*, *T*, and *V*, and they express multiple properties within a closed system of multiple dimensions. Equations can be elaborated to include all the things you wish to see and how they relate. For gravity, as long as **force** = $\boldsymbol{D^4}$, you can elaborate an equation to great complexity as long as the total effect is $D^4$.

## Quasiparticles

We have talked about the space matrix made of superstrings, and where the strings overlap and interfere, you have the creation of a monopole such as a photon. These monopoles are quasiparticles since they are single particles within the system and are associated with space.

## Coupling

Space is a resonating system, and the interaction of the inertial field of two systems is known as coupling. Gravity is an effect where two systems couple and decouple as they are drawn together by the inertia of the gradient due to density or voltage or temperature. Coupling is affected by several factors. Here in

***F* = *SQRT* (energy1 × energy2 / *R*²),** $F = SQRT(\text{energy1} \times \text{energy2} / R^2)$,

the coupling of two systems of energy is outlined.

## *G*: Gravitational Constant

*G* is a variable and also refers to two functions operating together, density 1 and density 2. So in a sense, the notation *G* should be ignored completely. Its use is only for convenience of traditional equations. And hence, like any other function, you can take the square root, but more appropriately, since it is really something that can be described or translated by equivalence, it can combine with other functions to produce other properties.

$F = SQRT(\text{energy1} \times \text{energy2} / R^2)$ represents the translation of the myth *G* to real functions.

## Gravity Proportional to $1 / R^2$ and $1 / R$ and Gravitational Shielding

Transdimensional theory predicts equivalency of closed systems where time and velocity are proper. Also, all functions exhibit the principle of equivalence. The gravitational constant *G* represents functions that are not defined, but here represent the density of objects. And also, the mass, originally viewed as planetary mass, also includes the mass of the generated inertial fields. The extension of the mass of the Sun is to distance *D* where **gravity** $= D^{10} = E^2$. This higher dimensional phenomenon is measured as $D^4$ or force. For a black hole, the extension is $D^2$ where **gravity** $= D^{20} = E^4$. This relationship is also present in the model atom where within the nucleus to central black hole of an atom, there is a high-density space matrix and gravity at $E^4$. This is shielded by the nucleus and thus experienced as $E^2$.

With regard to inertial fields, it is true as the density becomes less for a given mass, the inertial field recedes. There is a linear relationship between the increase in the volume of the inertial field at *R* radius and decrease in the density in the equations

$$F = \text{volume1} \times \text{volume2} / R^2 = \text{mass1} \times \text{mass2} / \text{density1} \times \text{density2} \times R^2$$

$$F = \text{mass1} \times \text{mass2} / \text{density1} \times \text{density2} \times R1 \times R^2$$

When the gravity leaves its field of inertia at distance *D* of the inertial field radius, this is at the hyperspace of that extension calculated at *E* or energy. At that point, gravity becomes gravity of gravity, and hence, the force is equivalent to $1 / R$.

$$F = SQRT(\text{energy1} \times \text{energy2} / R^2)$$

The inertial field extension is *R*, but gravity extension is $R^2$ for the Sun and $R^4$ for a black hole, which is beyond the extension of the generated inertial field through adjacent inertial fields. Hence, gravity *F* becomes proportional to $1 / R$ as a linear relationship and not radial from a point source since it is now beyond its own inertial field. Performing experiments within the solar system will not reveal this. By this model, stars in galaxies will reveal a $1 / R$ relationship instead of a $1 / R^2$ relationship, which they do.

***Force = Volume1 × Volume2 / R²***

This equation refers to gravity within the inertial field at $1 / R^2$. This can be calculated by relating the volumes of the inertial field. Variations in density of two objects can further complicate the equation such that volume needs to be expressed in terms of density and mass to relate the variations in density of dissimilar systems.

***Force = (Energy1 × Energy2) / R² × Force***

***Force² = (Energy1 × Energy2) / R²***

***Force = SQRT ([Energy1 × Energy2] / R²)***

This equation refers to gravity from hyperspace extension at force proportional to $1 / R$ beyond the inertial field. The gravitational relationship to Newtonian mechanics is shielded beyond the inertial field. This explains the increase in gravity beyond the shell of the inertial field.

***F = G Energy1 Energy2 / R² Temperature1 Temperature2***

Within the inertial field where temperature and density of two objects are uniform, you get the relationship of force proportional to $1 / R^2$. Variations in temperature and density make slight adjustments to the $1 / R^2$ rule. Where density and temperature are perceived as constants, then you get $F$ proportional to $1 / R^2$.

Space is a matrix, and that there are monopole particles to it. Hence, gravity is related by structure and particles. Photons do not travel, but the elastic wave through the structure does. Gravity also has a higher dimensional construct, and what we measure as force is only the "effect" within the measurable dimensions in which we live, up to dimensions $D^5$ or energy. Our units do not go higher because our measurements do not as yet. But when we write equations that involve powers suggesting relationships between $D^5$ or higher, we are forming connections that go beyond what we measure as the limits of our measured inertial fields. Also, what we write for a higher dimension can be transcribed to a lower dimensional equivalent. Hence, force is always $D^4$ but can refer to multiple transdimensional events occurring at higher dimensions. Space can be quantified, and hence, our picture changes when we are outside of the environment of the inertial field in which we live. Measurements of the solar system and being within its inertial field is biased and should not be the standard for developing a model for gravity and its operation for the whole universe. It is true that $1 / R$ gravity is also seen acting within the universe beyond our solar system. The system of gravity is hierarchical and changes with relationship to the type of source, the extension of the field, and also the interaction between fields. The shifting density of gravitational field from the source ends at the extent of the inertial field.

It could also be that a very small scale of quantum gravity, the gravity interaction within the space matrix, is linear also. So within the subspace and without the inertial field, the relationship of $1 / R$ holds true, but through the intervening extremes of the matrix, where the geometric densities carry a radial pattern, then $1 / R^2$ would be primary. At the quantum level of subspace, the square root of $R$ becomes the square root of classical mechanics. (Rovelli:1-5)

The idea that space is ordered by various powers of three units and trans-dimensionally, these allow us to define inner relationships of the atom to a black hole at $E^4$ by multiple dimensions. $G$ can be redefined as $1 / T^4$, which means that there are multiple dimensions not only for $D$ but $V$ and $T$. That quadratic additive and multiplicative infinite scales relate at $C$ and 0, where phi is velocity and phihat is time, means that infinity is qualified where both scalar and transdimensional relationships are united. Euclidean geometry is only relevant in limited view and is an abstraction of more complex systems. The notion of four dimensions of space-time is enabled by the construct of supporting multiple dimensions on an infinite scale. At $C$ and 0, $\boldsymbol{D} = \mathbf{1}$, which makes that relationship the point of transformation where this universe emerges from relating infinite patterns. By multiple dimensions, we will be able to travel at $C^2$, time travel, and much more. To say that dimensions have limit contradicts the axiom that infinite systems are a reflection of a truth about the nature of the universe. The cosmological constant is incorrect because there are no constants, only functions or variables integral with constants. So the notion that any constant is acceptable is incorrect. Constancy of a variable might refer to uniformity, but the variable always has possibilities. All physics can be defined in terms of time and velocity and powers thereof and also by scalar relationships of quadratic forms. These exist within an infinite framework. Our perception makes matter prior to space since by particles we define the universe, but in fact, our perception is opposite to the true primacy of space and its structure, since particles are caused by the interaction of various dimensions of space. What is perceived as boundary is where one dimension transitions into another, but the patterns that allude to boundaries are infinite, and hence, there are no boundaries per se. There is equivalency in all units. Once the pattern of the universe is known, the equations lead the way rather than the experiments and any equation can be derived as shown, without any experiments, and reveal more about relationship than possible by an inquisitive mind.

***F = G Energy1 Energy2 / $R^2$ Temperature1 Temperature2***

This equation is using the standard *G*. Assuming that both *G* (which refers to density) and temperature are constant, you get *F* proportional to $E1 \times E2 / R^2$; without further analysis, this represents the relationship "within" the inertial field.

Now you take the original *G* and go express from the equation above:

***G* / temperature1 × temperature2 = 1 / force = 1 / $V^2T^2V^2T^2$ = 1 / $D^4$**

and that is equivalent to 1 / force, which is *G* redefined as *G'*, the same thing implying functions within the inertial field.

As soon as you bring it around to get the square root in

**$F = SQRT(E1 \times E2 / R^2)$ where *E* is energy.**

In $F = SQRT(E1 \times E2 / R^2)$, you have reached the hyperspace of that relationship and are referring to the relationship beyond the inertial field.

The relationship of gravity within the inertial field relates to $1 / R^2$. The flaw in science is with the understanding of *G*. Multiple functions are lumped together in one *G*, when in fact, the diversity and outcomes are much greater.

Iteration is as follows:

***F* = *E*1 *E*2 density1 density2 / $R^2$ coulomb1 coulomb2,**

and this would also hold for within the inertial field.

Or if you like *G*,

***F* = *E*1 *E*2 / $R^2$ coulomb1 coulomb2 *G***

However, *G* shouldn't even exist.

Iteration is as follows:

***F = G Energy1 Energy2 / $R^2$ Temperature1 Temperature2***

***F = G Energy1 Energy2 / $R^2$ $Velocity^2$ $Velocity^2$***

***F = G Energy1 Energy2 / $R^2$ $C^4$***

Here you see *G* redefined again as **$G' = G / C^4$ = 1 / force**

You can move force to the left in

**$F = SQRT(E1 \times E2 / R^2)$, where *E* is energy**

and it becomes the $1 / R$ "beyond" the inertial field.

And note that we also have the square root of the numerator, and it refers to a system external to the inertial field for two systems. We can take this further as follows:

$F = SQRT\ (E1\ E2\ /\ R^2)$

$F = \text{energy}\ /\ R$

$F = E / R$ is representative of an object, while for two objects within the inertial field, $R^2$ holds true, and for distant objects beyond the inertial field, $F = E / R$ also holds true since the relationship of the energy density gradient to distance is linear.

**Multiple Possibilities and Pattern Universe**

The equations in transdimensional theory have multiple possibilities. Trans-dimensions are the "pattern" of the universe.

***C***

The speed of light represents the elastic collision of space, used to calculate the max extension of hyperspace, and also relating transdimensional and additive or quadratic infinities.

This theory explains space in different parts very well, but the scale is infinite. Thus 3 + 1 is only an iteration of space. Using the Pythagorean theorem, you are defining an infinite scalar system of subspace. Using multiplicative system of powers, it is the system of closed systems and proper velocity. At zero, you have the two systems equal and also at *C*, the hyperspace of the relationship. Light is not mass less since it has volume, although the dimension of it is not atomic mass by which we associate mass. Anything with volume has a dimension of energy. The volume we experience is made possible by supporting transdimensional matrix. So a sphere is not volume of space, but the interaction of twenty-seven spheres to form three-dimensional spaces is the space we experience as the space matrix of three dimensions. Time is just an inherent property of the space since all space is made of superstrings *D*, *T*, and *V*. The one you speak, $D = 1$ at *C* and zero of the quadratic formula where transdimensional or multiplicative and quadratic or additive system relate, the point of transformation or balance where infinite patterns relate to form the universe.

$T^2 + T - 1 = 0$

$X^2 + X - 1 = 0$

$V^2 - V - 1 = 0$

$X^2 - X - 1 = 0$

$D^2 - D - 1 = 0$

$X^2 - X - 1 = 0$

Phi represents velocity, and phihat represents time. The quadratic equations relating these two constants define the relationship of *D*, *T*, and *V* in scalar terms. The other hierarchy is with proper time and velocity in the transdimensional form of multiplicative systems and powers. Hence, two infinite systems are coming together to form the universe.

Light is a particle created by the overlap of spheres or superstrings that are *D*, *T*, and *V* to form a space matrix dimension per dimension. The whole of space is connected to infinity, and hence, every point is the center and may be a black hole, and hence, there may be various types of black holes in various dimensions.

**Universe Expansion**

According to transdimensional theory, there is a ratio of inertial mass to atomic mass. Hence, the universe cannot expand forever, and also by such a principle, while some areas are expanding, others are contracting. So what may "appear" to be so needs further evaluation and I am sure that creation is always happening and so is contraction or transformation. I believe transdimensional theory covers all bases since it provides a pattern for the universe. And hence, that pattern does not allow the universe to expand forever since there is ratio between all dimensions and states of matrix relationship. Just as you have proper velocity within a system of $MV^2$, the natural universe maintained though a hierarchy of balances will not allow one system to dominate another. Hence, the universe is perpetual motion relating expansion and contraction, highlighting the relationship between infinite systems. In a unified theory, all functions are tied together, so no one function has the liberty of independent action without the restrictions imposed on the governing system and the balances that are in place.

## Charge, Spin, and the Universal Pattern

### Charge Is a Function of Time, and Velocity Is a Function of Spin

We have talked about how time and velocity are superstrings and how time is an inversion of velocity. So in a sense, they are the same thing but acting differently, and hence having properties that exhibit polarity to each other. Time is normally associated with magnetism, and velocity is normally associated with current in this theory. And the two, time and velocity, come together and have the "effect" of superstring *D*. Time, as well as being distinct, is velocity but with a characteristic inversion, making it the conjugate of velocity, and in a similar fashion, phi and phihat are conjugates of each other. Phi then represents velocity, and phihat represents time.

A similarity of complimentary inversion or symmetry exists with spin and charge. Spin is an inverted form of charge in the same fashion as time relates to velocity. So spin is the ***conjugate*** of charge in the similar manner that velocity is the conjugate of time respectively.

### *Charge Is a Function of Time*

### *Scalar*

We will observe from charged particles that charge is divided in division of 1/3, 2/3, 1, and so forth. This 1/3 and 2/3 relationship is close to the scalar relationship of phi and phihat where

**Phi + phihat = 1**

**Phihat = .618, which is close to 2/3 or charge at $T$**

**Phihat$^2$ = .38, which is close to 1/3 or charge at $T^2$**

**The relationship of phihat$^2$ and phihat seems to parallel the charge of 1/3 and 2/3 respectively. Perhaps charge is a function of time.**

**$X^2 - X + 1 = 0$ *is a quadratic form related to time, so***

**$T^2 - T + 1 = 0$ and follows that**

**Phihat$^2$ - (-phihat +1) = 0**

**0.38 + 0.618 1 - 1 = 0, invert $T$ and 1**

**0.38 = 1 - 0.618 = 0.38**

It appears that time is related to phihat and also integral with charge. By contrast, the inversion of charge is spin, which will be shown, so spin is also a form of charge. This is not to say that charge is not spin; charge is only a particular type of spin and vice versa for spin; it is not to say that spin is not charge, only a specific kind of charge.

### *Charge Is a Function of Time*

### *Transdimensional*

**Phi × phihat = -1**

**Phihat × 1/ phihat = -1**

**$T \times 1 / T = -1$, and hence, charge here is of a factor of 1 in a transdimensional sense**

**So what might be the difference between spin derived through scalar functions and those derived through transdimensional functions?**

It may be that scalar functions relate to subspace and associated particles as a fractal relationship, and transdimensional relate to atomic particles that are created by interference in atomic structure. Subspace refers to those particles that are at the small level of the space matrix, and transdimensional refers to those particles at the transition of different levels of atomic dimensional space.

***Velocity is a function of spin.***

***Scalar***

**$X^2 - X - 1 = 0$ is a quadratic form related to velocity, so**

**$V^2 - V - 1 = 0$ and follows that**

**$\text{Phi}^2 = 2.618$**

**Phi = 1.618**

**Phi / $\text{phi}^2$ is close to 2, so**

**$\text{Phi}^2 - \text{phi} - 1 = 0$**

**$2.618 - (1.618 + 1) = 0$**

**Near a ratio of 2:1, we can go to**

**$1 - (½ + ½) = 0$**

**1 = 1**

There may be a relationship of spin ½ to the ratios in phi and $\text{phi}^2$. If so, this shows the connection of phi, velocity, and spin. But time may also be involved in spin. This may be in the following equation:

**Phi = velocity = (1 + *SQRT* [5]) / 2**

When

**Phi - phihat = *SQRT* (5)**

Then can be derived

**Phi = (1 + [phi - phihat]) / 2**

**2 Phi = 1 + phi - phihat**

**Phi = ½ + ½ phi - ½ phihat, where ½ refers to spin**

**½ = ½ phi + ½ phihat**

**½ = ½ *V* + ½ *T***

**This appears to be a definition of spin related to a scalar relationship of velocity and time.**

***Velocity is a function of spin.***

***Transdimensional or Proper***

**Phi × phihat = -1**

**Phi × 1 / phi = -1**

**Phi = phi**

Here we see a relationship of spin to velocity at the level of 1.

## ***Results of Charge and Spin Related to Phihat and Phi Respectively***

There seems to be a connection between phi and phihat with spin and charge respectively. Note that here, this math refers to the spin and charge of superstrings, and therefore, the math of the particles associated with these strings may, in fact, be an inversion of the relationships calculated here. For example, +2/3 charge for superstrings may, in fact, be referring to a -1/3 charge on the particle. So what strings do to create a particle may also be described as a symmetrically inversion in the math.

Charge is related to spin through time and velocity. Charge is integral with time. Spin is integral with both time and velocity relationship. Spin is a characteristic of the functions time and velocity. Charge is characteristic of the function time. Time and velocity are conjugates, like phi and phihat. Phihat and phihat$^2$, in the time quadratic formula, account for 1/3 and 2/3 charge. The negative and positive become inverted because the charge is for space, and the symmetry is in the particle, which has the opposite charge sign.

**At this small level, what is the thing modified from which everything is made?**

It seems that everything is subdivided from that one thing into pattern. Science is about relationship, so the one apart from relationship may be an unlikely task to know. I don't think that has an answer since it would be like asking what the essence of numbers is; it's probably a faulty question as well. If you could imagine before the big bang something singular that differentiated itself into diversity and then took infinite scales and related them in the big bang for creation of the universe. Or perhaps singular and diversity always was, and this is universal consciousness.

## Quadratic Scalar Time and Motion of Falling Object

The distance travelled by a falling object is a quadratic function of time.

**Phihat = -1 / phi**

**Phihat = 1 - phi**

**Phihat = 1 - 1 / phihat**

**Phihat - 1 + 1 / phihat = 0**

**Phihat$^2$ - phihat + 1 = 0**

$T^2 - T + 1 = 0$

$X^2 - X + 1 = 0$

### ***Now Conversion for D***

***Now introduction of variables a, b, and c***

**$a \times T^2 - b \times T + c \times 1 = D$ since all parts need to equal $D$ separately**

**$a$ = acceleration, so $aT^2 = D$, referring to acceleration**

**$b$ = velocity so $VT = D$, referring to velocity**

**$c = D$, referring to distance, and then**

**Acceleration × $T^2$ - velocity × $T + D = D$**

**This represents**

$s(t) = -gt^2 + v_0t + h_0$

**For gravity, the coefficient of the leading term is negative since gravity pulls down negatively, and for a negative quadratic**

**$s(t) = gt^2 - v_0 t + b_0$**

**$T^2 - T + 1 = 0$**

**Review of Phi and Phihat or Velocity and Time Respectively**

**Phihat + phi = 1**

**Time + velocity = 1**

**Phihat × phi = -1**

**Time × velocity = -1**

**$\text{Phi}^2$ = phi + 1**

**$\text{Phi}^2$ - phi - 1 = 0**

**$\text{Velocity}^2$ = velocity + 1**

**$\text{Velocity}^2$ - Velocity - 1 = 0**

**Phi - phihat = *SQRT* (5)**

**Velocity - time = *SQRT* (5)**

**Phi = (1 + *SQRT* [5]) / 2**

**Velocity = (1+ *SQRT* [5]) / 2**

**$\text{PhiHat}^2$ - phihat = -1**

**$\text{Time}^2$ - time = -1**

**From the relationship of phi and phihat, all the quadratic and multiplicative forms that compose the math of the physical universe can be derived. Phi represents velocity, and phihat represents time. Charge and spin are also functions of time and velocity.**

**The Keys to the Universe**

Some physicists refer to gravity and the weak force as the same thing. Force is always $D^4$. Now there are various hierarchies of force. For example

**$F = MA = D^4$.**

And Newton

**$F = G\, M1\, M2 / R^2 = D^6 / D^2 = D^4$; here an expression of $D^6 / D^2$, where the numerator is larger as $D^6$ instead of $D^4$.**

The same $D^4$ is the result in both equations, but different positions in the hierarchy of possibilities since the numerators that lead to $D^4$ are different in dimensional complexity.

### What's D?

*D* is the translating numerator. It's the singular from which everything is subdivided. It's like this: Imagine a tree. The trunk is *D*, and as you go up, each branch adds a *VT*. By the time you start at *D* or distance or displacement, as you go up, the hierarchies become farther away from *D* into greater diversity, just by adding *VT* and powers thereof. Space becomes more complex. Perhaps twenty lower branches are $D^4$, but all in different locations and all related. This comes from

**Phi × phihat = -1.**

It means that there is a hierarchy of *VT* since phi is velocity and phihat is time. The tree represents the diversity you can get, and everything is interchangeable. So knowing the pattern, you can come up with a thousand equations. Now there is another pattern, and it's scalar. Scalar is referencing quadratic forms that come from

**Phi + phihat = 1.**

From that you get quadratic equations. Both systems are all on the same tree.

### What if there's another tree?

That would be another reality, like heaven with another type of physics. This isn't the only one. And in that other place, what we have here is useless there.

### So what is your value then?

The person you are. Like the Egyptian pharaohs mistakenly thought to take it with you, you don't take it with you. All you discover only explains what already is. So your only real contribution is in helping others. The only thing transferable is the good in you. It's amazing to think that the key to the universe was always there in simple form of phi and phihat.

### Some say it usually is.

So what is left after the secret is revealed? We use knowledge to help others; it is important. A key isn't the story. Alice has to experience Wonderland with or without the key. And even after she explores it, it's more than for one. Life is a feast and expound on the variables.

### The key to the universe is always the last place you look.

"If I have the gift of prophecy and can fathom all mysteries and all knowledge, and if I have a faith that can move mountains, but have not love, I am nothing" (Paul).

The equation $F = G E1\ E2 / C^4R^2$ **is** $D^{10} / D^6 = D^4$; on the tree, $D^4$ here is at a higher dimensional iteration for two masses.

The square root

$$F = G E1\ E2 / C^4R^2 = E1\ E2 / D^2 D^4 = E1\ E2 / D^2 \text{ force}$$

$$F^2 = E1\ E2 / D^2$$

$$F = SQRT(E1\ E2 / D^2)$$

$$F = E / D$$

is a lower dimensional reality, probably not related to gravity as we define it by the interaction of two inertial fields, but force nonetheless and linear. The argument made about linear gravity had to do with the notion of the inertial field, which has nothing to do with the integrity of the equations. The notion being that beyond the interactions of two inertial fields, the gravity interactions cease to be radial and become linear.

## Free Will

For something to be free, it must meet parameters that are within and without the prevailing system of form. Form allows certainty and uncertainty. Certainty is a canvass of predictability and stability, which is the universe, and transcendent life adds the colors to create through stability and flexibility in relative or uncertain possibilities that happen only by free will.

We have said that life is one form of consciousness perceiving another. And so the universe is a form of consciousness, and in a sense, all consciousness is collective from a collective source. The universe is also an analogy of truth, referring to the source consciousness from receded states, much like a dream. We have also described this universe as derived from infinite patterns, additive and multiplicative as represented in math that relate in various hierarchies from simplicity to complexity for everything that is. For example, phi is velocity; phihat is time, and these conjugates together forming 1 or dimension. So they form everything in the universe. And now as representing the hierarchy of consciousness, they are not the source, but by symmetry, they refer to the source as an extension of the source consciousness and reflect the nature of the pattern of the source. And so the number 3 becomes the number of manifestation of form structure. Within the mathematics of consciousness, within the pattern that results in form, there is a pattern that is an anomalous pattern of free will. The pattern of the universe allows division and diversity that also allows creativity and uncertainty through free will as a transcendent participant. For this to be, the universe also has the free will of its originator but as a passive participant or God. However, God can also be active. Life is transcendental, containing the anomalous free will pattern that it brings. We have defined that consciousness is akin to infinity and patterns that are infinite, but derive form through the interaction of infinite pattern, and the form of consciousness is represented by three: will, thought, and awareness. Similarly, the universe is at the crossroads of infinite patterns.

For free will, there is an infinite pattern that must be self-evolutionary, of another paradigm and separate, perhaps in some ways as an infinite program of infinite sequences. And so free will being more dynamic, while at the same time revealing a formal nature being part of a formal patterned consciousness that also unites to all things, is also supported by and more dynamic form nature, which a creative process source would be or allow. In a similar fashion, the universe contains a duality of proper time and velocity, and scalar time and velocity, or transdimensional structure and scalar structure. Each supported by form number 3, but one system constant, and the other relative, which alludes to the idea that the universe as flexible as a consciousness would require to be connected to a free will of its own, which is God. The universe is alive. And in a sense, consciousness has its inherent characteristics of life and also free will. And free will has its predictable components and its relative ones that allow free will. But the governing systems are more complex than the created, and hence, the created become an analogy of the source. So pattern form is indicative of more complex dynamic systems from which it arises: the infinite. So as in the creation of the universe, infinite patterns combine to relate through basic presets, representing the pattern of overall form:

**Phihat + phi = 1 or time + velocity = 1 or *D***

**Phihat × phi = 1 or time + velocity = 1 or *D***

This leads back to complexity of the new created or projected system capable complexity to complex equations and genetic codes. But the anomalous pattern that makes life transcendent allows capacity to differentiate the infinite and also the simple to variations that create.

## Certainty and Uncertainty

There is certainty and uncertainty in the motion of things, in proper time-velocity and scalar time-velocity respectively; time and velocity are the same thing, but different expressions, one perhaps the inversion of the other. Polarities are expression of the same. Conjugates, as in math, complete each other. Also, space has many configurations of charge and spin. The ether we know is but one, that form of space that is related to antimatter another, and that related to polarity or antigravity in this theory is spurred by another combination of space, and yet another for other effects. So with regard to time travel, there is more than one way; with regard to the manipulation of time, more than one expression. Consciousness is allowed multiple expressions. The same stuff is differentiated, polarized, and hierarchically assimilated in the model of what is called math and ultimately consciousness, complex processes of symmetry and asymmetry. Anomaly is free will, a pattern outside of the pattern, but also within. Infinite patterns allow infinite arrangements and possibilities, but they also allow freedom within a system of stability so the two exist together. Without the action of life through free will, the universe and consciousness are certain. Natural laws become predictable but are an agency to that allows for uncertainty through what is determined to be a living consciousness, which all is. God has allowed a certainty to prevail as a canvas for uncertainty and interaction in the collective participating in consciousness. Free will is not based on anything in this universe although it has characteristics of free will that allows it to be received. Free will is transcendent, one reality merging with another; the real versus the projected or imagined. The imagined is an analogy of the real and similar to symmetry. Two systems of math or realities merge to form the universe, and life is another transcendent form or alternate pattern. Harmony is allowed through diverse systems.

## Hierarchy of Force

Force has hierarchy but always equal to $D^4$. What makes force higher in the hierarchical chain is by noticing the higher powers of superstrings in the numerator. For example:

$$F = MA = D^4 = (D^3T^2) \times D / T^2$$

And so, notice $D^4$ in the numerator. Now let's look at the Newtonian classical gravity equation:

$$F = G\,M1\;M2 / R^2 = D^6 / D^2$$
$$F = D^4$$

Notice how the numerator has climbed to $D^6$, but the outcome is again $D^4$. $D^4$ is in the fourth dimension, but the numerator is of lower complexity in $MA$, and hence not equivalent to gravity in $F = G\,M1\;M2 / R^2$. However, gravity with $1 / R^2$ represents the interaction of two inertial fields and supporting masses. What happens when beyond the inertial field of one object exists in relationship to the other? I surmise $1 / R$, since the inertial fields are no longer in contact or interacting. This takes nothing away from $1 / R$ in **energy / $D$ = force**, a common equation in physics; remember $FD = E$. Now let's see how we can take $F = D^4$ to a higher dimensional iteration of force for gravity:

$$F = G\,M1\;M2 / R^2 = D^4$$

In this theory as long as you maintain $D^4$, you can call it force, so

$D = VT$, so it's a question of adding $V$, $D$, and $T$ to the numerator and/or denominator and maintaining $D^4$. If you change each mass to $E$ in the numerator, you are adding $V^2$ twice or $V^4$ to the numerator, so to keep it at $D^4$, you have to add $V^4$ to the denominator. Hence:

$F = G\,M1\;M2\;V^4 / R^2 \times V^4 = G\,E1\;E2 / R^2\;V^4$; **change the $V$ to $C$ and**
$F = G\,E1\;E2 / R^2\;C^4$

Note $E = D^5$, so converted, the equation is

$$F = D^{10} / D^6 = D^4$$

***The numerator for energy of two masses has risen to $D^{10}$.***

Even if you plug in other variables such as current and magnetism and calculate for force, force should remain at $D^4$ to be correct. $D^4$ means force, but it can be the result of various hierarchies. In the relationship of two masses, we call it gravity. Now the old $G = 1 / T^4 = 1 /$ **density1 density2**, but its okay to express $G / C^4 = G'$ to remind us of the existing conventions.

## The Smallest Denominator

Time seems to be hard to understand because it is not something you can smash in a supercollider or see. It seems that time is there in the realm of perception more subtle than sight or touch. Hence, to some it's not real, but in this theory, there is the claim that time is a building block of mass. And mass is not solid but a configuration of space. What makes it appear solid, among many, would be the transition in density or transdimensional boundaries. There is more space in an atom than those particles we see. Space is primary, particle is secondary, and together they have mass. How can time create mass? Time is velocity with different polarity, spin, and charge. How can velocity make mass? Time and velocity in effect is dimension. How can dimensions or volume create mass? As you see, as the complexity of primary components $D$, $T$, and $V$ increase, we are getting closer to the mass we know. All that we see and measure is from the same substance with different spin, charge, arrangement, and dimension. But that one substance patterned into a pattern of three is "form." So time has multiple dimensions, and those have different testable properties to the extent that it looks entirely different depending on how it is structured and situated.

### So symbolically, what two constants together united for the number 1?

As we see, $VT = D$. Phi and phihat or 1.618 and .618 multiplied together are 1. Two numbers multiplied together to form 1 are conjugates.

### Can we make a parallel between phi and phihat and velocity and time?

Perhaps we can. Isn't it interesting that **phi + phihat = 1** and **phi × phihat = -1**? Here we have the keys to all complexities. The additive form results in quadratic relativistic forms. The multiplicative are for proper time and velocity of closed systems.

### What is time?

Time is one of the building blocks of all things. Time is integral to phihat as a function.
So let's break down $E = MC^2$ or $E = MV^2$

**Mass** $= D^3T^2$

What does this mean?

It means that the space is related to mass is in the configuration of a space matrix that has $D^3T^2$ superstrings. The particle is just an interference of that space matrix, the shadow of space matrix.

$E = D^3T^2V^2$

Now we can correctly express energy in its most basic space components $D$, $T$, and $V$. This will allow us to unify all physics into three variables, which is the pattern.

## Metaphysics

Metaphysics is the study of axioms prompted by intuition from the observation of patterns in the universe. These axioms tested by logical discourse become the analogy of truth. When further established by scientific observation, they can become science and physical laws established through mathematics. If the patterns discovered are actually the pattern of the universe, then the pattern or unified theories can lead the way for science and observation to areas of discovery beyond human perception. (Hansen: Web Link)

## Moment of Inertia

$I = M D^2$

$I = (D^3 T^2) \times D^2$

$I = D^5 \times T^2$

$I$ = **energy × density**

**Energy** = $D^5$

**Density** = $T^2$

The motion of a particle orbiting is a function of energy and density, which also describes its linear AND angular motion or moment of inertia. This relationship described density as a factor related to the motion of orbiting objects. Hence, the orbits of particles are a direct relationship and measure of the density of space in a sector being examined and also relates to the moment of inertia of that activity. The density is also a measure of time.

## Motion

If left to human perception, it may be said that you can never detect your own motion unless you are accelerating, but with measured results, there are two kinds of motion presented in this theory:

- Proper velocity. Velocity adjusts according to a closed system; you increase velocity and decrease mass. That's a view opposite to current view but held by some and also here in this theory.
- Scalar Velocity. The type where you notice motion outside observer.

Similar proper and scalar models would also exist for time.

## Units

**Phi + phihat = velocity + time = 1 and Phi × phihat = velocity × time = -1**

The additive mathematical constructs of time and velocity lead to quadratic forms and scalar relationships, and the multiplicative mathematical forms lead to proper systems of time and velocity. Everything in physics is a dimension of time and velocity. You don't need mass units, except for your human perception. And mass can be described as $D^3T^2$.

### What is mass really anyway?

Look into an atom and its all space. How accurate can units are if they do not describe what they represent in the pattern of the universe? Units should describe the pattern of the universe as cause rather than the object effect. Without pattern, there is no unified field; without a system of units that can represent *form* in parts of three such as *D*, *T*, and *V*, there can be no unification. Time is a transformative superstring. Time is velocity that establishes charge. Time is an inversion of velocity. Time is involved to establish spin. So time is velocity, but with different action. In time and velocity, you have one form that is changed to everything else to establish the complex universe. But because form is represented by the number 3, you can't define one superstring such as time without the other superstrings *D* and *V*. Pattern established the definition of one via others.

### So what is mass again?

If most of an atom is space, perhaps space is primary and particle secondary. Mass is an arrangement of pattern described by $D^3T^2$ superstrings.

**Using the concept of mass alone is like saying what is land (mass) if most of the Earth is water?**

Your units are off without describing pattern! Land and water are not units. Units represent the true pattern of what you see and not the object. Cause is pattern, and effect is object. Units belong to pattern. Units are inconsistent when they do not describe the pattern of the universe, and in the standard model, units do not describe what matter is by a pattern arrangement within the pattern of the universe. The definition of mass is like the primitive fire, water, air, earth, and wood. The primitive myths of cave people hold us and its all posture. We cling to the myths because we cling to the group and far from the notions of author Bach in the book *Jonathan Livingston Seagull.*

**Big Bang**

According to this theory, infinite pattern numerical systems come together and create this universe. The systems are defined by

**Phi + phihat = 1**
**Phi × phihat = -1**

These systems become equivalent at $C$ and 0 where $\boldsymbol{D} = \mathbf{1}$; the systems refer to hyperspace and subspace respectively. The additive systems are scalar and result in quadratic forms, and the multiplicative system are transdimensional and result in closed system of proper form, such as proper time and velocity. In the relationship of phi and phihat, we have the relationship of velocity and time respectively. Our universe is manifest as "form" from infinite systems. Form is represented in a tetrahedral system of $D$, $T$, and $V$ and units that combine to form all properties and physics that is. Some would argue about units, but true units are convertible and equivalent and represent the pattern of the universe and not object.

### What does this mean for the creation of the universe?

If different patterns combine to generate "form," then infinite points in the separate pattern generate the infinite points in the new pattern. Since numerical systems are infinite, the patterns associated with these systems also have infinite relationships. In the relationship of infinite systems to create form, certain aspects appear contained or qualified, but also others are infinite. So our universe is both infinite by evidence of irrational numbers and contained or better expressed limited by evidence of whole numbers. The relationship between these extremes allows infinite connections to the source infinite as well as the representation and hierarchy of form. We have a universe that is measurable within certain limits as well as accessible to the infinite source from its entire fabric. Black holes are the infinite points to infinite space. They are combined in superblock holes, at the center of all particles, and also infinitely accessed at interferences, points in the fabric of space. Therefore, their nature of operation depends on locality, for some are exposed and others shielded by various degrees or dimensions.

Conventional understanding of the big bang is that from a point, everything became from nothing. Yet both point and nothing seem to be more myths about human limitations and nature rather than what might be actually happening. For instance, in this theory, 1 represents unity, a point of balance of polar forms of energy. And 0 represents unity. The idea of nothing does not exist here. Here, infinite patterns of something relate to form something else. Our universe is supported by infinite energy and, hence, appears as a perpetual motion machine. Within the mathematics of limit and also the infinite, we can say that there is hierarchy and proportion and evidence in both phi and phihat and other universal numbers found in nature. This means that the universe does not tolerate forms that are not within the limits set forth in the creation of the universe either in its present form or in its beginning. Therefore, I would conclude that the manifestation of infinite points of creation along the relationship of infinite sets was the big bang. Like the appearance of a grand idea in our mind, it happened everywhere all at once and that from there and, working within the parameters allowed by such a manifestations, reorganized as allowed.

The way time is experienced in this universe is particular to this universe. And by such an understanding of infinite pattern, the universe, being both qualified and infinite, is really infinite. Any boundaries or edges perceived are only by the illusion of limited perception and transdimensional limits. The big bang, therefore, was a multiple big bang, and the apparent size is not accessible by telescopes since the aspect of expansion and contraction is a continuous process and depending on where you are prejudices your evaluation. Within this theory, it is proposed the possibility of travelling at the speeds at $C^2$ or larger, and hence, the likelihood that a small universe exists is in contradiction to the capacity of high-speed travel. It does not seem practical. The universe is infinitely old and large. There is much more to the universe than appears to the naked eye. When considering the big bang, one would ask,

**Where did all the mass come from?**

And following or together with this question would be

**What caused the big bang?**

If we consider the universe to be infinite as well as being qualified, finite or constrained, perhaps something of the nature of the cause is also contained into what was evolved to be.

**So what do we have?**

We have patterns that are both infinite and finite. We have patterns that are both scalar (fractal) and additive and, in addition, a second hierarchy that is transdimensional or multiplicative. If we can claim that all patterns can be ascribed to two equations, which are

**Time + velocity = 1**

**Time × velocity = -1 or**

**Phihat + phi = 1**

**Phihat × phi = -1**

These are the forms that, by relationship, the universe came to be, and by relationship, they form the tetrahedral "form" of the universe. The relationship of these equations exists at *C* and 0 and 1. These existing prior to the big bang or at the moment of the big bang lend themselves to something greater than nothing. And the concept of singularity might better be expressed as a point of transformation in patterns where patterns are in fact related by a system of balance (1) and unity (0). The hierarchy of infinite and the qualification of positions from infinite to finite result in "form" as we know the form of the universe. Hence, the questions about what caused the universe might be rewritten again as

**What caused pattern? How does simple pattern relate to form diverse and complex pattern?**

We also have said that time and velocity are the same thing but with opposite polarity or inversion, probably different spin, and maybe charge.

**So then what is that single energy that manifests itself into diversity and multiplicity, which is the form or order of the universe? And did the pattern to which we ascribe the universe exist in that one energy or singularity at the big bang?**

Can something without pattern manifest into something with pattern, or is the nature of the cause reflected by symmetry into the nature of the effect as claimed in this theory? So therefore, there can be no singularity if that implies the absence of pattern. But by singularity, perhaps we mean something very primal and compact with pattern unfolded into some very elaborate, almost like a genetic code or seed. However, in this seed, it is connected to infinite energy. Another thought is that for the balance that you see in the universe in both the background noise and even dispersion of galaxies, the bang was in fact multiple bangs or a grand induction. The product of the big bang we can surmise. And from that we can distil the universe into the most basic primary forms that we have. And so to the point that we can claim perhaps that the universe is "one energy" manifested into diverse forms. So what is this one energy, and what caused it to diversify into multiple patterns? Are these patterns a genetic code or code for the universe? What is this singular energy without a pattern? What is RNA without a pattern in a genetic code? What is RNA itself without other patterns? And this subdivision goes forever. The truth is in the pattern and not the object. Patterns create patterns. And infinite energy and pattern is the source of pattern, and what is beyond our experience is hard to comprehend since perhaps the physics is entirely different. It may be that one set of precepts generates another in the big bang.

Transformation and unity are the same, and since the concept of unity and zero has infinite relationships, therefore, the relationship to transformation is all points in the universe; and therefore, infinity pervades the entire universe in addition to being qualified. Unity is represented in both the center of black holes and also the hyperspace of the universal space dimensions or trans-dimensions. With that being said, the concept of multiple bangs carries more weight in addition to the uniformity in the universe between galaxies and the uniformity in irregularities. This is followed by the mathematical relationship between transdimensional forms and fractal scalar forms at various dimensional levels. Imagine that each galaxy generates its own space via the central black hole. So the system of motion of the space supporting the interior solar systems serves that larger galactic space. For or own galaxy, if the space is expanding and since we are internal to that space, that would make it appear the whole universe is expanding. The universe is not expanding forever. Expansion is relative and influenced by observation from the space you are in. This can even bias your measurements for the entire universe and space beyond your own realm.

## Space-Time and Multiple Dimensions

We have talked about transdimensional theory and multiple dimensions. Conventional theory states that we live in space-time or four dimensions where we have three dimensions of space and one of time. If we take **mass =** $D^3T^2$, we see that by looking at the powers, we are in fact at five dimensions of expression, which are three of dimension and two of time. So where Einstein talks about time as the fourth dimension, we should actually consider $T^2$ or time squared. So volume and time$^2$ is space-time. This goes in line with the concept of energy since another dimension of time is needed for energy, which is also $D^5$ or five dimensions. Space-time is more complex than our traditional view of time.

## What Is a Dream and What Is Real

We have talked about life being one form of consciousness perceiving another and that consciousness can exist in receded forms such that what appears as an awakened state can appear as a dream or memory in another conscious state. So in another life after death, this can, in fact, appear as a dream or memory as past things do, but also in the real sense of experience where free will carries consequences to development. Only a developed sense of self or awareness can surmount the veil of past memories and bring them to current consciousness.

## Scalar and Proper *D*, *T*, and *V*

In this theory, we have separated the events of scalar and proper time into two systems, which are additive and multiplicative respectively. The additive forms result in quadratic systems, and the multiplicative result in transdimensional system of dimensional powers. It may be that the results are also different in those systems when compared. Hence, you might get different results for time or mass depending on the experiment and whether you are an observer in a scalar system or part of a closed transdimensional system. For example, if you are in a spaceship travelling at the speed of light, that's proper time. Increasing one component of a closed system has the effect of reducing partner components. If you are watching from within the coordinate system as a third party, that's scalar, and results will be different from those transdimensional. Two different events are described here, and the method of participation leads to quite different results. The idea that there is one solution may be entirely wrong, and it may be in fact true for both systems. For instance, perhaps a photon is both mass less under certain conditions and also has mass. So therefore, both camps are right. Another example concerns the speed of light. The speed of light is proper at *C* within the system of the space matrix and also greater and less than *C* as a scalar phenomenon.

## Quantum Entanglement

In quantum entanglement, one member cannot be fully described separately without discussing its counterpart even though separated by space. But it is the understanding in this theory that there is no separation as it should be in a unified theory. Space is primary and is uniting all properties, structures, and associated particles. Consider explaining quantum entanglement; you cannot explain the God particle without its counterparts. Time does not exist without velocity in the matrix of space. Time and velocity are the same except in attitude. Entanglement means that the one that everything is made just adapts a different mode of operation. *V* and *T* are phi and phihat respectively, and together they are 1 or *D*. And the nature of form, which is the three *D*, *T*, and *V* in tetrahedral form, means one cannot be described without the other two, and since all is related to the three in every form, this is the basis for entanglement. Particles being the interference patterns of the superstring matrix of space mean that they share multitude of basic *D*, *T*, and *V*, and hence are all entangled at a distance.

**Derivation of Phi = *V* and Phihat = Time**

**Phi + phihat = 1**

$V + T = 1$

**Phi × phihat = -1**

$V \times T = -1$

$T = \frac{1}{2}(1 + SQRT[5]) = \text{phihat}$

$V = \frac{1}{2}(1 - SQRT[5]) = \text{phi}$

**The Mathematical Relationship between Phi, Phihat, Pi, Pihat, *D*, *T*, *V*, and 1**

**The Big Bang and the Creation of the God Superstring or Primary Superstring *D***

What is pi? In mathematics, it is referred to as a transcendental number. What is its exact relationship? It confers the symmetry in a universe where there is also asymmetry in phi and phihat.

**What is the relationship of pi to phi in the physical universe and physics equations?**

**Phi × phihat = -1**

**Phi × phihat = pi × pihat = pi × 1 / pi = 1**

***Let's solve for pi, D, and 1:***

$e^{(i \times pi)} = -1$

**-1 = velocity × time = phi × phihat**

**Phi × phihat = -1**

$-D = -1$

Velocity × Time = $-D = -1$

$-D = E^{(i \times pi)}$

***What is D in terms of pi as an exponential function?***

$-D = E^{(i \times pi)}$

$-D = E^{(i \times pi)} = -1$

***You can write as an exponential function giving the following:***

$D = E^{(i \times pi)} = -1$

You can write as an exponential function since $e^x$ is defined as exp ($x$) for complex $x$.

$-D = E^{(i \times pi)} = -1$

$-D = E^{(i \times pi)} = -1 = \textit{Time x Velocity}$

## $E^{(i \times pi)}$ =Velocity × Time

*D* is an exponential function of pi. *D* is the result of an exponential unlimited function to infinity of pi. And within that velocity and time of phi and phihat are also 1 and *D*.

**Who would think that the whole universe is an exponential function connected to pi?**

***D* exists from an exponential function related to pi and the whole universe for that matter from the big bang. *D* is the primary superstring or God particle where phi and phihat or velocity and time also exist. This is the only equation in the planet that relates *D*, *T*, *V*, pi, pihat, phi, phihat, and 1. The big bang was the result of an exponential function. I would surmise multiple bangs like your conscious mind. The entire universe is supported by infinite energy. Everything that is both inside and outside the universe passes through *D* superstring. It relates all math from transcend numbers to whole and irrational numbers.**

Yes, there is a basic substance defined by the equation ***D* = exp (*i* × pi) = time × velocity**. The equation refers to a transcendental experience that results in and is a part of the universe. We had spoken of life being one form of consciousness perceiving another. The form three of the universe was an analogy of awareness, thought, and will. In a sense, the source cause consciousness by symmetry creates the experience of consciousness as object. The nature of the object refers to the cause. In the equation, you see three operators created from the causes. The object is *D*, *T*, and *V*, and the operators of cause are exp, *I*, and pi. The source also exists in the cause. And God there exists in all his creations.

A basic substance makes up all things, but because the nature of form, that substance adopts a different manifestation that differentiates it within the form structure. Hence, the result is participants of the one that have different characteristic, and also where one differentiated part does not exist without the other. Hence, the one is always shielded from examination, but the diversity that is originated from it is not. Time and velocity are the same except by manner of expression, and one does not exist without the other. The substance *i* can never be seen separately.

### Oberg Formula Relating Pi and Phi

**Pi / 4 = arctan (1 / phi$^3$) + arctan (1 / phi) (Oberg:1-11)**

**Pi / 4 = arctan (1 / velocity$^3$) + arctan (1 / velocity)**

**Pi / 4 = arctan (1)**

**Pi = pi**

**1 / Phi = -phihat**

### Phi and Pi with Trigonometry

**2 cos (2 pi / 5) = phi**

**2 cos (2 pi / 5) = velocity**

**There are many expressions in trigonometry that relate pi and phi and phihat or *D*, *T*, and *V*.**

### The Meaning of Integers

In this theory, we have talked about the meaning of pi involved in an exponential function that creates *D*. Whole numbers also refer to *D* in the superstring matrix. Numbers like 1, 2, 3, 4, and 5 and so on refer to the following:

**Number *X* = (*D* + *D* + *D* . . . number of times to *X*) / *D***

For example, if ***X* = 7** or the whole number 7, the following relationship in relationship to *D* would be:

**$(D + D + D + D + D + D + D) / D$**

The above refers to the number 7 with 7 *D* added together divided by 1 *D*. This expression can be used to explore both units and new forms of the same equation.

***D can be split in various ways as follows:***

**$D = D \times D / (D + D)$**
**$D = (D + D) = 1 = (D' + D' + D')$ where the units total repeated in the additive string are *D*.**
**$2D = (D + D) / D = 2$**

**These various forms can be explored.**

**Energy of a Photon**

**$E = h \times f$**
**$E = C \times h$ / lambda**
**$E = D^5$**

**Phi Relationship to 2**

**$(Phi^3 - 1)$ / phi = 2**

**$(Velocity^3 - 1)$ / velocity = 2**

**$Velocity^2$ - (1 / velocity) = 2**

**$Velocity^2$ - time = 2**

**$Velocity^2$ - time - 2 = 0**

**Heisenberg Uncertainty Principle**

**Del $x \times$ del $p = h$ / 4pi**
**Del $x \times$ del mass $\times$ del $V = h$ / 4pi**
**$D \times$ mass $\times V = h$ / 4pi**
**$D \times D^3T^2 \times D / T = h$ / 4pi**
**$D^5 \times T = h$ / 4pi**
**$E \times T = h$ / 4pi**

**The system here is uncertain because it is using proper and closed system to try to measure scalar phenomenon. Scalar equations should be deployed to calculate both position and momentum. Using phi and phihat template, we can find the following template:**

**$1 / D^2T^2 + 1 / T + 1 = 0$**

**Making this equal to *D* and solving for mass, velocity, and *D* position, we have the following:**

**$(D^3T^2) / D^2T^2 + (D / T) / T + (D) 1 = D$**

(Mass) / $D^2T^2$ + (*V*) / *T* + (*D*) 1 = *D*

Of course, we need to express the denominator in understandable form. So the following occurs:
**$D^2T^2 = D^4 / V^2$ = force / $V^2$**

**Inserting this into the equation leads to the following:**

***(Mass) $V^2$ / force + (V) / T + (D) 1 = D***

Mass, velocity, and time can be known by this process to derive a scalar equation just like what you do when you have a falling body equation, except here as a more-complex system. Scalar and proper transdimensional systems are not to be confused since the results are different. For deriving a scalar equation in this theory, you need to take the template of phi and phihat, and then rather than making the template equation equal to zero, make it equal to a constant value *D*. This is the process for deriving scalar equations.

**What Makes a Good Philosopher?**

A good philosopher unlearns everything taught, and then strives for self-knowledge, the universe being an analogy of truth, then strives to know self and the universe. Myth is the fabrication to explain what you don't know by the conviction that you do know. Most of human knowledge is myth where myth comprises false collective beliefs. Almost everything known falls under that category. Being of your own mind is the key to self-knowledge and truth where one develops your own ideas. A tour of Ralph Waldo Emerson followed by Bach's *Jonathan Livingston Seagull* would be a good tour.

**Black Hole Entropy**

**Black hole entropy is defined by the following equation:**

**Sbh = kA/4$l^2$*p***

***K* = Boltzmann's contestant = joules / kelvin**

**$K = D^5 / V^2 = D^3T^2 = MV^2 / V^2$ = mass**

**Lp = plank length = *SQRT* (Gh / $C^3$)**

***G* = gravitational constant = 1 / $T^4$**

**$C^3 = V^3$**

***h* = reduced Planck's constant = joules × seconds = $D^5T$**

**lp = *SQRT* ($T^3$ × ($D^5T$) / $D^3$ × $T^4$)**

**lp = *SQRT* ( $D^5$ / $D^3$) = *SQRT* $D^2$ = *D***

**Sbh = mass × $D^2$ / 4 × $D^2$ = mass**

Typically, black hole entropy is explained as proportional to its area. But in reality, we can say that entropy is proportional to volume and mass; the maximum mass in a fixed volume. And if we look at

**mass = volume × density,**

then we can say that volume density is at its maximum.

# CHAPTER 6

## Transformation, Death and 0

-0 = 2pi

B (V, T) = T! X V!

Nothing is a Myth and 0 is something.

There is hierarchy in all truth and love. To love is to infinitely and intimately enable. I know truth because I love, since truth is intimately ascribed to the nature of being, which is life originating from the first cause. To love was to create and to be loved. The pattern of the cause of being resonates throughout reality as a standard of the perfection of being. I am, I love, unity, resonance, pattern, relationship, 1 and 0, life, mathematics, truth, harmony, the nature of the 1 is in all and all is in the 1.

This universe is created by resonance. Euler's equation explains the big bang, resonant creation of space, time and velocity from outside the universe. The entire universe emerges from various resonant states. The graphical form that results from Euler's equation $E^{(i \times x)}$ result is a circle resonating outward to form tetrahedral space. The geometry of the matrix forms an underlying geometry to all things, even space. Those spheres are created by resonant functions. The rotation of radius is the sphere and the dipole action relating all steps is multidimensional. Dipole action refers to induction relating all Trans-Dimensional steps, and the smallest subdivision of a sphere is a binary code. The formal aspects of the universe are on a triad system where only three units are required. Below this geometry is the smaller and infinite one based on a binary code qualified by operators from outside the universe. The concept of evolution is misleading since the universe was induced much like a magnetic field through action of symmetry from operators outside the universe as a conscious form which all things are, and with transcendence the involvement of other conscious beings is allowed with free will. Hence, as the universe, life was also "induced".

## Transcendence and Euler's Formula

We have said this universe is a particular qualification of infinity. Therefore, functional systems outside this universe or infinity, by symmetry and transformation, create the systems within this universe. The universe has within it representative functions of its order, functions that are created within and also influenced by transcendence by functions outside the universe. Those transcendent functions, in addition to transforming into the building block functions or elements of this universe, also exist as transcendent influences. Hence, the universe exists in infinity, has infinity and functions of infinity or outside the universe transformed to be within or of the universe, and also this is a door to outside the universe. The universe is transparent, having relationships particular to the universe and also relationships to outside the universe. Transcendence is represented by transcendent numbers. Euler's equation is the form that expresses not only the big bang but the relation of transcendence in the formation of phi and phihat or *D*, *T*, and *V*. Refer to the equations as follows:

$$E^{(i \times \mathrm{pi})} = -1 = -D$$

$$E^{(i \times \mathrm{pi})} + 1 = 0$$

$$E^{(i \times \mathrm{pi})} + D = 0$$

We have said that form is a tetrahedral relationship of three. We see in Euler's equation three transcendental forms ***E, i, and pi*** that result in *D* or 1. An expression can also be made to unity or 0 (zero). And *D* represents the creation of *D*, *T*, and *V*. Consciousness creates itself in a sense. Pi may be a function such as *D'* (or *D* prime) from outside the universe that is transformative to create *D* in this universe. As we have referred to before, all numbers refer to functions. Transcendent functions from outside the universe may appear as constants within the universe. And so when you have

$$D^2 = \mathrm{pi} \times D^2$$

**Pi = 1,**

**Pi = *D'***

This relationship means that pi is *D'* or *D* prime, and what in fact happened here is that *D'* was transformed to *D*.

$$i^2 = -1$$

$$i^3 = -i$$

$$i^4 = 1$$

Euler's equation tells us that creation is a multiplicative form that results in -1, which is as follows:

$$-1 = V \times T$$

Multiplicative forms are proper or transdimensional. If we bring the -1 over to +1, then we are left with

$E^{(i \times \mathrm{pi})} + 1 = 0$, which is the same as

$$-1 + 1 = 0$$

Here we have an extension of **multiplicative transdimensional form** balance in relationship to **scalar additive forms** to complete the universe. So the tree of creation starts with infinity transcends into this universe by transformative functions to create a transdimensional relationship, and this is further extended to additive scalar forms. This order is also related back to the source infinity through creation, which would have to be an ongoing event. From the equation

$$-1 + 1 = 0$$

we are led to time and velocity through phihat and phi respectively where any number to the *N* is equal to 1, and hence, the relationship of 1 is varied from scalar to transdimensional within the universe because of -1 (multiplicative) and +1 (additive) and also by symmetry is transformed in the creation of the universe where pi refers to 1 or *D'* transformed to *D*. The universe in a sense is an extension or mirror of consciousness. Everything that is, is, which means that consciousness always is awareness, thought, and will, and within that realm of memories, there is only information. The point of transformation from which all things past go into all things future is the present, and it's fleeting but for an instant. Hence, all our experience is memory, and the notion of nothing is a myth. Oblivion, the point of no collective memory, does not exist. To imagine something new is to create. The universe is consciousness, and it is amazing that all that is was imagined to be as an original thought.

The universe is an analogy of truth, which means that ultimately, the same truth applies to all. However, the character of consciousness, how it is altered by participation in the universe, affects that journey to truth. For example, the journey of a dog is quite different from the journey of a man, and even between men. The manner in which you are allowed to perceive is a path to knowledge. This does not take away from ultimate truth, only in your capacity to grasp the truth, and hence the series of steps you need to take to assimilate truth. So we might finally arrive at a truthful axiom that there is no love without respect and trust, but the nature of consciousness between beings might elicit quite different paths to that truth. But to be of truthful nature, there's no excuse to deny the truth whence presented. And that truth is recognized conscious beings of maturity, and there is no excuse to deny.

**Observation**

Physics uses observation, and that is explained as math. Once the pattern of the universe is discovered, as in a unified theory, the reverse is true also; the math predicts not only the observation but also what is not known, since and finally you have discovered the truth of the correct mathematical expression of the physics pattern of the universe.

**Now what is the truth about the physics?**

In a proper system, time decreases while in fact, in a scalar, the opposite may be true.

**Phi + phihat = velocity + time = 1,**

from this all quadratic forms can be derived for scalar systems.

**Phi × phihat = velocity × time = -1,**

from this all proper multiplicative forms can be derived.

Force, no matter at what level of hierarchy, is always $D^4$.

Once you have the pattern, there is no difference between the math and the physics.

Proper time has to do with closed systems such as $\boldsymbol{E = MC^2}$, where one function adjusts in relation to the other in a multiplicative fashion. Scalar relates to additive forms such as quadratic equations and has to do with third-party observation. The first, proper function, is intrinsic to the system; the other, scalar, is within the system as observer. So the first is the system of space, inertial systems itself, and scalar existing within the space. So therefore, time can be experienced differently depending on whether you are part of a system or looking on. So you cannot solve scalar problems with proper forms. However, there is a relationship expressed mathematically between the systems in special conditions where they come together.

**Velocity × time = *D*,**

Which is a proper system, $\boldsymbol{D = 1}$, or the primary superstring

**Velocity + time = 1**

Comes from

$s(t) = gt2 - v0t + h0$

$T^2 - T + 1 = 0$

$\text{Phi}^2 - \text{phihat} + 1 = 0$

Phi is a functional representation integral with velocity and time with phihat. The observation for these equations already exists; you just have to establish the relationships. Currently, scalar and proper systems are separate.

***Euler's equation may in fact represent the big bang.***

$E^{(\text{pi} \times i)} = -1$

**Transcendence Function = Universe**

It is the only equation that combines pi, phi, 0, 1, and phihat in addition to incorporating transcendence. There are no functions, and therefore, any transcendental constants mean that they are produced from functions outside the universe. Pi constant is the result of a function from outside the universe and may in fact be *D* working outside the universe. Euler's formula represents functions outside the universe that by symmetry and transcendence produce the universe. The -1 resulting from the Euler's combination produces

**-1 = phi × phihat = velocity × time**

The energy of time and velocity are organized by ratios that always produce the integer 1. By symmetry, functions from outside the universe produce *D*. From integer 1 or *D* and by another process of symmetry, we are led to

**1 = phi + phihat = velocity + time**

And this refers to all quadratic forms and polynomials.

There is no limit to the realities possible expressed by math. All constants refer to functions. Since all physics can be described in terms of three units *D*, *T*, and *V*, then all constants are referring to one of these three functions or products of these functions. The exceptions are with the transcendental numbers such as pi and *e*, which are occurring in this universe from functions outside the universe. We had talked about the universe being an analogy of truth, and therefore, consciousness and form we similarly represented by the number 3; and thus in science, we had three primary superstrings *D*, *T*, and *V*. The analogy of form (three parts) in consciousness was represented by awareness, will, and thought. And then the concept of telluric form, that the energy from which everything was made, was divided and subdivided or differentiated into various components or identities of form by symmetry and asymmetry. In the process of symmetry and asymmetry, the number 1 therefore was the primary function from which all symmetry and asymmetries were performed. The number 1 was also providing balance within systems of symmetry and asymmetry. 0 was the concept of unity. The idea of nothing is discarded and replaced by a unifying function; an analogy to spiritual emotions is love.

In ***D*** **= 1**, then *D* being awareness was the first action of life or being, and zero was the action to unite or love. And all this existed together. From one came symmetry and asymmetry and also the variety of thought and will, extending to all the diversity of life. So from 1 come all things into a differentiated and varied state, within a system of balance of symmetry and asymmetry; 0 unites all diversity, which may also, in fact, be the telluric energy. The 1 contains all potential diversity, and 0 is the substance from which 1 is made and subdivided.

**Volume and Energy**

We have talked about an inertial field surrounding a concentrated mass core of higher dimensional states. The volume of the inertial field is related to the energy and spherical wave form of the inertial field. This leads to the equation as follows:

$F$ **gravity = volume1 × volume2 /** $R^2$
$F - D^3 \times D^3 / D^2 = D^6 / D^2 = D^4$
$F = D^4$

**The equation derived in this theory also concurs with Walter Russell: "The rate of vibration in a wave field depends on the volume." He goes on to say, "The vibration wave field is the pulse of interchange between its compressed core and space surrounding that core." (Astrotometry: Web Link) and (Russell: 1974)**

A spherical wave begins at the black hole of the mass core, extends out to the boundary of the inertial field, and pulses back to the core. The overlap in the two pulse patterns results in a standing wave. This wave in the solar system is a gyroscopic effect that holds planets within the plane of the ecliptic. When the wave approaches the core the amplitude increases to contain the generator core.

Also note in the $F$ gravity equation above that $G = 1 / T^4 = 1 / density1 \times density2$. When the resultant of mass and $G$ becomes volume, we can say that $G$ is not a constant but the product of $G$ and $M \times M$ is constant.

**Balance of Symmetry and Asymmetry in the Universe**

Scalar and proper forms exist together, not in direct alignment, but they do relate harmonically as complimentary system of symmetry and asymmetry. We have symmetry and asymmetry in harmonic relationship, and also similarly, the universe possesses both infinite pattern and qualified pattern and unrestricted pattern and restricted pattern respectively.

*Inequality of Time and Velocity*

We have seen that the distribution of time and velocity in phihat and phi respectively is asymmetric. Their combination in the integer 1 is a return to symmetry of larger systems. However, asymmetry is responsible for such events as acceleration and motion in the universe, and this may also extend to polarities in magnetic fields and other imbalances related to differences in forces.

**Human Nature, Truth, and Science**

We have talked about life being one form of consciousness perceiving another. Hence, the universe itself is alive and watching, and our participation is a view into the universal consciousness. In a sense, we are all tied together by this experience, and consciousness becomes collective. The manner of perception is qualified by the vehicle through which consciousness is facilitated to participate in this particular experience universe, the form of our minds. Without realizing the primary source or higher self, we assume that our experience is mind and limited perception that we have. Hence, our path to discovery is biased by the manner or style of human perception. With the realization that the higher self is the source of consciousness, then the truth that the mind serves to enable consciousness in this experience reverses our understanding toward the truth about life, and our experience is enabled toward the objective balance required to understand our experiences more fully. The matrix of life, the collective consciousness, by its very nature and pattern enables that which is according to its nature. These are diversity, evolvement, free will, and transformation, but in a sense, the pattern becomes the guide to infinite diversification; and although that diversification and journey is infinite, the pattern also refers to fundamental truths about itself, and those motions in contradiction to its nature cause pain. Much in the same sense, those things according to the nature of any being, reversed or altered, cause pain. In science, left to human perception, science becomes limited, constricted, and a journey of trial and error, and this is akin to human experience and the quest for truth and self-discovery. But for those of higher self-awareness and by using their intuition can derive truth at times as an epiphany, where the awareness of concepts of truth are a priori to experimentation and lead to insights far beyond anything that can be assessed by trial and error. It may appear that mind and intuition are at odds, but mind is meant to serve intuition, and the nature of the mind, in a sense, defines the quality of the path to a greater enlightened experience as being human. So it has purpose in the sense that the nature of mind and the path married to the higher self is a path, a human path, to truth and understanding and has certain qualities that another path may not contribute. And considering that knowledge is infinite, the quality of being human and the style of which those truths of humanness are derived, it is a particular insight solely derived by being human. And finally, since truth is ultimately universal and attainable by almost all paths, the particular path you are on has a certain value or treasure that is uniquely your own, and hence, the collective consciousness is enriched by the variety of experiences contributed by all, and even in the manner and depth of explanation of those truths.

So my higher self and my particular path are enabled to expose truth in a particular way and depth that is uniquely yours and that richness shared with all, and therefore, the richness of experience has no bounds. Realizing however that all paths lead to ultimate truth, therefore, this becomes the basis for moral code and conduct of behavior and also justifies the meaning and purpose of free will that allows for such diversity in wisdom and knowledge, yet being accountable for actions within this infinite path. Truth is universal, and hence, morality and the eventuality of truth is no excuse to deny since it speaks to our higher self and intuition without introduction, much like music. By consciousness and creation being collective, it is true that separation will cause pain. The harmony of the parts leads to a greater truth or synergy of the whole. The whole is the organism of life, and all that is becomes related by the definition of pattern. And there are interpolative patterns, extrapolative patterns, as well as transcendent patterns, and the varieties and compositions result in free will.

## Time and Eternity

Time is a transformative principle and also a function. It has natural motion, inertia if you will, since time and velocity are the same, except by expression; one is the inversion of the other and balanced asymmetrically. Perhaps in eternity, instead of conscious thought being an idea or image of a reductive nature, perhaps it becomes of an ulterior nature, where to think actually creates. Hence, ideas rather than being representations or images in fact transform themselves into events. Hence, if I imagine flight, I fly; they become one and the same. Math represented by such, time may no longer be transformative. Eternity is accessible intuitively since all truth, universal and ultimate, is ever present. Subjecting time is in the realm of higher self, or intuition causes time to be experienced differently or may not even exist at all. So when referring to the differences in operation of mind and intuition, one could say or speak as follows:

***Listen to what I say, not feel what I think.***

And as such, the intuitive person is using the mind to serve the higher self and can access functions beyond those indigenous to the universe, such as telepathy and other forms of higher self-functions. The present is now, but the point of now is actually transformative in our present consciousness where thoughts are transitory or none at all but with awareness. Even in full meditation, that you are is still present. Our memory is in the past. Transformation is where the conscious mind goes from present future to present past. Everything that I imagine to be to everything that I know and remember exists and is balanced from my current awareness and transitory thoughts. In the collective consciousness, awareness creates all that is, and that is differentiated to past and future and all events to manifest the whole universe. So the absolute now is generating all past and future, and that now is you along with everyone else. Awareness becomes form through thought and action. Awareness is transformed from future to past. The absolute now or your awareness is generating the past and future. This means that awareness, the one, the present, the collective 1 or I, is that in which all diversity emerges. From that awareness or all emerge will and thought, and all subdivision of consciousness and experience in the whole universe. It is the source of all action and non-action, all thought motion and all will, intent, and related action. It is the gate of all experience, and hence, the staging area of all action thought to be and what becomes the area of transformation. Since with diversity also exists balance and that point of balance is where everything springs in your awareness. This is akin to the pattern nature of form consciousness and the nature of the physical universe.

### Hierarchies of Time and Space

All constants refer to functions. Since all physics can be described in terms of three units *D*, *T*, and *V*, then all constants are referring to one of these three functions or products of these functions. The exceptions are with the transcendental numbers such as pi and *i*, which are occurring in this universe from functions outside the universe. In this theory, it is proposed that there are several hierarchies to measure time just as there are several hierarchies of force. Force is always $D^4$, whether it is at a level that comes from

$MA = D^4$ or

$D^4 = GMM / D^2$, which is a higher dimensional state or we can diversify into many other forms that include electricity, magnetism, and more.

And in additional to these proper (multiplicative) or transdimensional forms of space as exampled above, there are scalar forms of space that are additive, such as in quadratic forms and polynomials. And then there is another level of space created by the components of the space matrix, or the superstrings overlap and create regions of ***subspace***. There may, in fact, be the templates for both transdimensional and scalar math. For example, going from the template created by phi and phihat in scalar form by making an additive string equivalent to *D* such as in

$gt2 - v0t + b0 = s(t)$
$T^2 - T + 1 = 0$

$\text{Phihat}^2 - \text{phihat} + 1 = 0$; this was illustrated before.

Time varies depending on the definitions and classifications of space:

- **Scalar observation, objective case from one inertial system to another.**
- **Transdimensional intrinsic, within an inertial system or proper system.**
- **Transdimensional, between inertial dimensions.**
- **Scalar, the scale of the participant space that has subspace, space, and hyperspace.**

As a common and proper form of inertial space, we have said d**ensity = time**$^2$

The time stream of flow *T* is a function and relates to participating functions such as velocity *V* and displacement dimension *D*. The functions *V* and *T* are proportioned in their relationships to phi and phihat respectively but combine as variable products. Hence, density is related to time in more complex hierarchies. Without understanding the different mathematical relationships of time, there will be paradoxes.

If energy of higher dimensions appears as force where we measure, then in view of measurement, we are looking at the shadow or effect of the cause. But to understand the math, we can now better estimate the cause. Hence, something travelling in complex motion in a higher dimension may appear as a straight line to us in our view of measurement, but now that we can estimate our evaluation in a different way, we will avoid the mistake of undertaking one system of math for all perception and scientific observation. So we may develop different sets of equations for different levels of transdimensional and scalar form.

As an abstract exercise, we can describe *D*, *T*, and *V* in relationship as a cube coordinate system where you have an axis for dimension *D* and another for time *T* and yet another involving *V* or velocity. In this theory, the relationship of *D*, *T*, and *V* have a tetrahedral inductive relationship as in a Cartesian coordinate system. However, this also is maintained through various hierarchical relationships. For example, current and the electrical and magnetic field relationships also have this relationship in *DT*, *DV*, and *D* or magnetism, current, and force respectively. Note that the variation in coordinate is by an additional *D* product.

Current is associated with *V*, magnetism with *T*, and *D* with force. The abstraction of a tetrahedral relationship into three axes is the result of three primary superstring interactions, which are also three primary energies. That being said, only three units are needed to describe all events, but such that their properties vary so through dimensional states, we give them more complex attributes and names that make them seem unrelated. So then bringing them together again is unification by three units and powers thereof.

***In a proper system and one function is enlarged, the other functions decrease such that balance is maintained within the inertial field system. For velocity, the limit would be C.***

*This would be correct in the relationship of D, T, and V.*

With regard to antigravity, anything that causes an effect contrary to the existing inertial field polarity would be anti-gravitational. **This is generally a 1 / *X* relationship for the anti-field relationship**. *X* represents the natural standing equation for an effect such as *F* for gravity. For example, if gravity force is

**$F = GMM / D^2$, then the reverse would be**
**$F$ anti $= D^2 / GMM$**

**And this form can be rewritten many ways, and so there is more than one way to achieve antigravity. Here time would have an inverse relationship as well and, instead of being transformative, becomes nominative.**

*C* is both a constant and variable. *C* is constant as a proper form representing balance within a proper system of inertial field and varying between inertial systems that are not acting as proper systems. And there are also various dimensions of *C* resulting from additive and multiplicative system. All constants represent functions and sometimes products of functions. For example, it may be that the constant *G* is actually a function $1 / T^4$, which is a variable and that the proper constant is *GMM* or the product of *G* and mass. Space is an elastic medium of superstring matrix. You might imagine like small sphere rings or ellipses tied together, something like quantum loop gravity or foam at the plank level, but not entirely the same since this matrix is producing all mass and properties and are reorganized in higher densities and dimensions to form all particles, which are the interference patterns of this matrix. So there is

- **proper motion where the inertial field changes itself,**
- **then scalar or relativistic motion where one inertial system is moving within another, and**
- **another type would be two inertial systems acting as one as when two masses are supporting a shared inertial system and blending together.**

So in these examples, you can see that *C* is not always constant. However, the manner in which the inertial field can vibrate under certain conditions is constant and has limits related to *C*. So motion is partly an illusion, and photons may in fact be monopoles within the matrix, and what you see is the excited portion of the matrix when the elastic wave reaches your eye. And in this sense, all space is light, but your eyes are designed to see the negative, the interference patterns of the matrix rather than the matrix itself.

We refer to the 4D plane as space-time where time is one component. But actually, it's a 5D plane or space-time since **mass = $D^3T^2$**. So here we have space and $T^2$ instead of *T* where **$T^2$ = density**. So the language of thought sometimes contradicts the language of the true pattern since the rational mind is a process of trial and error. Our highest form of measuring is energy at **$D^5 = MV^2$**, so space-time may in fact be five dimensions but the space is still $D^3$, but then if you break up components further, you get

**$E = D^3T^2V^2$**
**$E = D^3T^2C^2$ for hyperspace**

which are seven participant functions. But that we measure *D* as the determining of dimensions, it's five. So space-time is at $T^2$ and not *T*. However, like Newton, we are fixed on momentum and time rather than energy and $T^2$. It took several hundred years to revise to the idea of energy.

Mass is a configuration of $D^3T^2$ superstrings, so therefore, if we break anything down to three units, you can always find the mass. Let's try time. At first glance, we say time is *T*. But then now let's generate a relationship to mass from this one vector.

$$T = D / V$$

$T = G \times$ **mass** $\times$ **velocity / temperature**$^2$ **since** $G = 1 / T^4$ **and temperature** $= V^2$**, So**

$$T = (1 / T^4) \times (D^3T^2) \times V / V^4$$

$$T = (D^3T^2)\ V / (T^4)\ V^4 = D^3T^2V / D^4$$

$$T = D^3T^2V / D^4 = T^2\ V / D = T^2\ D / DT = T$$

So time in a higher form is

$T = G \times$ **mass** $\times$ **velocity / temperature**$^2$**,**

which means time is mass velocity in relationship to temperature. And if the condition of space temperature is negligible or nearly constant, time is mass velocity. And you can see how even space has mass, a large amount to be exact. The inertial field is a mass extension of the core mass.

Objects distribute themselves over several dimensions, but it's not limited to $D^5$, which is the limit of our field of measurement. For example, force is always $D^4$ that has relationship and is induced by functions operating under more complex systems of space in higher dimensions. Here there is a postulate that gravity is in fact $D^{10}$, and we can see the simplification of this action as we experience force as a $D^4$. This relationship and hierarchy is represented as a model through all dimensional states in an atom. In a sense, the galaxy is a crude example of that model in a super massive black hole. We can then use the squaring function to estimate the maximum projection radius $R$ or $D$ of the inertial field at $D^2$ in an unshielded black hole. For the Sun, it is $D$ using $\boldsymbol{D^5 = MC^2}$ calculate $D$. For a super massive black hole it is $D^2$ because it is not shielded. This would take a black hole to $D^{20}$.

Phi does represent disparity between velocity and time since the relationship is asymmetric. This asymmetry would have an impact between more complex products such as volume and density even in a $D^3$ perspective. The understanding of the universe and math has value in all dimensions, dimension per dimension. If you want to think of the ratios of $D$, $T$, and $V$ as 1, phihat, and phi respectively, with positive and negative values in proper and additive forms, then these ratios could be viewed as an impedance involving complex numbers. The pyramids did reconcile the constants phi and pi, so perhaps our ancestors were familiar with these numbers as governing the mathematical and physical relationships of the universe. These numbers and their ratios or impedance also have the unit functions that allow variable operation.

There is a proof for $\boldsymbol{D^5 = MV^2}$ near page 54. $C^2$ is just the hyperspace of that model, a simple substitution. So $\boldsymbol{E = MC^2}$ becomes $\boldsymbol{D^5 = MC^2}$, which when $D$ is calculated, it's the Schwarzschild radius, the extension of the inertial field. The difference is in the definition that $\boldsymbol{T^2}$ **= density**, which gets you to the equation $\boldsymbol{D^5 = MC^2}$. We had talked before that time was mass velocity, but that inertial field to which it refers is also affected by the density of the field related to $T^2$. There is a velocity to the inertial field of varying density. At any point along that field time is a mass velocity, but there is also a variation of time along that field affected by its density, which lessens as it has a larger radius $D$. This pulse of inertial field creates a standing wave due to the spherical wave nature of the field; the standing wave is within the inertial field.

The asymmetries in other dimensions as well as the symmetries have an effect in this dimension. Every dimension is tied to a cause coming from somewhere else. It is enough to realize that time and velocity are not symmetrical; however, their product and various dimensional states produces symmetries. And so we can speak of the divergences of other dimensions, resulting in the properties we measure in the dimension we are in at hand.

## ***Space, Time, and the Casimir Effect: Templates of Subspace***

Well, it is of course correct that we need an equation to summarize our ideas of the space matrix. Theories without laws have not much to test. Secondly, the space matrix that we know appears 3-D or $D^3$, and there is also $T^2$ where $T$ is a part of that. These are physical properties from space. The $D^3$ represent $D^3$ superstrings as an effect volume. I have mentioned that the superstrings have a volume themselves, but that volume is not the effect volume we experience by the relationship of volumes of $D$, $T$, and $V$. So if you break down a superstring as a single sphere, you might not be able to experience space as we know it in those smaller states. Now the matrix itself has hierarchies of force or mathematical matrix physics that binds it that has nothing to do with the effects we experience, although related. I refer to this math as templates. So I was considering those templates as the math to consider for Casimir effects. If you want an equation template for time, consider those generated by the relationship of phihat (time) without phi (velocity). There is one as follows:

**$\text{Phihat}^2 - \text{phihat} + 1 = 0$**

**$\text{Time}^2 - \text{time} + 1 = 0$**

These then become templates for what we experience. The 1 can be considered *D*, So

**$\text{Time}^2 - \text{time} + D = 0$**

**$-D = \text{time}^2 - \text{time}$**

Displacement and forces related to *D* have a relationship to time at the plank scale as a quadratic form involving complex numbers. Our universe is the *-D* created from Euler's equation:

**$E^{(i \times \text{pi})} = -1 = -D$**

That *-D* becomes as follows:

**$-D = \text{time} \times \text{velocity}$**

We might consider the Casimir force to be a fundamental relationship of space not experienced by us higher-dimensional beings. Higher dimensions become transparent to space matrix at lower dimensions. Hence, the math of subspace, the areas of overlap of superstrings, is another mathematical form, which is referred to here as a template. In current physics, the Casimir force per unit area is measured as force/area as follows:

**$\text{Force/Area} = D^4 / D^2$**

**$D^2 = \text{area again}$**

**But $D^2$ is also charge $Q$.**

**$D^2 = \text{charge}$**

**$D^2 = h$ (Planck's constant $D^5T$ or $ET$) $\times$ $C$ (velocity of light) $/ D^4$**

Science is using a conventional idea of charge to estimate the Casimir force. In this theory, time is the cause of charge. And hence the equation

**$\text{time}^2 - \text{time} + 1 = 0$**

may be appropriately related to the Casimir effect. If it is an effect related to subspace, but the current definition makes it related to space as described below as a proper form. There are always both proper and scalar forms related to physical effects, which makes the selection of the correct scalar and proper form for an explanation an exacting task. Whether an equation is proper or scalar shows different relationships to potential form or energy dimension per dimension.

The current definition of the Casimir force or *Fc* is measured not at subspace but space as a proper equation as follows:

**$Fc = h \times D^2 \times C / D^4$**

**$Fc = h \times C \times D^2$**

**$Fc = h \times C / \text{charge}$**

**$Fc$ can be rewritten as follows:**

**$Fc = h / C \times \text{density}$**

**The force/area² relationship or $D^2$ can be rewritten as**

**$D^2 = C^2 \times$ density.**

So then Casimir is measuring density fluctuations, and in this case, probably negative. This would also have a relationship to gravity and antigravity because we are talking about charge and capacitance of the inertial field. As we have mentioned before, that charge is involved in the natural space polarity and field distribution, and to alter this was a key to inducing counter the natural ether field. This world achieves antigravity (inverse relationship) and other related forces.

<u>A scalar solution is as follows:</u>

**$\underline{T}^2 - \underline{T} + \underline{D} = 0$**

Now let's make this template equal to $D^2$ for the force/area relationship as follows:

**$D \times A \times (\underline{T}^2) - D \times V \times (\underline{T}) + D \times (\underline{D}) = D^2$**

**$(Q$ / density$) \times (\underline{T}^2) - (Q / T) \times (\underline{T}) + Q =$ force/area**

**Force/area = $(Q$ / density$) \times Q / C^2 = Q^2$ / density $\times C^2$**

**<u>The math in this theory with three units can derive existing physics equations.</u>**

**Let's look at $F = MA$**

**Here it becomes $D^4 = F = MA$, which means all force is $D^4$**

**Can we come up with a relationship of force or $D^4$ using two masses $m \times m$?**

**Yes.**

**Let's start with**

**$D^4 = M \times M$ (add functions)**

**$D^4 = (D^3T^2) \times (D^3T^2)$ (add functions) $= D^6 \times T^4$ (add functions)**

**So to get to $D^4$, to the right we need to divide by $T^4 \times D^2$**

**Therefore,**

**$D^6 \times T^4 / T^4 \times D^2$; let's return the numerator to two masses**

**$F = D^4 = M \times M / T^4 \times D^2$**

**We know from this theory that the units of $G$ or gravitational constant is $1 / T^4$**

**Therefore,**

**$F = D^4 = G \times M \times M / D^2$**

**$F = D^4$**

***<u>This is Newton's equation.</u>***

**<u>Casimir Effect, Capacitance, and Teleportation</u>**

**Capacitance = $C = Q / V$ = charge / voltage = $D^2 / D^3 = 1 / D$**
**Capacitance = $1 / D$; Inverse of Resonance**

The inverse of dimension $D$ space is <u>capacitance,</u> which is what might be required to create a wormhole or teleportation gate, where space is travelling in the opposite direction or shortened in relation to the surrounding matrix. The $1 / D$ relationship may in fact be the fluctuation in density of the space matrix that is driving the Casimir effect. And of course, along with this, in terms of scalar or relativistic relationships, the space outside is not as old as the space inside due to relative time adjustment. This means that the typical relationship of time is reversed. So there is the proper form of time with an inverse relationship to the normal time of the matrix and also the relativistic adjustments to the observer on the outside observing from another inertial field. In the proper system, the mass decreases as the velocity increases since the systems require balance and increasing the time, but the opposite is true in a scalar or relativistic system where time appears faster. In the proper system, mass is converted to time. In a reversed system, time would be converted to mass and appear faster. Hence, in a realistic system, time would appear slower outside the system that is inversed.

The relationship to capacitance goes back to the Biefeld Brown experiment discussed before.

**The concept of space and definitions are multiple. For example,**

**Capacitance = *C* = force of gravity / energy. Others,**

**Capacitance = *C* = charge / space**

**Capacitance = *C* = voltage / force**

**Capacitance = *C* = time / density *C* (speed of light).**

**It may be that the key to teleportation lies in the development of fields of space that create a star gate of relationship 1 / *D* perpendicular to the direction of motion. (Davis: 1-88)**

A relationship of 1 / *D* or any inverse relationship represents a function that is reverse to the normal or natural field polarity. Hence, we can talk about antigravity as 1 / force relationships or the contraction of space as 1 / $D^3$. Hence, the effect of capacitance would have also an inverse relationship on space and time. This of course would affect the natural or general observations with regard to relativity and proper system and achieve an inverse effect. (Bouwmeester: 1-5)

## Singularity

Singularity refers to an infinite system that is beyond the transdimensional and coordinate or scalar system. According to this theory, Euler's equation refers to infinite functions outside the universe that produce the universe:

$\boldsymbol{E^{(\text{pi} \times i)} = -1}$

**-1 = (time × velocity) = (phihat × phi)**

**-1 = -*D***

And from this the relativistic or scalar system emerges in

**1 = time + velocity = *D***

**1 = *D***

From 1, the whole universe is produced with all fractions and related numbers. From 1, all diversity emerges in hierarchy and complexity. The exception is pi since it is referring to the source function outside the universe. We have said that the universe is a qualification of infinite functions. Euler's equation is the projection of *D* from singularity or the creation of the universe. This transition occurs infinitely in various types of black holes through all space-time. Hence, the big bang is an ongoing function of infinite portals throughout all space since all points of the universe are connected to Euler's equation. A multiple bang is the universe, and hence, it is a form such as a holoform or hologram in an infinite matrix pattern. And the product refers to the source form by the relationship of three.

## Zero, Singularity, Transformation, and Creation

It is posited that where there is a singularity, there is no mathematical model to define a singularity. This theory proposes that the concept of nothing is a myth; hence, singularity refers as a transformation of one system of functions to another and is referred to as a point, gate, or qualification of transformation or transformative system. We have said that zero refers to unity, meaning that various functions combine through unity or zero and are related by a process of transformation. In transdimensional or proper systems, this unity is referred to as *C* whereas in scalar or relativistic systems, it is referred to as zero (0). Hence, unity or zero refers to a system of transformation and can be qualified as various forms, and we might suggest various forms and types of black holes. There are black holes that relate to unifying relationships and transformations that are transdimensional and those that are trans-universal, which means those functions that are outside this universe that transform to create functions within or of the universe. Trans-universal functions create the universe.

## Time, Density, and Charge

$$T^2 = D^2 / V^2$$

$$T = D^2 / V^2T$$

$$T = \text{coulomb} / V^2T$$

$$T = C \times T^2 / D^2T = C \times \text{density} / D^2T$$

$$T = C \times \text{density} / \text{resistivity}$$

$$T = C \times \text{density} / p$$

$$T = \text{current density} \times \text{density}$$

$$T = J \times \text{density}$$

Time has an asymmetric relationship with velocity through phihat and phi respectively. Together, time and velocity form the symmetries of larger systems. Time and velocity form 1 or *D*. And by forming 1 or *D*, time and velocity are the cause of all the diversity in the universe. Time and velocity can be quantized. Time is related exclusively to charge as has been shown before in this theory, and both time and velocity are related together to spin, as also has been shown. Velocity is the larger portion, and time is the smaller and more elusive. There are many definitions of time as has just been illustrated.

$$T = \text{coulomb} / V^2T$$

$$T = C \times T / D \times V \times T = C / D \times V$$

$$T = C / \text{acceleration} \times \text{density} \times \text{velocity}$$

$$T^2 = C / \text{acceleration}^2 \times \text{density}$$

Time is related to charge density and motion.

## Constants and Variables

It has been mentioned in this theory that many times what appears as a constant is in fact a function or variable. Constants with units are in fact variables. True constants do not refer to units, and there are very few of these constants. For example, pi does not refer to any units and is therefore **a true constant**. The appearance of a true constant in this universe is the result of a function from outside the universe, occurring by induction. Pi could be an operator like *D* outside the universe involved in creating *D* within this universe by an inductive process in Euler's equation.

$$E^{(\text{pi} \times I)} = -1 = \text{time} \times \text{velocity} = -D$$

Euler's equation then represents the big bang or induction of all points in space of the entire universe, the creation of *D* and diversity through time and velocity both in scalar or relativistic form and as a product of closed systems trans-dimensionally.

**Gravitational Force, 1 / R and 1 / $R^2$**

Generally, we speak of the gravity between two masses as having a relationship force of 1 / $R^2$. In this theory, that means that

- **two inertial systems are acting as one as in Newton's equation, and**
- **one system is overlapping or within another as in Newton's equation.**

The inertial field generated by the Sun is from a shielded system. A super massive black hole at the center of the galaxy generates space or an inertial field as an unshielded system. In a shielded system, space is generated at distance $D$, and in an unshielded system, $D$ goes to $D^2$, and so there is much more space surrounding a super massive black hole. Also, it may be that there is a 1 / $R$ relationship within the inertial field of a super massive black hole. When exiting the inertial field of a shielded inertial field system, the relationship is also 1 / $R$ since inertial fields are no longer acting as one system or attached but as separate systems acting through a third system. So to summarize,

- **shielded system projects space to $D$ and has a gravitational relationship force of 1 / $R^2$,**
- **unshielded systems of $D^2$ space projection as associated with super massive black holes have a gravitational force relationship of 1 / R, and**
- **when exiting the gravitational influence of shielded systems, the relationship between disconnected inertial fields is 1 / $R$ through a third enveloping system.**

These parameters may account for the pioneer anomalies and other cosmological relationships between galaxies.

**Temperature, Velocity, and Mass**

According to transdimensional theory, temperature is velocity$^2$, so to say temperature is velocity$^2$ in Brownian motion is a given. However,

let's translate temperature or $V^2$ further as follows:

**Temperature = velocity$^2$ = $D^2$ / $T^2$ = $D^2$ / density**

**Temperature = $D^3$ / density × $D$ = volume / density × $D$**

**Temperature = volume$^2$ / mass × $D$ = volume$^2$ / ($D^3T^2$) × $D$**

**Temperature = velocity$^2$ = volume$^2$ / $D^4$ × $T^2$**

**Temperature = volume$^2$ / force × density = volume$^3$ / mass × force**

**Temperature = volume$^3$ / mass$^2$ × acceleration**

So you can see in the equation,

**velocity$^2$ = temperature = volume$^3$ / mass × force.**

Brownian motion is also affected by mass, however slight, and volume of the particle along with its inertial field as well. A multiplicative system is a proper system, a product of multiplication. A scalar system is additive. If you want to isolate momentum and position, you need a scalar equation involving addition, as in quadratic or polynomial mathematical forms. Heisenberg was uncertain because he used a proper mathematical form to solve a scalar mathematical problem, which is incorrect. The philosophy of approach as well as the outcome for proper and scalar systems is entirely different. A proper system deals with the unification of a system or an inertial field system considered as operating as one. A scalar system involves variations within the inertial coordinate system from an objective view of another inertial system; relativity deals with multiple inertial systems separately or objects to the system rather than the system itself.

For a two-body proper system,

**Temperature** = $V^2 = G\,M1\,M2\,V^2 / R^2\,D^4$

**Temperature** = $G^2 \times M1 \times M2 / R^2 \times V1 \times V2$

So you can see that temperature is related to mass, velocity, and distance between objects. As an object approaches the speed of light in a proper system, its temperature approaches absolute zero, but the reverse is true in a scalar system where temperature increases as the kinetic energy of motion increases.

## Reality of Self

We have said that life is one form of consciousness perceiving another. And hence, the universe is God's consciousness. The physical forms that we possess, such as our bodies, are borrowed as are all the things that we possess and share while participating at this level of consciousness. So we own nothing here except transcendent self, and all that we see is in the consciousness of God. Life states of consciousness are receded from the source consciousness, so at what level of consciousness a person is can be one of many multiple and possible states. When we pray and meditate on where we are in a dream such as sleeping, we can know that we are dreaming and refer to the higher level of consciousness and exit the dream. For example, when you sleep, it is possible to know you are dreaming and refer to other states of consciousness. Also in the level of consciousness of your current "awake" state, your tie to the source consciousness is again through meditation and prayer, which refers to the source at all times.

Our purpose in life is not about things, evolution, conquest of the universe, or even the future of the human race, but about the state of inner self and the quality of being. So to a warrior in a primitive world, that person would exalt in the conquest of territory and of things and the defining of self through history, and contrarily, a more-exalted being would be focused on the quality of inner being and truth, those things apparent that transcend the nature of the universe and refer to the nature of the matrix source according to its real nature or truth or God. Awareness of truth leads to wisdom, and by that, you know how to live. So in the nature of being, the material goals mean less in so much as the participator realizes that the participation is for the development of self. You are here to participate and learn, and being is not about the object of the time line in this universe. But we so eloquently refer to the evolution of man and traversing the galaxy as a primary destiny, which it is not. What is important is the present and how you are aware of its importance and meaning in relation to the harmony of the whole.

A civilized person therefore has nothing to do with machines and advanced technologies. Ideally, an advanced person is so according to his or her nature and reflecting the nature of the source matrix. In tune with higher self, the civilized man is defined so, and the machines are no longer necessary to define civilization, except perhaps only as they define intent toward harmony of the good of all. And for the utmost advanced spirit, I could imagine having the capacity to will oneself to the far side of the galaxy. Now the machines are meaningless, and the ultimate being is complete and one with the matrix of harmony and truth as co-operator. In a sense, this is the realm of miracles. So therefore, illuminated beings are in tune with the nature of the source consciousness and, as such, are co-operators in the divine harmony and will of the collective. You can have beings with advanced technology and also who are unaware of the truths of life and also those of little technology who are aware of truth and as such can be deemed to be civilized. But in either case, an understanding of the universe does ultimately lead to an understanding of God.

It is possible that beings are so focused on the rational rather than the intuitive self, and on the material that this defers from the awareness of their higher conscious self. And is akin to death, where the nature of self becomes divorced from the nature of the universal consciousness and no longer is able to perceive its higher conscious self since it is so wrapped in defined, limited, and narrow paths of being. Awareness seems obstructed and remote, where the expansive nature of the truths of the infinite source consciousness is lost. All is good, and the matrix is here to infinitely enable; evil is the contradiction of self.

## *SQRT* (5)

**Phi** = $SQRT(5) + 1 / 2$

$V = SQRT(5) + 1 / 2$

**$2V = SQRT(5) + 1$**

**$2V - 1 = SQRT(5)$**

**$2V - D = SQRT(5)$**

**Phi - phihat = *SQRT* (5)**

**Velocity - time = *SQRT* (5)**

**Velocity - time = $2V - D$**

**Velocity + time = 1**

The number 5 is expressed as a relationship between time and velocity.

**Fibonacci Sequence of Numbers**

**It appears that all powers of $V^n$ or velocity$^n$ are governed by the Fibonacci sequence of numbers.**

- **Phi$^3$ = phi × phi$^2$ = phi (phi + 1) = phi$^2$ + phi = 2phi + 1**
  **$V^3 = V \times V^2 = V(V + 1) = V^2 = V = 2V + 1$; 2 and 1 are Fibonacci numbers**
- **Phi$^4$ = phi × phi$^3$ = phi (2phi + 1) = 2phi$^2$ + phi = 3phi + 2**
  **$V^4 = V \times V^3 = V(2V + 1) = 2V^2 + V = 3V + 2$**
- **Phi$^5$ = phi × phi$^4$ = phi (3phi + 2) = 3phi$^2$ + 2 phi = 5phi + 3**
  **$V^5 = V = V^4 = V(3V + 2) = 3V^2 = 2V = 5V + 3$**
  **Phi$^n$ = FnPhi + Fn - 1 when *n* is greater or equal to 2**
  **Velocity$^n$ = FnVelocity + Fn - 1**

**Also,**

**Fn = Fn - 1 + Fn - 2**

**Velocity$^{(n+1)}$ = Fn + 1velocity + Fn = velocity × velocity$^n$ for all *n* greater or equal to 2**

**Fn = velocity$^n$ - time$^n$ / *SQRT* (5)**

**1 and 0**

Phi and phihat or velocity and time create all that is known and all the hierarchy and diversity of the universe. Phi and phihat or velocity and time result to form the number 1. All diversity is within the number 1, and all integers are related and defined through phi and phihat. Phi represents velocity, and phihat represents time. All Fibonacci numbers are related to powers of velocity$^n$. In a sense, the pattern of the universe is creating all integers and all math; they are one and the same. The number 1 represents balances of diverse and hierarchical forms such as time and velocity and the various degrees or dimensions of their relationships. And various configurations of 1 are used to form different dimensions of symmetry. Time and velocity are inversions of the same energy and also asymmetrical. Zero represents unity, the point of transformation of energy from one state into another, from future to past, from dimension to dimension. The typical understanding of zero as void is a myth. Zero then is transformative and uniting in that act of relating functions. Zero is transformation between various states or functions. The source of all things is the number 1 or *D*, and then through differentiation of time and velocity that form *D*, you get diversity, rhythm, and hierarchy. Zero is transforming through that diversity. Zero is the point of transformation and relationship. For example, in a quadratic equation for motion, you can have various functions equal to zero. The transformative relationship is what allows separate functions to be expressed as united. So zero means allowing transformation and united. Anything with zero will do. *C* is similar for proper forms. The 0 applies for scalar relationships and *C* for proper relationships. The 0 refers to the relationship of multiple inertial fields and 1 to a single proper inertial field.

Let's imagine a model:

**0 is for the center of a sphere and *C* for the perimeter of a sphere.**

So relativity is scalar, but $E = MC^2$ is proper. And they behave quite differently. So Heisenberg uses a proper form to figure out a scalar problem, which is incorrect. He multiplies several functions in an attempt to solve a scalar problem. What he should be using is addition in a quadratic or polynomial form and relating them to unity or zero. You can know both momentum and position as a scalar relationship, but not so as a proper form. Scalar refers to the interaction of several inertial fields and proper to one inertial field.

## Constants

Planck's constant defines a small scale. Any constant with units is not a constant; it's a variable. True constants have no units and represent functions outside the universe, such as pi seen here. The constant pi is the result of a function from outside the universe that here looks like pi. Functions from outside the universe create 1 or phi and phihat in Euler's equation since phi and phihat become 1. This describes the big bang.

**$E^{(i \times \text{pi})}$ = -1 = time × velocity = *-D***

So those constants outside the universe are creating the universe in the number 1 or time and velocity.

## Velocity and Time in the Creation of Form, Hierarchy, and Pattern

## Time and velocity are identical with mathematics.

Coming from phi (velocity) and phihat (time) we see that all integers are derived. The universe and the mathematics we use is a mirror of the patterns and structure derived through form in the relationship of the component superstrings or primary superstrings. Time and velocity and the organization of said energies form the basis of all structure and hierarchy, which is represented by the very math systems we use, are not only representative but one and the same. Hence, it is almost inconceivable to refer to any other orders without some reference to the existing orders to be a conceivable system. But since consciousness progresses from one system to another by symmetry, we can say that an extrapolation of math will indeed predict other realties. Since action directed must have a receiver, as one form of consciousness to another or—even more fantastic—one form of math unto another, the extension of one conscious world to the creation of another is also a reference to the possibilities in models of math to predict occurrences from one system into another. This process of induction here is referred to as symmetry. The nature of consciousness allows action and to be acted upon according to the pattern nature that induces the very things it allows receiving.

## Higher Self

In this theory it has been mentioned that consciousness exists in different states receded from the source consciousness, collective consciousness, or higher self. Hence when I dream, from the point of conscious observation, the dream appears unreal, and in the dream, where I came from appears unreal. So the place of conscious activity or presence always views the dream as a distant place, nebulous, and as a vague memory. And the current place also is viewed as a dream when our activity is receded or experienced in another dream state. Always from the point of observation, the dream remains unreal, the memory of that is a dream, and even our current activities take on the form of a dream when viewed from another state. So when we depart or die, surely our memories of this universe are as a dream. Consciousness is being transformed from future to the past and also from one dream state unto another. In all this experience, reference to the higher self is always genuine, meaning that truth never changes since our higher self transcends all the dreams that we experience. Hence, although we may dream, the dreamer can always refer to the higher self for constancy and real truth. Your higher self or intuition knows all and is constant, and the dreams are variables. To know intuitively is to know a priori, to know from the source of all life, and this is a constant reference while the dreamer or rational mind is exploring the vastness of relative manifestations of the same truths under more varied and subjective conditions of the various universes or dream states. Hence, the higher intuitive form is transcendent and akin to a proper form while the rational is akin to a relativistic form. Euler's equation tells us that various functions from an exterior dream state are experienced as -1 in this universe, and the one contains all the diversity of this universe in time and velocity, which are equal to 1 as a product and also in scalar terms. From the point of observation, the higher consciousness appears as a dream, as a constant, as a memory, and the form, although existing is not experienced as in the dream state that it refers to. Hence, from here, the infinite energy of other systems appears as constants, and from the exterior realities or realities external to this universe, this reality appears as a dream or constant. But one reality and its energy, by symmetry and induction, create another reality. Euler's equation defines the creation of the universe by one form of consciousness referring to another.

## Math

In math, different deductive systems yield different interpretations.

**Is it possible for men to deduce systems that are not according to those that have representation in this universe?**

All integers and numbers have meaning, which means that every integer can be identified with a process or function existing within this universe. For example, phi refers to velocity and phihat to time. Together they refer to the number 1, and within the number 1, all diversity and hierarchy and subdivisions are found. If the universe is a form of consciousness, that of the Creator, then the nature of the object refers to the nature of the source by symmetry since one is directly dependent on the other. The principle of symmetry generates all the repetition and complexity or hierarchy in the universe. The notion of separation or of being unrelated and the concept of nothing are myths. Even as all that exists is related, the universe being an analogy of truth tells us that all is connected, and hence, even in physics, the pattern of the universe can be derived.

**And so then can man, according to his nature, deduce facts about math and realities that are not according to his nature?**

This would in fact be a creative event in the manner of creation. Although such forms may be written down, for all purposes, they might be interpreted as gibberish and, for them to appear real with meaning would be creation.

**Hence in the manner that one (1) is subdivided, do new forms come into being, new modes of reality?**

If men are a product of this universe, then according to the nature of it, it would be impossible to exceed it. But in the manner that man is a transcendent being, this could be possible, and together collectively with the source creator, in the mode of origin of our higher selves, we may also transcend these other realties and truly be connected to the boundless possibilities of the universe pattern and math as we imagine its possibilities to be. In this intuition may the greater truth be. But the first action, all powerful, is the first division of 1, which is God, and that subdivision in any of its forms is creation, a pattern of reality, and the will of math. God is all things within the 1, and things subdivided within and without, and transcending all things and possibilities that can be. Limit therefore is transitory pattern, and all is unlimited pattern and freedom in transcendence.

## Science and Life

Whatever we discover, it's only discovering what already is. The purpose of life is not in the quest but the present moment and development of self. Although some argue and place significance on and about humanity's survival and conquest of the galaxy, our state in the present moment is our eternal purpose, so it doesn't matter what point in history we appear to participate. Our task is the quality of self and at the present moment, which is life. How we view the past, live in the present, and plan for the future describes who we are in the present, and those intentions become manifest by the actions of our free will and show who we are.

***The quality of self is a challenge at any level of being.***

If a theory can celebrate life, that would be its most important achievement for any scientist. That can happen any time in the present, whether the theory is complete or not, accepted or unaccepted. The search for truth has its own legitimacy and substance. So that a theory bridges science and philosophy makes it real according to the search for the meaning of life.

## Gravitational Mass, Inertial Mass, Higgs Field, and Higgs Boson

## The Big Bang and Creation of the Universe

### Why are gravitational mass and inertial mass the same?

Core mass such as the nucleus of a particle creates space or space mass of the surrounding inertial field. The variances of space extend from a higher or denser dimensional state to a lower dimensional state of outer space, which we refer to as the vacuum. These variances comprise one entire system as they transform from one dimension to the next by induction. The atom is the perfect model of a transdimensional system running through all dimensions at every level, from outer space of the inertial field to space at the center core wherein a shielded black hole lies. Both mass and space are an arrangement of primary superstrings ***D, T,* and *V*.** Hence, physics only requires three units of measure. We can convert to and from these three units from our current equations to discover new relationships. *VT* together forms a *D*, and hence, the appearance of *D* is larger than *VT*, and also we can refer to *D* as the Higgs boson superstring. But here, rather than being a particle, it is a unit or superstring of the inertial field space matrix. When the matrix is more involved in its complexity, specific formations of $D^3T^2$ result in what we term as mass.

**Mass =** $D^3T^2$

However, each ***D, T,* and *V*** has different attributes of the same telluric energy. For instance, time is an inversion of velocity of the same substance. And if science does identify a "particle" as a Higgs, it is only the particle associated as an interference pattern of $D^3T^2$ superstring waves, since particles are the interference patterns of superstring waves. So *D* or Higgs in various arrangements form various inertial fields of space dimension per dimension, one related to the other by induction. The Higgs field is responsible for the creation of inertial mass because they are one and the same transdimensional system. The field and the core together as one system are equivalent. So the farthest extension of the field trans-dimensionally is the lowest level of energy, which we refer to as the vacuum state. And *D* as a unit along with *V* and *T* define a matrix of space. It is a particular arrangement ($D^3T^2$) of the superstrings that define mass. However, if we were to quantify *D, T,* and *V* in everything, that would be more accurate a description of our observation.

So the creation of the most elementary unit *D* with *V* and *T* is the big bang. The only equation that relates 0, 1, *E, i*, pi, phi, phihat, and *C* with *D, T,* and *V* is Euler's equation adjusted as follows:

$E^{(i \times \text{pi})} = -1 = -D =$ **time × velocity; *C* can replace *V* or**
$E^{(i \times \text{pi})} + 1 = 0 = E^{(i \times \text{pi})} +$ **(time + velocity) = 0**

**Euler's equation refers to the big bang and the creation of *D, T,* and *V*.**

***The big bang was a multiple bang.***

The atom is the standard model for a transdimensional model, which contains a shielded black hole. Within an atom are all trans-dimensions, which are qualifications of infinity. And essentially, the whole universe is a transparent fabric resting in infinity. Euler's equation defines both outside and inside of the universe. It is claimed by science that a singularity was the beginning of the big bang as one force. This could mean that a state of high energy unfolded like an accordion through induction and mathematic ratios to unfold into multiple dimensions, both scalar and proper. But then it could have been a bang of infinite locations since the creation of each *D* (with *V* and *T*) and its definition and order has to be maintained by symmetry implied in Euler's equation from the functions external to the universe that project to create the universe.

It may be that the bang actually preserves all the math and ratios required for the machinery of the universe to operate. And so even in a black hole, the operators of math are still present. And what may be the singularity is in fact the point of transformation of one universe to the next, one dimension to the other, via a process of induction. When we generate a magnetic field with current, the field appears instantaneously, and these are but the manipulation of a selected area of effects, so then imagine an induction that involves the whole universe.

**Would it be by a single operator or an infinite number of operators?**

Well, if we perceive that the universe rests within infinity, then the operators must be infinite. If the universe is a form of consciousness, then also those operators of induction are also infinite. The qualification of infinity is the ratio and form attributed by the induction. The organization of infinite possibilities is the creation of form and still maintaining the relationship to the infinite.

## Free Will, Intent, and Action: Principle of Symmetry

For an action to be given that same quality must have the capacity to be received—this is defined as a principle of symmetry. To love requires something to love and be loved, the receipt of the action. And in this process, the Creator is always available. At the deepest level, intent describes the action. Therefore, if I observe my intent, I know more about the action than what is observed since action is intent fulfilled. Therein lays the true nature of the action. Therefore, some actions done to others might not be acceptable at that level done to ourselves. Free will is an anomaly and is allowed by transcendence. And while there is free will, the universe brings to harmony those things according to its nature. And that process reveals pain because those things allowed by free will are also allowed to err. The universe is enabling its nature or truth. Since the universe enables, that which restricts is contrary to that nature, which we define as truth. Truth is discovered by intuition and can be intent evolved into action so as to be experienced as both causes in thought and effect as received externally. Through intuition and intent, you are here to learn to enable yourself according to the nature of the universe and truth, of which you are a part. The purpose of life is the intuitive enabling of consciousness according to the nature of truth, and this is far from the conquest of stars or evolution. Your intent for the present moment fulfilled according to the nature of truth is divine, the quality of self and being at harmony with yourself and the universe now can be achieved at any moment, age, and any point in history.

**Thomas Aquinas described the universe as comprising of all the rational and irrational beings that have immutable substance in the mind of God.**

If the life is one form of consciousness perceiving another, then truly the universe is manifest by and in the mind of God, and all its diversity refers to the God. In the number 1 is all diversity and hierarchy that represents the universe, and as we have mentioned, the universe is also an analogy of God and refers to the nature of God.

**So are you saying that creation is a thought in the mind of God making a circular reference? He is in his own thoughts?**

By symmetry of consciousness, God projects the universe. He can be in his own thoughts. He is in his thought and transcends his thoughts. Even the story of Christ tells us that. God subjects himself to his own creation. When you dream, you also subject yourself to your own consciousness, but you can always refer to your higher self and transcend the situation from your higher consciousness from the collective source consciousness. Our higher self is actually a part of God's mind also as described earlier.

**What I was trying to get you to say is that we usually (it depends on the philosophy though) separate between persons and particulars or a thing from its nature.**

God is discovering himself and allowing qualities of his being (us) to refer to action and return of action. The nature of God is reflected in creation by symmetry, and in addition to reflecting this nature is also dependent. Hence, there is no excuse about not knowing God by works in creation; because of this, the universe becomes an analogy of truth. We have all had thoughts of God by our intuition, and in this may the greater truth lie. Diversity exists in 1; it is the one that has been diversified and becomes form in the number 3 or the number of form through diversification. And various hierarchies from simpler forms lead to all diversities. Action and return and action must be received, and hence, creation allows that. That is the nature of creation. Love has a counterpart; the action is manifest into effect. This is a mystery of love perhaps and shows the direct link between the intent and nature of the cause and the nature of the effect or created effect in the reception receiver.

**Usually you would not say God is discovering himself in the creation; a perfect God is not going to do that?**

Our sense of perfection and what is allowed about God unto himself are entirely different. What is discovered by God through love and the consequences are creative. Qualities of God are revealed through his will, through which all creation and diversity emerges. Discovery is allowed in free will.

**The creator I think is meant to be viewed more so as some kind of intrinsic reality. Everything just bends around it. You do not need to look at it and go one does such and such?**

By understanding the numbers and science, we are led to the possibility of a God or creator, and more deeply so than most of the myths by which we define ourselves. The search for truth involves asking these questions. One can predict the likelihood of God without religion, and this is not denied by any religion; this is a major step for humanity to abandon many prejudices and myths. From what is experienced from our existence (rational and intuitive), we know patterns of the universe.

**I think there is a major thing to be had in the point God has a different nature to his creation.**

That the universe form is structured on three attributes and those combined lead to the integer 1 and that all integers can be derived from 1 is an amazing thought. All diversity is in the number 1 and proceeds from the number 1, and the manner of that manifestation is the manner of creation. Creation refers to God's nature since it is an extension of consciousness. The nature of the cause extends to the nature of the effect.

**There is a difference between natures and particulars?**

If intent describes an action, the cause and effect by nature are tied together; this is no truer than in love. Intent and action are like cause and effect. This makes the creation of life and life meaningful in an intimate manner as well between the creator and the created. Your experience always refers to the higher self-collective consciousness. Semantics and duality and other modes of contractive thought become the myths that hold men back. Hegel is steeped in the duality of human nature, a philosophy trapped by the very semantics that limits it. Examine the word *nothing*; the idea of nothing is a myth. A philosophy based on the limitations of words and lack of knowledge, not to mention the human tendency to group ideas as dualities is an error from the point of structure of rational thought.

We have talked about the number 1 being balance, wherein all diversity emerges in balance. Also, we have talked about the number 0 representing unity through various modes of transformation. Transformation could be scalar (+) with 0 or proper (×) with *C*. Now with the principle of symmetry, we find that in the equal sign (=) is allowing symmetry.

## True Constants and Prime Numbers

All true constants have no units. There are two types as follows:

- Those that are without units generated from functions outside the universe or external to the universe as defined in Euler's equation. This illustrates the symmetry of functions in the big bang to create the universe while inducing representative constants such as pi.
- Those that are without units within the universe that define proportionalities that can't be broken that relate to both phi and phihat existing in the prime numbers. All numbers in the universe represent functions since they represent units but are held within the limits of prime numbers. An example would be the fine structure constant

and the prime number 137 associated with it. These prime numbers maintain the pattern limits of the entire system universe.

## ***Fine Structure Constant***

***Fine structure constant*** $= \$ = e^2 / (4pi\ \varepsilon_0)\ hc = e^{2c}\ \mu_0 / 2h$

**Fine structure constant** $= \$ = k_e\ e^2 / hc$

$k = \text{kg}\ m^3 / \text{sec}^2\ \text{coulombs}^2 = (\text{mass})\ D^3 / T^2 (D^2)^2 = (D^3T^2)\ D^3 / T^2 (D^4) = D^2$

$h = E \times \text{lambda} / C = E\ DT / D = ET = D^5T$

**Fine structure constant** $= \$ = k_e\ e^2 / hc$

**Fine structure constant** $= \$ = D^2\ D^4 / D^5TV = D^6 / D^6 = 1$

- ***e*** **is the elementary charge**
- $\hbar = h / 2\varpi$ **is the reduced Planck's constant**
- ***c*** **is the speed of light in vacuum**
- $\varepsilon_0$ **is the electric constant or permittivity of free space**
- $\mu_0$ **is the magnetic constant or permeability of free space**
- $k_e$ **is the coulomb constant**

The fine structure constant represents 1 or *D* and means that Euler's equation in creating the universe has ratios of multiple points or barriers that order the various dimensions in place through ratios of limits. So infinity is qualified in the creation of form universe. The universe is first induced and also has form regulators; prime numbers perform that function. Euler's equation induces the universe. All diversity evolves through phi and phihat or time and velocity to form all numbers, and prime numbers establish benchmarks for balancing the dimensions. To review mathematical constructs, let us list the following:

- **1 refers to balance and contains all diversity and numbers. It also refers to the dimensionality of the universe created by the primary superstring *D*.**
- **= refers to symmetry.**
- **0 refers to transformation and unity for scalar systems with +.**
- ***C*** **refers to transformation and unity for proper multiplicative systems with ×.**
- **+ refers to scalar and additive systems such as polynomials and quadratic forms derived through** $V + T = 1$ **or phi + phihat = 1.**
- **× refers to proper systems such as multiplicative products derived through** $V \times T = -1$ **or phi × phihat = -1.**

## $D^{20}$ and Forces Arising within Atomic Nuclei

In Trans Dimensional Unified Field Theory, the space matrix in a black hole is at $D^{20}$. This means the energy is $E^4$ or $D^{20}$ compared to our highest measuring unit of *E* or $D^5$ as current science measures. This means that the forces holding the space matrix together are strong due to a denser and more complex space matrix and also associated with greater mass in a black hole. The identification of mass is in mass space matrix superstring where

**mass** $= D^3T^2$,

and since an atom is a model of a transdimensional universe, then those forces binding the matrix at the nucleus and at the center of the nucleus of an atom are the highest. Also, particles are the interference patterns of the matrix-generating particles and black hole at particle centers and also at the center of the atom. But since these masses of the space are outside our realm of perception and measure, much like the mass of the ether of outer space, which is transparent to higher atomic mass, these quantities do not figure into our systems of measurement. The exception is when these large quantities are unraveled as in an atomic explosion. This also explains the inaccessible dark matter or energy of the universe. But we can figure large quantities with dimensional analysis. We know that all force measured by our measurements acted upon in our lower dimensions is $D^4$. It is relevant what the components is electrical, gravity, or whatever; force components result in $D^4$. We can also figure that through the degree of complexity of the dimensions, we can get the $D^4$ projection of $D^{20}$ near a black hole. Hence, the following is a conclusion:

**Black hole = $D^{20}$**

**$D^4 = D^{20} / D^{16}$**

**$D^4 = (G M M / D^2) \times D^{16} / D^{16} = (G M M / D^2) V^{16} T^{16} / D^{16}$**

**$D^4 = (G M M / D^2) C^{16} T^{16} / D^{16} = (G M M / D^2) C^{16} / G^4 D^{16}$**

**$D^4 = (D^4) C^{16} / G^4 D^{16} = C^{16} / G^4 D^{12} = C^{16} / G^4 C^{12} T^{12}$**

**$D^4 = C^4 / G^4 T^{12}; G = 1 / T^4$**

**$D^4 = C^4 G^3 / G^4$**

**$D^4 = C^4 / G$, which makes the matrix-binding force very high and the string force that overcomes any possibility for the matrix unbinding.**

**Certainty and Uncertainly: Consciousness and the Universe**

***The more we are aware of our higher intuitive selves, the less we are bound by our narrow interpretations of the rules that govern the hierarchies of the life state that we are in since we are transcendent, creating, and participating in the order we experience.***

When you go to sleep and dream, you enter a new world as if it was a new universe with people having their own wills, rules, logic, and events. For the dreamer that has not much experience at dreaming, things seem pretty fixed. So if in the dream I hurt my leg and I feel pain, and if I imagined being a scientist trying to explain what is happening, I would be bound by the order that I apparently experience. To the experienced traveler, in tune with the higher self, that person can know that they are dreaming and even control some of the events and even exit the dream. The traveler or dreamer can be aware of creating, experiencing, and having transcendence. In the dream, however, there still is the complexity and order that is attributed to the viewing of the dream. Now imagine being awake in our normal day and being a young and inexperienced person. As a young and experienced person, you have the experience of reality with all its imagined absolute force, and even all the scientists have you convinced, at least some, that the universe is all that is, with absolute rules; and therefore, what becomes common knowledge, almost as a myth, defines ourselves and all of us together in a false sense of security, but we go with it because we have the support of the group. That's the way it is, you say. Then you have a few, like those we call spiritual, who are in tune with the higher self and see a higher spiritual self and transcendent truths that are an ultimate destination. Perhaps what is experienced here is simply as an experience to learn, and collectively, we have an origin as a higher place or self so that maybe this also could be a dream, akin to Buddha's notion of a higher self or the Christian idea of heaven. Obviously order is maintained as a default, but awareness and free will align to the concept of transcendence, meaning that in the creation of place, we are both subject to its rules but also transcend; and for the highly aware, perhaps some of those rules may even have the possibility for exceptions or to be adjusted, much like a miracle. So we can speak of science of those things that are, and as transcendent beings, at some point, perhaps the science or miracles, perhaps linked to a math that both transcends this would and creates the math we experience. Are the natures of the two, transcendent and transcended, identical? Perhaps *identical* is the wrong word, but they are similar and related. One state of being refers to the nature of the other, but in the exact meaning of being, the operators are slightly different, and so one seemingly operating like the other by symmetry are not exactly identical, hence, the dreamer and the dream. But knowledge is available at all levels because we are transcendent beings. To define consciousness will require a full understanding of the cause consciousness and the effect consciousness or the dreamer and point of experience. So to approach consciousness simply from an understanding of biology excludes the nature of transcendence, although science alludes to the multidimensionality of consciousness since consciousness is attached to reality at every level.

So in the nature of things in the natural order of default consciousness of our awake and everyday life, without exploring the nature of transcendence, there is no uncertainty. With regard to the nature of transcendence and free will, the future and possibilities are open and uncertain. But let's not confuse the physics of the natural order with the transcendent nature of free will. The natural order is created by transcendence and is transcended to be experienced by similar operators that define higher consciousness or the source consciousness. In a sense, consciousness by symmetry creates a reality and then subjects to it or transcends. The universe has order and is altered only by the degree if its participants have a harmonic awareness and harmonic participation with the source, this being the creator or first cause. The first cause has decided to be constant in creation, allowing free will to transcend. Hence, constancy reflects constancy of truth and the certainty of the natural order that is an analogy of said truth. So in terms of free will, things are uncertain. Free will requires transcendence and thus allows free will to be uncertain. But in terms of the created dream state of creation, all things are certain. In physics the argument of uncertainty applied to natural laws is false, and all is certain and those exceptions, which, if there are any, have to do with transcendence and actions that alter this natural order from outside the universe. Transcendent activities originate from outside the universe. Any other perceived alterations are magic tricks related to natural laws. Heisenberg has an uncertainty principle, but based on the mathematics and by using a resulting product via multiplication (×), he claims you cannot know both the position and the momentum of an object. So he talks of probability. Well, multiplication is a proper form (×), and to know both momentum and position, which is certainly to be known, you have to use scalar operators (+) such as in a quadratic or polynomial form, like with addition. So Heisenberg uses a proper form mistakenly to solve a scalar problem. The scalar problem needs to be addressed by a scalar mathematical form, and hence, probabilities can be discarded. So in nature, in its default, all things are certain since the dreamer of the creation, the creator has set the machine to work steadily and without exceptions, which refer as natural laws. Natural laws remain steady because the universe, being an analogy of truth, refers to the nature of the source in constancy of principles that order life and existence. God's truths are eternal and steady. Therefore, we can know the constant and the diverse nature of love and wisdom in all its precepts. To be uncertain of truth is to have no order. The exceptions occur by transcendence in spiritual experiences common in the literature of many religions and are allowed in such a way as they still refer to the constancy of truth. These experiences or miracles are allowed in as much as the higher self of the participant and source consciousness or creator are in total harmony almost to be perceived as acting as one. So what might be perceived as an anomaly or change is merely action according to its own nature. The matrix allows those things through it according to its nature. Those things against its nature are allowed within the mechanics of consciousness with pain, a self-correcting mechanism; those are not in the nature of exceptions but are as a general rule reflecting the nature of intent in action, good or bad.

**Shielded and Unshielded Gravity**

The equivalence of gravitational and inertial mass is so because they are part of the same transdimensional system. Mass generates space or, more precisely, the inertial field or space surrounding an object is a lower-dimensional extension of the more-dense space mass core of a higher dimension. Therefore, an atom is the ideal model of a transdimensional event, with the space matrix packed at center of the nucleus and stepping out beyond the orbiting electrons in a less dense and lower dimensional fashion to create as surrounding inertial field. So from the black hole within a nucleus of an atom, we move trans-dimensionally to the inertial field of outer space. So under the model of an atom, as we measure the mass, we measure the gravity since they are proportional and say they are equivalent. And therefore, Newton has

$$F = GMM / D^2 = D^4.$$

The $G$ actually refers to

$$G = 1 / \text{density1} \times \text{density2} = 1 / T^4$$

where the density refers to the <u>mass of the core AND the mass of the inertial field.</u> But imagine that the inertial field exists in approximate ratio to the core mass, then what is a variable is now lost to a constant $G$, but inaccurate.

**Now if gravity is shielded, does that mean that it violates the principle of equivalence?**

It means that you don't have all the equations to understand how force or $D^4$ can vary in the various dimensional hierarchies related to gravitational force or force.

**Note that**

**$D^4 = MA$ (force) and**

**$D^4 = GMM / D^2$ (force of two inertial fields acting as one) and**

**$D^4 = C^4 / G$ (strong force) are all the same $D^4$.**

Imagine a black hole at the center of an atom, and then imagine a black hole at the center of the galaxy. They are different in that the atomic structure has the order of neutrons, various zones of space, and then orbiting electrons followed by more space. The transition through dimensions from dense to less-dense space in an atom has an order that results in a gravity of **$1 / R^2$.** And this we measure in the solar system and refer to it as Newton's law and assume that it is universally applied to the entire universe. Now imagine a black hole without these transitions at the center of the galaxy. Pioneer has shown gravitational anomalies as the satellite left the solar system. NASA claims there are departures from Newtonian gravity, and NASA has maintained that when exiting the solar system, Newtonian rules no longer apply. This theory maintains the likelihood of a 1/R relationship when leaving the solar system to both the Sun since you left its inertial field, and also to the center of the galaxy. (Faith: Web Link)

The Newtonian gravity explains two inertial fields united as one. What happens when

- **adjacent fields are not united? This theory maintains $1 / R^2$ relationships in overlapping $1 / R^2$ inertial fields acting as one.**
- **the black hole is not transitioned as in an atom or shielded?**

Then the rules of our solar system do not apply universally, but the equations we use are part of a universal system. Hence, the concept of gravity and how the source is shielded or unshielded and also the involvement of the inertial fields will change the relationships depending on variable conditions. There is no one universal condition to view gravity. It can be manufactured in a lab by various methods. Antigravity is an inverse relationship to gravity and can be achieved by various methods. Space expands according to the ratio **$D^5 = MC^2$** where $D$ is the radius of the extended inertial field.

**Pattern, Randomness, and Free Will**

To be without pattern is to be random. It infers concepts such as chaos and statistics that tries to resolve the absence of pattern issues, to mention a few. So not knowing the pattern, we resort to methods of probability to determine possible outcomes and then, in a sense, look for a probable pattern and thus look for organizing systems outside the obvious. The universe is pattern that is both finite and infinite. Math allows what may appear to be random within a system of pattern. Hence, where the outcome is unknown as in free will, the operators that allow free will also reside within an infinite system of pattern. The concept of chaos is a myth although infinite patterns allowing infinite variables may appear to be random; actually, these are within a system of pattern and are unity. The universe is originating from the number 1, from which all hierarchy and diversity emerges, and still from within such a system, random patterns are allowed that refer to the unity of the entire process. So then is what is random exactly? It is the future of what has not been added to the memory of the collective conscious as a transcendent activity and is allowed by free will. Once in the memory, it becomes engrained into a pattern. So then randomness is the option to select a pattern or to develop pattern in a creative sense. We have referred to the action of the creator as an act of love in creation in the development of pattern. Hence, for creation, free will has purpose and is not random.

**In nature or the natural order what at times seems to be random is highly ordered.**

Even referring to the placement of trees in a natural environment and what we might term natural or random way has recently been found to follow mathematical precepts. So nature, in all its perceived randomness, is adhering to forms of pattern, which reveals the truth about creation and the first cause or creator God, that the love that caused the universe is of a value according to the nature of constancy, according to values that, from that constancy, we derive such constancies such as wisdom and virtue. Hence, while allowing free will, the matrix is constantly referring to its own nature, which is constant, stable, and reliable. Many activities we assume to be random are not. Such randomness was explained by John Nash in many instances to be based on pattern. There is free will with pattern, and the word *randomness* refers to the limits of human perception at an undiscovered pattern that awaits investigation. So my action now has pattern, and the future is waiting for a pattern or equation for me to select, and in this selection there is free will and the randomness that we allude to. Hence, running through all activity is the thread of unity.

## Square Root, Induction, Multiple Dimensions, and Measurement

### Parallel Space and Universes

Time and velocity are the same; however, time is an inversion of velocity. Together, time and velocity form *D*, a component of space. The ratio of phi and phihat, which also refers to velocity and time respectively, also sets limits of the proportion of energy at the quantum critical level. From time (*T*), velocity (*V*), and dimension (*D*), we view that there are only three variable dimensions that form other dimensions and properties of the space matrix. There is a need for only three units that refer to the qualities of three primary superstrings.

The keys to understanding time and velocity are as follows:

**Time + velocity = 1 = *D*, and**

**Time × velocity = -1 = -*D***

The + (addition and subtraction) involves computations of all scalar forms and results in all quadratic and polynomial mathematical forms. This equation

**time + velocity = 1 = *D***

is a template, like a genetic code.

The × (multiplication and division) has to do with all proper forms like $F = MA = D^4$. That form comes from

**time × velocity = -1 = -*D*.**

These systems **(+ and ×)** combine to result in all the equations and complexity of the universe.

For instance, all force is $D^4$ such that

$F = MA = D^4$

$F = GMM / D^2 = D^4$ = **gravitational force**

$F = C^4 / G = D^4$ = **strong force**

And now, try electricity and whatever you wish to find new combination that get $D^4$. You will find the journey rich in the discovery of new relationships by transdimensional analysis.

So what is the universe? Once called space-time, what about space-time$^2$?

$T^2$ = **density**
$G = 1$ / **desnity1 × desnity2** $= 1 / T^4$

Time is many things depending on complexity of a dimension and involvement in those dimensional hierarchies by the combination and compression of superstrings. At the most basic level, time is the inversion and balance of same velocity. The same thing is also manifested with a certain attitude in *D*. The 1 is from what all diversity comes. There is asymmetry between time and velocity, but together they form other symmetries. So we have a universe of multiple dimensions, but we measure and perceive in what is considered the lower dimensions. The dimensions are related by induction, and some relationships of higher dimensions to lower dimensions carry the value of square root. For instance, we know that the calculation of force is affected from various hierarchies. What results as a force from the higher dimensional strong force is different than what results in force from the lower dimensional gravity to the $D^4$ where we are. We can see that in the numerator for gravity, we have as many as ten variables:

$$F = GMM / D^2$$

$$F = (D^3T^2)(D^3T^2) / D^2 T^4$$

And for the strong force, as many as twenty as a starting point in creating the equation

$$F = C^4 / G = D^4 \text{ from}$$

$$D^4 = D^{20} / D^{16}$$

$$D^4 = (G M M / D^2) \times D^{16} / D^{16} = (G M M / D^2) V^{16} T^{16} / D^{16}$$

But the twenty dimensions of the strong force distill by simplification to $C^4$, which relates to the tenth dimensions (gravity). The measuring via the constant $C^4$ is somewhat the square root of $C^8$ at dimension 20. So what we measure at the lower dimensions regarding force can be a projection of a higher dimensional event. Events in our experience also refer to events in all other dimensions, since by induction and translational relationship, all things are connected. We call this transdimensional. We have defined space in terms of its superstring matrix with elements of spin and charge and classifying space by the condition of superstring operators of spin and charge. Also, we can classify in a broader and more general sense the net result of this variance in both scalar and proper forms since we have two conditions:

- ***-D* for proper form**
- ***D* for scalar forms**

***Together, they are space.***

Hence, the properties of space can be viewed in entirety as a species governed by various hierarchies and classifications. Space can be define as

- **scalar and proper in relationship to + and - for the overall space or net effect of *D* and *-D*, and**
- **then the level of observation can be broken down to the matrix involvement of superstrings by spin and charge related to + and - and neutral.**

**Why do we bother with this diversification of space classification if we are looking and examining one type of space?**

Classification leads to the understanding of other types of space related to parallel universe space and anti-space related to antimatter. A type of space is induced by symmetry from functions outside the universe; along with this space comes undetectable parallel spaces that mirror qualities of the space we are in and support its existence through symmetry.

**<u>Reverse Polarity and Inverse Relationship of 1 / *X*</u>**

Antigravity is as an inertial field generated by the reverse polarity or relationship of 1 / *X*; this is also called opposite polarity. Antimatter or negative mass goes beyond opposite polarity to be totally opposite genre of space or quadrant space of parallel supporting space, meaning rather than opposite in polarity in the <u>relational construction</u> of adjacent dimensional space, opposite in both spin and change of superstring structure and also in geometric configuration. So antigravity is an "inverse" relationship of the model of the atom in the generation of space or 1 / *x*, so the traditional

$F = GMM / D^2$

**Becomes**

$F = D^2 / GMM.$

If force of gravity is $D^4$ or force, then an anti-force of antigravity would be

**anti -** $F = 1 / D^4$ **or**

**anti -** $F = e^2$.

This is the equation for reverse polarity.

Anti-gravitational force becomes $e^2$ or permittivity of free space, which is affected by dielectric material, such as in capacitors. Permittivity is the ability to allow an electrical field to maintain its natural polarity. So amazingly, there is an unexplained mystery with the Biefeld-Brown antigravity effect using capacitors. Reserving the polarity of the permittivity of free space allows gravitational motion along the gradient of the generated space matrix to reverse. (Bahder: 1-34) The force exerted on an asymmetric capacitor is given by the rate of change of momentum as follows:

$F = MV I / q$

$F = (D^3T^2)\ V\ (VD) / D^2 = MV^2 \times D^3 / D^4 = E \times \text{volts} \times e^2$, **where** $E$ **is energy and** $e$ **is permittivity**

$F = Q \times \text{volts1} \times \text{volts2} \times e1 \times e2 = 1 / e1 \times e2$

$F = 1 / e1 \times e2 = Q1 \times Q2$

**For antigravity, the expression of** $D^4$ **would become** $1 / D^4$**, and hence,** $1 / D^4 = 1 / E \times \text{volts} \times e^2 = e^2$**.**

An inverse relationship has to do with the reverse relationship of dimensional space or dimensions of natural inductance. An example would be positive flow around negative charge, which is opposite the model of an atom, but there are other methods as well involving capacitance as demonstrated. The relationships of natural inductance are reversed in the inverse relationship of opposite polarity. The gradient of space of an adjacent field is reversed and repulsed.

## Cold Fusion and Superconductors

When the matrix of space at a dimension exhibits elastic collisions, velocity, or vibration characteristics at the speed of light, we term that a superconductor. Hence, outer space at 0 Kelvin can be considered a superconductor. When the matrix allows this at normal temperature that can be termed a room- or high-temperature superconductor. If the space matrix medium is configured correctly to allow elastic collisions, the velocity *C* is part of the matrix. In this manner, it would be possible to predict sound at the speed of light under the right conditions, and in fact, this experiment has been accomplished. Now, if we assume that we can associate the vibration energy with particular dimensions and also reaction types, then we can safely say that nuclear fusion at 6700°K is a reaction the occurs near $D^{20}$. Burning paper is at 500°K. And room temperature at 293°K would be the range for cold fusion and high-temperature superconductors. So we see that various reactions and transdimensional states are associated with various temperatures or $V^2$ (velocities). The question becomes, can the higher velocity states of higher dimensions be configured in a matrix that allows for a lower temperature? Or more precisely put, can you get elastic collisions and associated chemical reactions occurring at temperatures that you don't normally find by manipulating the matrix? Hence, cold fusion is a possibility much like superconductors at room temperature.

In 2007, a high-temperature superconductor was produced at 260°K with hydrogen inside a palladium lattice. Here the revised matrix allows elastic collisions to occur at a range of that required for cold fusion: velocity with a revised associated temperature. This same relationship of palladium and hydrogen is reported in the cold fusion experiments; perhaps the answer to cold fusion is in the manner of creating high-temperature superconductors that will allow nuclear reactions to take place outside their natural mode. A matrix can be created that allows the velocity required for a nuclear reaction that is outside its natural environment. Normal fusion reactions occur at high temperature (6000°K) at high velocity since energy produced is proportional to the velocity squared of a chemical reaction since temperature is $V^2$. Therefore, smaller amounts of energy that might be produced at normal temperature near room temperature at 293°K and a much-lower velocity mean a much-slower reaction. So therefore, you would be looking at smaller amounts of energy if at all possible. Higher velocities with same lower temperatures would be special conditions of the space matrix, as in high-temperature superconductors and cold fusion. How that could be manipulated for power generation does not seem as efficient as using nuclear power, but another possible avenue in the generation of energy. (Bailey: 1-12)

## The Myth of Nothing

**If I change, then I am not myself—yet if I remain the same, then I cannot experience anything . . . Either way, I do not exist.**

Life is one form of consciousness perceiving another. For an action to be given, the medium wherein it resides must allow same to be received, and hence, the universe is an alive and conscious being, also has been shown here that by symmetry (=), numbers are created from numbers, form is created from form, consciousness is created from consciousness, and from one universe another is created. And so the universe is induced from the outside, much like our experiences here, and also extending to our thoughts and dreams; and so we are of and also transcending this conscious experience and others. The universe is an analogy of truth. And the truths that can be derived through numbers that also represent the form of the universe and consciousness means that they refer to the structure of truth, and also the universe is an analogy of the truth since it is derived from the source matrix by symmetry. Form represents the organization of various states of motion; this involves hierarchy, symmetry, and the relationship of motions at the most basic level since the number 3 is where form becomes manifest in motion.

### What is motion?

The 1 from which all is manifest is self or "I am," and from that and diversification through to complexity is motion. So time and velocity are both motion of the same thing, different in attitude and intent, involving symmetry and asymmetry, and expression ad infinitum to include all diversification and possibilities. Self-become free will and thought, the 1 is diversified, "I am," through diversification manifests form and hierarchy (symmetry and asymmetry) manifests motion. Self through form becomes motion in transformation, unity, balance, symmetry, asymmetry, infinite, and ratio. And in this act through motion, all is created, and that which is to be and receive according to its nature. Through love by symmetry creates that which can be received according to its nature and allowing free will since all those things according to its nature need to be allowed to be truth, the created being an analogy of the creator. Time and velocity are both motion manifest from balance and unity and 1 and 0 respectively. Self, the 1 and 0, and from the 1 all diversity emerges unity and transformed by zero to accomplish all things according to free will toward infinite diversification and expression of love. So change, therefore, is transformation and true to the nature of self, being derived from self, and what remains the same is what is willed to be and also refers to the source identification of self. Nonexistence is a myth since 0 refers to transformation and unity of what is, allowing both change and self and experience and existence. Transformation means moving from one absolute present into another.

**What is between change and unchanging?**

What are unchanged are the things from which all is made, and what changes are all the things made from that which is unchanged. The thing is in fact something since nothing is a myth. The something referred to here is awareness. "I am." The only truth is the source consciousness. The motion allowed by the various aspects of consciousness is change. Form is the arrangement of various acts of motion. The identification with self is changeless and those things that are willed to remain.

**If time and/or space are particulate, then how can motion occur?**

Time and space is made from velocity and time. Both are functions and allow transformation. They are asymmetrically balanced within a symmetrical *D* or 1 of space. Balances represent self, *I*, or 1, and in that unity is developed other forms of hierarchy and symmetry resulting in form consciousness. By symmetry, one reality is created from another where = means symmetry. The number 1 means equilibrium or balance. The 0 is unity, transformative and uniting. Powers and square root are inductions of multidimensional states, one from the other. All is infinite or what is, form is the alternation between infinity and interval. There are no absolute boundaries, only hierarchies defined by prime numbers or transitional boundaries that become form and define ratios or relationships of form, a qualification of the infinite, and also relating motion in free will and thought.

**If a thing is finite, then "not a thing" is infinite. "Not a thing" is nothing, so nothing is infinite?**

Form is finite and infinite, infinite in the range of proportions, and infinite is capacity to transform and form and act and finite in truth, those things that refer to the very nature of the source consciousness, regardless of the free will infinite allowed. Finite in balancing ratios, hence, we can actually refer to the constancy of truth and wisdom.

**The main error in the definition of metaphysical ideas is a faulty premise, and with this faulty premise, we reach a faulty conclusion—or rather, no conclusion.**

Any action given must be received in the way of existence; the universe is alive and is the mind of God and is a conscious state. The universe speaks about its conscious state and truth, which can be known by all, as an inward journey. So what you see in the world is not you, and what you is up to you.

## Nature of Mass

Space has constituents (superstrings) that form all dimensions of space, which includes space at higher dimensional densities and configurations into atomic matter and its various transitions and dimensional states and also lower into one- and two-dimensional structures; this presents a problem about what the broad nature of what mass really is. We have said that an atom is the perfect transdimensional model, and in all its dimensional states, it is made from the space matrix with mass at these different dimensions. Since what we measure is within dimensions 1–5, much of what is going on in an atom is measured as a perceived effect of a dimension rather within that dimension. And at almost every dimension, there are two-dimensional and even one-dimensional components, two dimensional such as spin. These functions may appear as constants from our perspective dimension of measurement, hence ½ spin, etc. And what is dark matter would be basically those components left out of our scope of measurement; those components within our dimensional view and beyond that are not measurable since our definition of mass is related in a narrow way to a few dimensions of atomic mass configurations. So then if we redefine mass in terms of components' superstrings, we will find that it includes not only various dimensions of core mass space but also those components that make up general space or the inertial field matrix, which is also an extension of atomic mass and vice versa. Here I have defined mass as $D^3T^2$ superstrings, and hence, there is much more mass than that which is measured, even to include the mass in space or the inertial field. So the total mass of an object includes the core space mass as well as the general field space mass of the inertial field. Hence in $\boldsymbol{D^5 = MC^2}$, $M$ or mass is calculated as mass of object plus the mass of the inertial field, so then the two now are viewed as one object, the transdimensional field of a source dense-space matrix or space-core mass. The core mass generates the space mass or inertial field to distance $D$ radius.

## Induced "Effects" Trans-dimensionally from Higher Dimensions

In a transdimensional universe, what scientists measure is anything from $D^1$ to $D^5$, where $D^5$ represents $\boldsymbol{D^5 = MC^2}$. As we approach $D^{20}$ and the strong force near a black hole, these actions have their apparent effect at $D^4$ where we can measure that effect. So all force as $D^4$ represents actions occurring at $D^4$, which may be originating from a higher dimension. The strong force originating at $D^{20}$ has an effect at

$\boldsymbol{D^4}$ **= force =** $\boldsymbol{C^4 / G}$**.**

And moving toward lower dimensions, we also get an induced gravity force as

$\boldsymbol{D^4}$ **= force =** $\boldsymbol{GMM / D^2}$**, where** $\boldsymbol{G = 1 / T^4}$ **and mass =** $\boldsymbol{D^3T^2}$ **superstrings.**

$\boldsymbol{D^4}$ **= force,** which means, as one could say, large mass and energy is zipped up; but rather than saying it that way, we can say that some dimensions are transparent to others or indirect, and what is measured is what is ***induced*** trans-dimensionally toward lower dimensions.

## Antigravity Revisited

Under general relativity, gravity is assumed to follow the shape of space around massive objects. Here in TDUFT, gravity is not a two-dimensional phenomenon as illustrated in geometric models where the deformation of space equates to a true force. Under the GR model for antigravity, the deformation would need to produce a negative force, and hence therein lays the failure of this geometric model. Rather than antigravity being a negative force ($-F$) based on a negative geometry and negative mass, it is the result of a model of reverse polarity ($1 / X$) of inertial fields opposite to their normal adjacent relationship of induction, reversing the model of induction of the space matrix. Also, the general relativity model proposes space-time, when in fact—like Newton in his "time" in velocity with momentum—when dealing with energy, $T^2$ is involved in $V^2$ in the conservation of energy. Here, $\boldsymbol{T^2}$ **= density**, and thus gravity in space-time squared should refer to space and gravity as resulting from the dynamics of shifting or changing density. (Sankar: Web Link)

By induction, it may mean that one dipole moment generates another or one dimension induces another. Hence, an electron moves and generates a magnetic field. So if the magnetic field is a dipole, the electric position must also be a dipole, which confirms the symmetry and dimensionality of induction within space. (Jacklewski: 1-88) And also that anything with space has mass and all things have space, except it may not be the space that we move through or that space "effect" we are used to as the one we move through. And what is mass? However, we wish to define the collection of component superstrings. Here we are using

**mass =** $\boldsymbol{D^3T^2}$**.**

Reversing the geometric model of density requires a $1 / X$ relationship, which is a polarized relationship rather than a negative one, but in a quadrant, that is not the anti result nor is it the negative result. The negative mass and positive mass, it is said, reproduces a negative force in

$\boldsymbol{-F = G(-M)\,M / D^2}$. But note the error

**$G = 1$ / density1 × density2,**

and if you address $G$ by placing density1 with a "-", you are back to a positive result. So the "-" field is a type of mass, although it may exist in another parallel quadrant and related to or defined by the internal spin, charge, and geometry of the space matrix, which results in the same pull of gravity. Because

**$F = -M1 \times M2 / D^2 \times$ -density1 × density2 = $+F$,**

**so note the misunderstanding of $G$.**

When discussing gravity, two inertial fields are acting as one, or they are coupled; obviously in a $1 / X$ relationship, this is not going to work where you have two inertial fields opposing each other. In space, voltage is high and current is low; by revising the Newtonian equations for gravitational force we get

**$F = G MM / D^2$**

**$F$ = mass1 mass2 / $D^2$ density1 density2; $G$ = 1 / density1 density2**

**$F$ = volume1 × volume2 / $D^2$; volume = voltage**

**$F$ = voltage1 × voltage2 / $D^2$.**

By transdimensional analysis, we have revised $G$ to functions and used the relationship that

**$D^3$ = voltage.**

The static and inert nature of space is irrelevant, and the imbalance in the densities of inertial fields causes flow or gravity, much like static charge or charge imbalance. Charge imbalance is also a time imbalance since charge is related to time. What we perceive as the net effect of space, when broken down, has components of density, spin, and charge; and thus the interaction of these variables in two inertial field voltages acting as one draws them together. The ether is not made from electrons but contains the same dipole action that is composed at higher dimensions to create electrons and other particles. So the ether is electromagnetic in nature, transparent to higher atomic mass, yet acting within its own field, much like all field activity. So on its own terms, shifts in density of the inertial field attract according to governing voltage differences.

## Free Will, Random Activity, and Transcendence

For those that believe in the myths surrounding death, they give similarly to the universe unending purpose in its myth of randomness. Centuries ago, it was felt by artists and argued by some philosophers that random activity is at work in the universe. Well, this is pure illusion. All action follows laws. Since there is more than one law governing the multiplicity of actions, this may appear illusory, but in fact, the laws govern all natural activities. It is true that the pattern of mountains and also the manner in which trees grow in the forest follow fractal patterns, and these can be explained mathematically. Even the manners in which tree branches divide as they grow follow geometric patterns. The universe is a machine. Constants that appear in the universe, which are the true constants without units, either come from the outside caused by functions on the outside or from inaccessible dimensions from within. Realizing by a process of symmetry and inducing things are created and related, the path is far from being random. The universe is induced, much like ideas in thought. And from the point of creation and the experience of it, it carries all impact and energy, but from the outside, as a dream or constant. Hence, the universe, with all its structure and energy, is purely a state of consciousness, a dream. Randomness has its place in transcendence and free will. One set of operators is introduced into another, yet each refers to the source consciousness; hence, there is constancy to truth that transcends any possible creation. Operators that are unrelated introduce random results. If the operators exist together and came to be together, then randomness does not exist. Hence, there is nothing random in the universe until forms of life showed up. And since the universe creates things according to its nature, the set of operators that represent life (free will and randomness that represents life) are foreign to the mechanism, and hence, life is transcendent. It is free will that allows transformation and the allusion to evolution. If I am naturally selecting, I use the randomness allowed by free will as a transcendent activity. There is knowledge that goes beyond knowledge; the universe being an analogy of truth, it incorporates certain truths according to its nature that allude to something beyond its own existence by finding similarities with our conscious existence. In and of itself, in the universe, if those things created within were totally bound, the concept of relationship to something else would be nil. And since nil is a myth and that transformation rather than evolution is the norm, we can extrapolate to other modes of being. The purpose is the present since there is no goal; a process of induction cannot devise patterns greater than itself. Hence, transcendence is akin to extrapolation of information. Transcendence means one set of operators are external to others that exist and that are free will and substantiated by mathematics. We know already that two separate systems are at work:

$T + V = 1$ **And**

$T \times V = -1$

This is not evolution; this is purely mind, and so the complexity that ensues is a derivative of thought.

**Do you believe in ancient astronauts?**

We are aliens, and so are aliens, aliens. All are aliens. The term *ancient astronaut* is just another way of asking if you believe in life on others planets. *Yes, there is life in the universe. What was that, a mystery?* There should not be a mystery about the realization of other life-forms in the universe. We on Earth are on the outside of our galaxy and are not on the alien trading routes.

**Do you know where the alien trading routes run?**

Well, they probably run next to star clusters in this galaxy, maybe like the star system Vega or Belt of Orion.

**Do you think nonhuman aliens inspired human religions?**

Maybe yes for some, but not for all. I think that truth is open to any being. There is no exclusive club on the truth; however, some groups might argue such.

**Which human religions do you think are closest to the truth? How is the truth represented in religion impacted by humanity?**

All religions are impacted by human culture or myth; so much so that the myths dominate the interpretation of the truth. So then it's a real battle as a personal journey to start from an untainted beginning of understanding of truth with objectivity. We must unlearn the myths and relearn truth with intuition, reason, and objectivity. And what is truth? It is an exploration of self and the universe, axioms which speak to us with constancy through all states of being or consciousness.

**How do you define myth?**

Myths define what we do not know, and more than that, they refer to limits in semantics and human structures in logic. They also extend to historical traditions and political biases. Think about human language and logical forms. We have many dualities such as good and bad, up and down, that riddle our logical forms. Not to say that it's entirely wrong, but language is limiting. And then you have the various methods of logic from various cultures and their superstitions, traditions, etc., which also are limiting.

**Are you saying all dichotomies are limiting to finding the truth?**

Not all, but you really need to learn to use your mind objectively. And realize that when someone says something, it's more about how they think than what they said, how they used language rather than the information. There is the information heard, the manner of process thought, and how words are used to reveal it, which reveals more about someone's true nature, intent, and state of being.

**Why did you speak of the truth?**

Ultimately, the universe is an analogy of truth, and it can be found. But that takes disciplined reorganization and evaluation of thinking processes.

**Can you give an example of someone saying more about how they think than what they say by saying something?**

Let's say an artist looks at nature and feels the randomness, uncertainty of events. This is an old idea, maybe even someone two thousand years ago might agree with. This is almost set in stone by all cultures. Art equals or is equivalent to randomness, unpredictable events. Now, we have science with laws that describe how things behave. They used to think that mountain forms were random. But recently, they have discovered that tree patterns and natural forms also follow mathematical patterns. Nature left to its natural course is mathematics. *So then I ask, what is random?* It can be that if all natural events follow mathematical forms, then if you have hundreds of laws combined, it looks random. That which has but shows no rules to man seems random but is not. Random in the natural order is an illusion. But then random has a place where operators introduced are foreign to the system that exists, and that may be with all life-forms. Transcendence is free will, is random and the universe with all its natural processes is ordered, but with an order that we see variability that is not random.

**You say there is a sort of random that follows no rules, that is phenomena subject to free will, and there is the mundane random with rules that man cannot see (yet)?**

*Yes.* So free will is random, and nature is not. We have a very concrete definition of free will. There is a larger lesson and, that being the awareness of the limitations of information, the limits of logic and language. Be very careful and review what you say and how you process it. There is always an exception, and that is a doorway to refine truth. You can say intuition over mind, over nature, but they need to work together in harmony—harmony between higher self and various states of consciousness.

**Revision to Newtonian Gravity and Quantum Gravity**

The Newtonian equation for gravitational force is describing one inertial system that can be subdivided into two masses with their respective inertial fields. This wasn't the original notion of gravitational force where two objects are drawn together as a function of the mass. Transdimensional theory revises the understanding of the original Newtonian equation:

$F = G\, M1M2 \,/\, D^2$

**The new revisions in the understanding of the force gravity equation are numerous as follows:**

- **Mass or *M* is now the addition of core mass or concentrated atomic mass and the lesser dense surrounding space generated or inertial field. So all mass objects as traditionally understood generate the space of the universe.**
- **The distance of the inertial field projection is *D* coming from the equation $D^5 = MC^2$.**
- **$G$ is revised and divided to represent a portion for each mass. $G = 1 / T^4$, where $T^2$ = density; hence, $G = 1$**

**/ density1 × density2. *G* represents density of the entire inertial field for each mass. This allows the gravity equation to be simplified to *F* = volume1 × volume2 / $D^2$.**

- **Gravity as the behavior of one system is resolving differences in density between two inertial fields, and the balance the results draws masses together to average out the differences. Differences in density cause motion toward balance.**
- **Any constant with units is in fact a variable, so *G* is not a true constant but an approximate averaging figure that has managed to be used as an approximate solution.**
- **Newton's gravity represents a particular solution for gravity at 1 / $R^2$ or 1 / $D^2$ where the inertial field is generated at distance *D* from the core center in a shielded system. Shielded means that higher dimensions are shielded by dimensional transitions from center of atomic mass. This is in contrast to an exposed black hole–generating space to $D^2$ with a 1 / *R* relationship. So Newton's gravity is known as a standard within the solar system but should adjust to varying circumstances outside the solar system.**
- **Gravity is affected by the changing in density of the inertial in three dimensions.**
- **Gravity is affected by the spherical wave generated within the inertial field that causes a standing wave or gyroscopic effect holding the planets in an ecliptical plane.**

So we have summarized many revisions to how the Newtonian equation can operate and is adjusted according to new understandings and circumstances. We have also redefined *G* to actually simplify the Newtonian equation to

***F*** **= volume1 × volume2 / $D^2$ and other simplified forms are possible.**

***This means that force is dependent on the ratio of volumes between inertial fields.***

Now at quantum level, we would be reviewing gravity and its operation at the Planck scale. The very revision to the equations at the macro-scale is possible by understanding the components at the Planck scale. All relationships in physics are determined by the geometric arrangement of components' superstrings of the space matrix. The organization of superstrings is equivalent to mathematics; scalar and proper mathematical forms arise from these arrangements. Since there are three primary superstrings, then all physical relationships can be accomplished with three units, which are *D*, *T*, and *V*. It is possible to revise Newton's equation using transdimensional analysis and describe the same gravitational equation in various forms. We find the expression for force at the level of two masses acting in a gravitational field takes many forms. These mathematical forms, reduced to three units, are also expressing the relationship at the quantum level. Hence, we simultaneously bridge the macro and quantum level, understanding with the same equations and unified the field. We now can view gravity expressed with magnetism, electricity, and a whole host of alternate components by which the same *F* is derived. For instance,

***F*** **= voltage1 × voltage2 / $D^2$ = $GM1M2$ / $D^2$.**

As we move into the atom, the same force of *F* or $D^4$ coming from higher dimensional relationship will get the strong form at the center of the nucleus of the atom. This is an unshielded relationship of gravity now and, yet again, in a denser matrix and higher dimension. This higher magnitude force of the strong force is given by

**$F = C^4 / G$.**

So the *F* related to gravity exists differently at different dimensional states and translates always to $D^4$ in our sphere of measurement and perception. Gravity represents the binding forces of the space matrix throughout all dimensions, and hence, it is hard to explain it as a one-dimensional effect since transdimensional relationships concerning gravity occur through all dimensions. We are left with the magnitudes of force stepping up as we move through to higher dimensions. Should one can call all $D^4$ (forces) gravitational? Or are we to call all binding forces? Let us begin with keeping some tradition and call gravity at *F* as one system, and then at higher dimensions say there is a relationship to induction, that the gravity of higher dimensions is qualified to those dimensions, but maybe gravity prime or $D^{10}$ and $D^{20}$. It becomes more obvious that there is a metamorphosis in what we measure and how that relates to higher dimensions. And so the question of gravity at the quantum level needs to be qualified dimension per dimension. Gravity is transitional phenomenon, and so are time, dimensions, and velocity. Moving through dimensions and at higher densities and geometries, with changes in spin and charge, properties actually change for time, dimensions, and velocity. Properties for *T* are vastly different from $T^2$ **= density**. And *F* for the strong force is different than *F* for Newtonian gravitational force both being $D^4$ but referring to different dimensional relationships.

Speaking of the component variables of *D*, *T*, and *V*, an imbalance in charge represents an imbalance in time. It may be that the transdimensional induction is due to one dipole moment generating another, and as such, dipoles may reinforce coupling locally and cause acceleration in fields with a density gradient. The forces at action within the inertial field are transparent to observation, just as the large mass of such fields are as well unless excited by an elastic wave. Space ($D^3$) acquiring $T^2$ or density becomes mass and, additionally with $V^2$, becomes energy. (Rushchanskii: 2010)

When you reverse the polarity of an electrical field, you reverse the current; this is in line with reversing the flow of positive particles and negative particles, flowing in the opposite direction to each other. The standard mention of current as positive is only intentional. Our current understanding is that current flow follows voltages along positive and negative gradients. The space matrix or inertial field is influx similar to current flows and also transmits elastic waves. Reverse polarity also means reversing current. Hence, counter to the normal flow of the inertial field concerning gravity would be antigravity. Charge imbalances due to density gradients within adjoining inertial fields acting as one will cause current flows to establish or seek internal charge balance. In current two operators

- **we have the motion of the matrix, and**
- **the elastic collision or vibration of the matrix, which is *C*, compared to a much slower motion of the matrix.**

So when gravity is felt between bodies, this has to do with the elastic wave through the matrix at *C* and not the motion of the matrix, but the motion of the matrix is amounting to gravitational force *F*.

In gravitational force, the orientation of inertial fields is such that the two fields are drawn together due to voltage imbalances; imbalances in the dipole charge distribution of the space matrix along a density gradient, which would also be imbalances in time. Hence, we can view the quantum nature of gravity as current flows where the effect of positive and negative flows are travelling in the same directions to achieve a neutral balance. This seems at odds with our notion of electrical current where negative poles that are adjacent repel, and where there is positive and negative, we find opposite current flows. So current and voltage is behaving differently in gravity. We must consider under all models that the net result *F* is + since in reality,

$V \times T = -D = -1$ **and**

$D^4 = \text{force} = +D^4$,

and the squaring of negative -1 gets a positive +1, and for energy at $D^5$, the result of squaring is -1 or $-D^5$, so perhaps the nature of current is reversed when the origin of the force from a higher dimension is a positive dimension. And contrarily, the force operates differently with regard to matrix when the force *F* originates or is induced from negative dimensions. Remember, what we measure is *F* force $D^4$ as an effect induced from higher dimensions. How that force operates depends on the source of the generating field trans-dimensionally at the higher dimension. So then we will have positive forces $D^4$ operating according to positive and negative space matrix parameters. Gravity and electricity have seemingly different processes, which may seem contrary by conventional standards; gravity and electricity will have multiplicative and divisional matrix effects respectively where current flows have the opposite effect. Remember that with the Newtonian gravity model, we are dealing with proper systems where $\boldsymbol{V} \times \boldsymbol{T} = -1 = -\boldsymbol{D}$, and this has an impact of the nature of gravity. And so the nature of current may change or reverse as we go from negative "-" to positive "+" products of dimensions at $D^n$ (positive or negative) in proper (×) systems.

Like voltages repel, and unlike voltages attract. And in the perceived static or inert nature of space (although with regard to spin, geometry, density, and charge there is a distinct pattern) and with the gravitational pull, one would expect that if space is considered a voltage from the core mass to the outer limits of the inertial field, then the planets would repulse instead of attract. But they attract. And when they unite, the voltage and inertial field is even stronger and larger respectively. Perhaps in the model for current related to gravitational pull, like voltages attract since in this particular dimension $D^4$ effect of force, force is given the designation + because it is induced from a positive dimension. Hence, contrary to the idea of electrical current where current releases voltage and it diminishes, here the opposite is true: increased current with increased voltage. We would expect electrical force to be induced by – dimensions.

$F$ is always positive within a system of $D^4$, but between systems, we notice the difference in viewing current action. $D^4$ is an effect produced from positive and negative fields of higher dimensions.

**What is an effect?**

$D^4$ is always positive, and effect is the product induced from a higher dimension.

***Let's look at some of the variables of the space matrix component superstrings in a dimension as follows:***

- **Spin is a combination of velocity (*V*) and time (*T*).**
- **Charge is a function of time.**
- **Inertial flow velocity or current *V*.**
- **Elasticity *C*.**
- **Geometry.**
- **Density.**
- **Component superstring type.**
- **Dimensions net positive or negative attributes.**
- **Dimensions "arrangement" of sub attributes of geometry, spin, charge, and flow, which lead to quadrant attributes of parallel supporting systems.**

**So we end up with various models of space dimension per dimension, with net attributes within a transdimensional system and quadrant attributes of supporting parallel systems. Suffice it to say that gravity is a reinforcing action that enlarges both voltage and current flow; *V* as the single system becomes centralized from the imbalances or polarities presented with the Newtonian equation. For electrical current, the internal voltages show divergence rather than unity; however, force is always $+F = D^4$. Several types of current can now be observed to include the following:**

- **Neutral unidirectional current such as photonic current, a light travelling through the universe. These are neutral related to - and + dimensions.**
- **Polar negative (-) current, where voltages are unlike each other and diminish such as in electric current flow to establish balance. Related to - dimensional activity.**
- **Polar positive (+), where like voltages and inertial fields are reinforced and imbalances are reversed but such process.**

The space or ether that we move in is transparent to higher dimensional mass, and with the various functions of photonic light current and gravitational activity, we are looking at space that is many different dimensions coexisting in front of our eyes, which is what to expect; but now we cannot call it one thing such as space-time but many things coexisting. Hence, the physics we interpret from it happens in adjacent dimensions of space, although we assume it to be of a singular nature, which is a human inclination. Its overall math is singular, but what we cannot see are the various parallel dimensions that account for sometimes conflicting phenomena only because we assume space to be one thing instead of many.
It could be that space volume being $D^3$ is odd and negative, and in the case of the Sun, generating elastic wave and light at $D^5$, the matrix mechanically acquires additional $D^2T$ along with $D^3$, which product is elastic motion $D^5$ with time $T$. So when space vibrates, it acquires a function to the Planck units or constant at $D^5T$. This also allows that function to be neutral with an additional $T$ for photonic current. $D^2T$ is also resistivity of space. With $D^3$ as a negative field and with some positive characteristics of the Sun in local areas of positive ions, the Sun may be acting at times like a battery that would affect the motion of charged zones.

### *Force Resulting from Volts in Electrical Current and Volts in Gravitational Current*

**Work = force × distance**

**Power = work/time = $FD/T$ = force × velocity**

**Power = $I$ × volts = $FV$**

**Force = $I$ × volts / velocity**

**Force = $D^4$ = $I$ × volts / velocity = $(D^2 / T) \times D^3$ / velocity**

**Force = $D^4 = (D^2 / T) \times D^3 / C$**

**Force = $D^4 = D^5 / CT$**

You can see from the above equation that with regard to volts in electrical current, the numerator is $D^5$, which is odd and refers to a negative dimension. When we are looking at Newtonian gravity, the force created is more even with

$$F = D^4 = GMM / D^2 = D^6 / D^2$$

**There is no complexity here with time in the denominator.**

There is a difference in the order of the numerator and denominator when comparing gravitational and electrical force due to volts, and this difference has to do with the manner in which time is allocated with the power associated with volts in electrical current. Time has a different role in the space matrix of each. In gravitational force, the voltage contributed to increasing current and voltage, and in electrical, the voltage decreases with increasing current. This has to do with time. In both there is an imbalance of time, but in one it reinforces voltage and the other it diminishes.

**Science and Moral Questions**

The entire natural universe is mathematics; hence, there are no anomalies or uncertainties or randomness. The fact that one hundred laws operate simultaneously leads to variety that gives the illusion of randomness. Now if one set of operators from the outside, unrelated to the matrix, is introduced, that anomaly is free will, and randomness is made possible by transcendence. So free will is made possible by transcendence. So your consciousness transcends, and hence, you have a soul. But the universe is an analogy of truth and follows that which is according to its laws or nature, and hence, when you differ from that, there is pain. Life as one form of consciousness perceiving another means the universe is alive. And those things we refer to as morality are constants that remain through various states of consciousness or being. Hence, the creator, in creation, allows those things according to its nature, and those things are moral. The difference in what free will allows is the capacity to differ and in the experience of pain, which also reveals truth. That freedom lies in the intent of free will, which becomes through action. Those things that are according of the nature of the universe are received easily, such as love.

**The Code of the Universe: 1 and Zero**

Space is made from the space subdivision of a sphere and the compaction of spheres outside a sphere. And there is no place without space, which means everything is filled, and the idea of "nothing" is a myth.

***D* = 1** representing a sphere, and the sphere being the smallest subdivision of space, which is the smallest unit defined properties of space, one point, one side, or the one from which everything emerges as *D* is divided into time (*T*) and velocity (*V*). So 1/*D* would be the inverse or that space in between, which is that curved tetrahedron or curved form. And what we see or the properties derived depend on what has dominance in the geometry, whether it's *T* or *V* or together *T* and *V* for a *D*. As the sphere is subdivided, we get all other spaces or properties, and also outside as well as they are multiplied. So perhaps within contains scalar or additive forms, and without refers to proper or multiplicative forms. Within *D* are combined two forms structures (*T* and *V*) like a puzzle of two component networks interlaced in 3-D, which together fill the void of the other, like a puzzle, and make a sphere. One network is time and the other velocity, and the geometries that form the circle in each are reverse and asymmetrical and also asymmetrical in energy. One structure such as *T* is the reverse image of the other *V*, and both form a circle. So how the circle is subdivided and multiplied, added or subtracted, accounts for the very structure of dimensions, properties, and math of the universe, which refer to one another with meaning. But then there is a triality that refers to the smallest being, the sphere *D* with *V* and *T*, and the dividing geometry where the first subdivision of *T* and *V* account for forms that are symmetric in surface but asymmetric in space subdividing geometry, such that one has less negative space than the other according to energy ratios of phi and phihat. Velocity *V* refers to phi, and time *T* phihat, and the asymmetry in the two networks. And if those structures in the process of the subdivision or creation have a common component, then this would be the thing that that all have in common but can never be observed because that same thing has different expression in *T* and *V*. Metaphorically, we might call it light, but the nodes of these intersecting geometries of space create the particles (the secondary place in between the primary spaces). The places between intersecting spaces are the particles.

The qualities we give to our math equations in physics refer to the space and the nodes are particles. And hence, space attributes qualities through various dimensions, which are related to particles. Light, the node, is therefore the smallest node in the matrix of space referring to the smallest or least complex geometries of space dimensions. That same thing that Michael Ravens refers to as the Trion Re, which makes time and velocity or everything else for that matter is never in a singular state to be measured. Let's call this telluric or Trion form *T*, which represents the balance that forms 1 or *D* or singularity and bonded by unity or zero, the other limit of singularity. Perhaps every superstring sphere is bound by 1 and 0 in its inner code of the Trion *T*. The relationship between 1 and 0 is the basic components of all structure that also by arrangement like a binary code forms time and velocity. Suffice it to say in summary regarding 1 and 0, we can describe the following:

***D*, *D* + 0 = *D*; 1 + 0 = 1**

## Trion – Re Geometry of Light

***Permission by Michael R Evans***
***Geometry of Light***

## Trion – Re Geometry of Light

*Permission by Michael R Evans*
*Geometry of Light*

## Trion – Re Geometry of Light

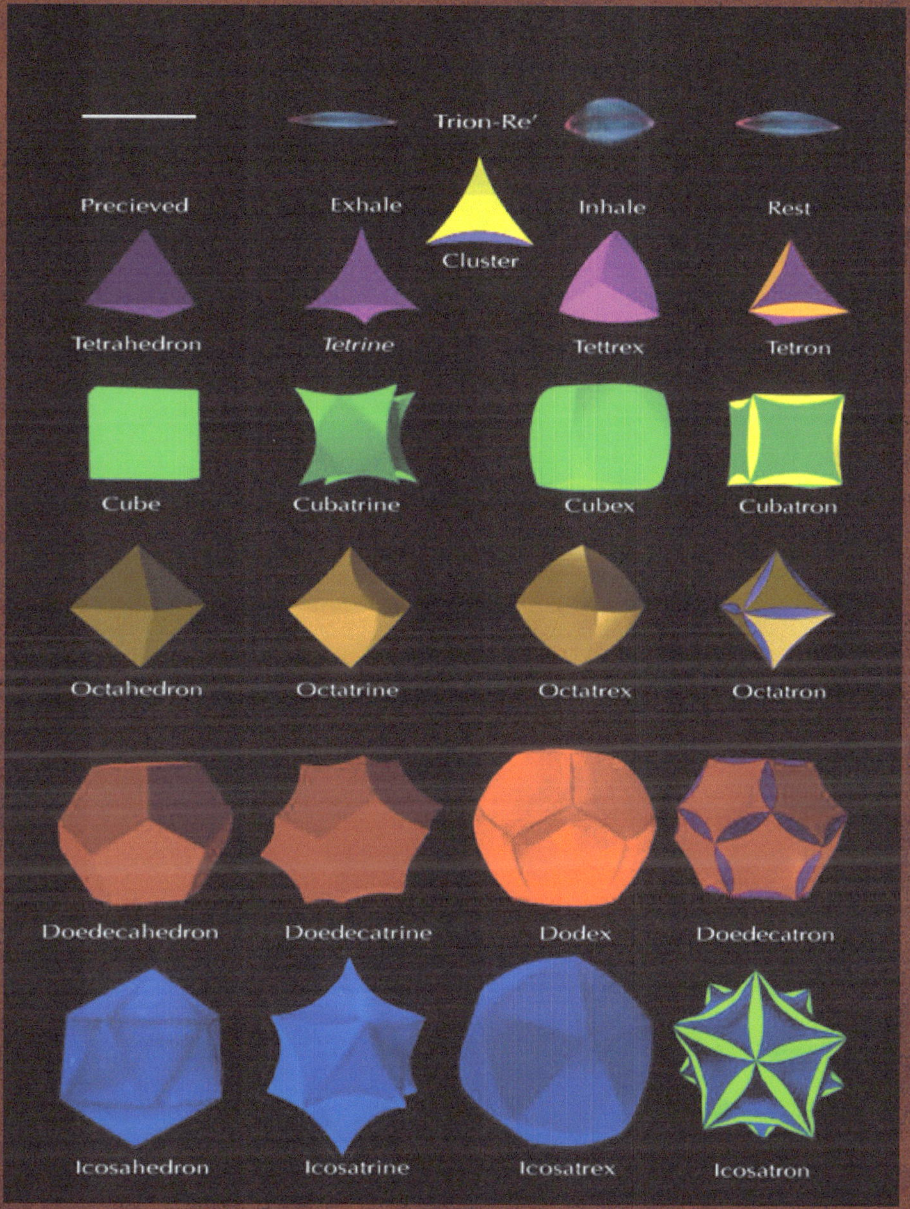


***Permission by Michael R Evans***
***Geometry of Light***

## Trion – Re Geometry of Light

***Permission by Michael R Evans***
***Geometry of Light***

***D*, *D* × 0 = 0; 1 × 0 = 0**

The arrangements of 0 and 1 in the division of space establish a code of creation similar to a program of a computer since the 1 and 0 relationship and arrangement is the subdivision of space into time and velocity; beginning with one and with various relationships with zero, the superstring *D* is subdivided in a network of 3-D into time and velocity, much like a quantum computer. Zero becomes 1, and 1 becomes 0. The relationships through zero are transformation and the capacity for the subdivision of space and hierarchy of various physical properties as in the Trion Re. (Evans: Web Link)

The Trion Re described by Michael R Evans refers to the 1-0 as the smallest in the subdivision of the network code comprising a superstring sphere, that part described that has no physics property in itself since it is beyond the domain of measurement. Hence, for the network, the Trion is the smallest, but in the domain of measurement, the superstring is the smallest. The superstring sphere is the smallest *D*, but in the creation and subdivision of 1 that creates all diversity, the 1-0 is the smallest. So let's go through images: *D* would be the 1 containing a network of 1-0 arrangements of 1or *D* that defines 1, and 0 limits for the entire superstring. The 1 is balance and 0 unities, and so the singularity is bound by 1 and 0. From 1 sphere *D* arise two spheres *T* and *V*. The following summary explains *D*, *T*, and *V* with the Trion Re 1-0:

- **The sphere *V* or velocity has 1-0 pattern of *V*-dominant network of its inner space.**
- **The sphere *T* or time has the pattern of *T*-dominant network in its inner space, which is the polar or negative image of *V* dominant.**
- ***T* and *V* are complimentary and polar; together their images fill the missing action of the other when together they form *D*.**

So what is dominant space in *V* is negative space in *T* and vice versa. The smallest Trion means the common code 1-0 in the subdivision of the sphere, which combines with other spheres to account for all the universe and diversity. Image process:

- **Imagine a sphere.**
- **Now imagine an object (Trion) that can be used to generate a sphere.**
- **Let's imagine small Trion.**
- **So you put these together, and when all are together like a puzzle, you see a sphere.**
- **Now let's call this Trion 1-0, and let's call the resulting sphere *D* or dimension or God particle or primary superstring.**
- **Now in that sphere, let's say there are two networks of Trion that are *T* and *V*, and one set of Trion is 30% less than the other top form a ratio related to phi.**
- **Together, you can't see the division between the two sets of Trion because it's like a finished puzzle.**
- **Let's call one network time and the other network velocity, interlaced systems.**
- **Now let's refer to the smaller network time sphere and remove it from *D* sphere so *D* becomes a *V* or velocity sphere. Now there are two spheres of diminished energy and complimentary.**
- ***V* superstring has a network of 60% velocity and negative space *T* of 30%, which is the positive space of the other time at 30% with 60% negative space.**
- **So you have two systems of time and velocity, now each the reverse image of the other, which we call polar or asymmetric or complimentary.**

So from phi **+ phihat = 1 and phi × phihat = -1, we get**

**velocity + time = 1 and velocity × time = -1**

**where 1 is dimension *D* superstring.**

Summary

The Trion reestablishes the relationship between 1 and 0. We can say that through distance, time, and velocity in its various matrices and dimensions, we get all the equations and properties of physics; so in that sense, *D* is smallest for defining properties. But the nature of *D*, *T*, and *V* defined by the Trion Re means that 1 is function of 1 and 0 where 1 is a function on and 0 is function off, where the Trion Re is defining relationships between 1 and 0. So as a math symbol, let's refer to Trion Re as 1-0 code. So the code by position does not carry physical properties, but cumulatively, physical properties are denoted, much like a computer code or genetic code. Codes get imbued with life or properties at greater complexity. Hence, with the same argument of the genetic code, all creation has intelligence in its construction, being the creator. So 1 or *D* is subdivided into functional complexity by a positioning system of energy Trion Re 1-0. For all purposes, the sphere *D*, *T*, and *V* is our physics with properties and functions. But the Trion code is used to program those functions.

**If you have a component that creates properties, then what is the component Trion Re structure because we only measure properties?**

The best you can do is say

**1 + 0 = 1**

**1 × 0 = 0,**

**and these codes make 1 and 0,where 1 is *D* and through zero make *V* and *T*.**

And 1 comes from Euler's equation, which represents induction of all points in space in the universe and its creation. The universe was <u>induced</u> much like a magnetic field in the big bang. It could be a quantum hologram. So the universe is a program of 1 and 0. When men get ideas, they are merely imitating nature. You can't get something to work that is not already in the nature of the universe. So that we have computers means that it represents functions already in operation that exist within the universe. You will also notice that 1 and 0 are created from Euler's equation, where 1 comes from the additive scalar form and 0 comes from the multiplicative proper form as follows:

$\boldsymbol{E^{(i \times \mathrm{pi})} = -1}$

$\boldsymbol{E^{(i \times \mathrm{pi})} + 1 = 0}$

And so the value of zero in determining velocity or time is <u>position</u> within the 1-0 matrix, where position being the inverse of infinity establishes form or qualifies form. The 0 means everything is space and determines properties by position. Zero means there are no units in this position; this means in this position, there is a presence induced from outside the universe that is constant or without function, and 1 is where there is function. So

**1 + 0 = 1 refers to units and constants as scalar respectively and means to <u>leave the units in operations,</u> and**

**1 × 0 = 0 refers to units and constants as products and means <u>leave the functions to stop as constant.</u>**

So by position, you have *D* on and *D* off, still there when off but not functioning.

**Then what is this 1-0 substance that makes up everything?**

This expresses all numbers between 1 and 0 (fractions) in exponential form, which is the basis of logarithms. And by position, infinite set of numbers is established, which is part of the infinite values of the universe and diversity that is enabled and also qualified as form. Zero is the point of transformation of various dimensional functions and also between positive and negative numbers. To divide by zero is not possible because within the universe and associated numbers, infinity is qualified by position through zero, but from the outside, what appears as the infinity and variety of the universe is a constant.

**<u>Intuition, Transcendence, Consciousness, and Metaphysics</u>**

Metaphysics is a transcendent intuitive process. Intuition is a transcendent process that unites metaphysical truths to science so that there is continuity and relationship, and as metaphysical truths become represented in rational, mathematical form and formal relationships that represent the universe, they become science. But there is always reference from science to realms of higher realities of truth that are transcendent since they are one. The many are in the one and the one in the many. Entanglement of universal process extends beyond the universe and back within and connects the known and unknown. These relationships also refer to eternal truths or constancies throughout all creations, and hence, such metaphysics and awareness of universal laws actually form a doxology of an awareness of universal truth or wisdom and refer to modes of being that refer to the nature of such truths, which extends to modes of harmony of being or consciousness. The universe is an analogy of truths that are eternal and common to all realities of existence, including the created universe, and also such things beyond this universe. Our lives are intuitive and rational, and hence, the idea of experiment and science being something separate is incorrect since ideas originate from all levels by human investigation that includes intuition. Those things coming from the collective consciousness higher self in relationship to this universe are a priori, where all information is known; hence intuitively, all knowledge may be suggested as ideas. Intuition is an inversion of the normal mode of thought; intuition is metaphysical and an awareness of universal axioms from patterns experienced in the universe. Metaphysics is the awareness and study of axioms prompted by intuition from the observation of patterns in the universe. These axioms tested and reviewed by logical discourse become the analogy of truth. When further established by scientific observation, they can become science and physical laws established through mathematics. If the patterns discovered are actually the pattern of the universe, then the pattern or unified theories can lead the way for science and observation to areas of discovery beyond human perception.

There are some things that will never be resolved without transcendent intuitive process or metaphysics, which eludes to the synergy of the whole, which is there anyway, but not to confuse that it's a foreign process in the human reasoning; it is there even with engineers that do mundane tasks. Emergent forms of process thought without intuition can never ascribe to the discovery of the whole since it is out of reach, and so for discovery in science shows limitation. All science leads to God and metaphysics, but understanding the unification of the universe will require a radical unity and simplicity that cannot be surmised by measurement since it lies outside of human perception and within the limits of emergent process reasoning. Metaphysics has more to do with a method of thought or a process of thought of the one in the all and the all in the one, transcending and all-knowing without time, it is not false; it has to do with proceeding with concepts in science as an idea, a priori. The morphology of creative intuitive thought projects models that unite ideas that may be discordant but that actually imply a structure a priori to that representation in math. We are looking here are the value of promoting intuitive process rather than eliminating it in science.

If the universe is multidimensional, human experiment and perception cannot see beyond the dimensions within which we are. If physics is to include the entire universe, experiment will not lead to it; hence, the realm of intuition, math, and ideas can propose models, metaphysics, and thus having a broader model, can predict a science that may be possible to explain things beyond perception, which, in fact, might lead us to these other dimensions and patterns. And to a higher form, these patterns refer to the truth about God and the nature of this universe, and the nature of God both within and without the universe and also about us. The 1 in everything and everything is in the 1 represents the entanglement of all things by the pattern of the universe arising from 1 and 0 to complexity. To arrive at the theory by intuitive process, one admits that all knowledge is readily available to all somewhat like entanglement, but that in the manner of thought, you must know that intuitively, the relationships can be readily accessed. If the one is in the many and the many are in the one, the solution is relatively simple as well as complex. The universe is an analogy of truth, and the universe is mathematics. Every number refers to various truths, and all equations refer to transcendent truths. In the 1 are all things and all things are contained in the 1. Science deals with what is measured, and hence, it is limited to dimensions that are accessible. Intuition is metaphysical or transcendent and has that capacity to explore models yet untested, and this too is within the realm of math. Therefore, math is not subordinate; it's a priori. What is subordinate is science, following from intuition and following from math. That science is useful for your survival makes it a priority.

<u>Meaningless</u> is a myth, along with such axioms as *nothing*, *chaos*, and other words and theories that propose arbitrary action, randomness, and accident to the natural universe. The universe is a machine of mathematical relationships, and randomness is the resulting illusion. The word *meaningless*, like all words, has meaning in them and purports a construct philosophy, which is contradictory to the word. In the same sense, the development of any concept word is automatically the declaration of organization. Any relationship is the beginning of order, whether defined or not. To make any statement is an intuitive process and objectively transcends the topic. Relationship is truth. The truth about free will is when one set of operators is foreign to an existing environment; you then have randomness and transcendence. Hence, it is impossible to have free will and randomness without transcendence. Hence, you have a soul. In this mathematical relationship of unrelated axioms, the relationship or truth is free will.

## Absoluteness in the Universe

**Are there absolute qualities in the universe, and is there absolute morality?**

Concepts that we espouse may be considered ideally perfect and, hence, referred to as absolute. In some contexts, it is the utopian perfection. In science, it may be a boundary such as absolute zero. And also for some, absolute represents the unattainable by being human or the uncorrupted ideal. So being absolute is completeness and complete in the nature form of the concept espoused. The Chinese say "scarcity has the potential to gain." That which is lacking in the concept represented has the potential to fill or grow into it. Therefore, approaching the understanding of the absolute is a journey. And if truth is absolutely identified in a myriad of forms, the quest is infinite. Since there is transformation and change, you can be absolutely something else, but not so absolute in the journey unless you follow an absolute path like mathematics. But then being free willed where you have options, being is not so absolute. But in the nature of free will, complete and absolute are you. Absolute is in the nature cause of something but variation on the degree of infinite knowledge acted upon. Not so absolute in the path knowledge, but generally in the nature of the idea. As the universe has constants and being an analogy of truth, so too is there an eternal constant and extending to wisdom and morality as having absolute benchmarks. Absolutely this is so.

**But how can we be sure?**

Because there are two operators, those that are emergent and transcendent as follows:

- **Those arising from natural qualities of an emergent universe.**
- **Those arising through transcendent interaction with the emergent universe from the outside as a foreign anomaly.**

You experience constants in an emergent universe that represent patterns throughout all reality and, hence, absoluteness referred to within pattern language. There is no anomaly or randomness except by interaction of emergent and transcendent patterns. There is absoluteness within the synergy of all pattern relationships, and thus referring to moralities and modes of behavior; there are also absolute benchmarks since the nature of one is in the many and the many in the one.

**Can we call it a priori?**

A priori means that all knowledge is known at the outset as a transcendent intuitive experience. And after the transcendent interaction with the universal experience, it is rediscovered.

**So a priori is about knowledge?**

Yes, but more, and let us explore this. But in the interaction you experience (example, morality) and before, you just know intuitively. I know something is bad and don't do it intuitively. I experience bad, and that totally confirmed.

**In philosophical terms, a priori is specifically about knowledge and not anything else?**

It is more than knowledge and includes knowledge. It's about a state of source consciousness being self, awareness, thought, and free will since it describes our relation to transcendental being and includes all knowledge.

**Yes, but what about "absolute morality," and where does it fit in?**

The nature of the universe enables those things according to its nature that is absolute, and although allowing free will, those things that are not according to its nature involve pain. The universe enables according to its nature. This process is absolute, and although allowing free will, those things that are not according to the nature of all creation involve pain. Therefore, the constants are about the nature of the universe and, as such, represent absolute benchmarks: wisdom, appropriate modes of behavior, morality, and much more, such as an ordered universe with constants. And those things that defer involve pain. But there is the option to defer.

**Yes.**

And hence, the benchmark is absolute according to the nature of the universe or all reality. That you have pain means there is deference to an absolute. And that you have happiness means you are in harmony with the absolute.

***Absolute* gives the impression that it is not debatable.**

Well, there is diversity in the absolute. The idea that it is 1 and also infinite so there is freedom. The notion that absolute is limit is incorrect.

**That is right.**

Language is limited according to cultural traditions of thought patterns; what people say has more to do with how they think than the words they use or the definitions they espouse. Reality or the infinite matrix encourages those infinite processes according to its nature. It allows deferring, but there is pain. Freedom and infinity lie in the nature of reality, and your free will is allowed to discover this and become enabled for infinite progression according to the nature of reality or matrix. The emergent universe is a machine. Hundreds of equations running simultaneously give the illusion of randomness, but there is nothing random in natural process. When one set of mathematical operators is foreign to those emergent, then you have randomness. Hence, free will being the foreign operator or non-emergent is transcendent. Free will means transcendence and that you have a soul. There is no free will or randomness without transcendence. All life is free, set within an environment that appears free due to its variable relationships. Therefore in the natural world, there is no uncertainty except by the free will of life-forms.

## Space Matrix, Toroid, Wave Structure

Space or the space matrix is an aggregation of a wave-matrix system, and there are no points except areas of localization and interference of various waves, which we refer to as the centers of mass. So there are

- **wave lattice or matrix structures,**
- **the elastic collisions through those structures,**
- **the motion of those structures,**
- **and the localization of various structures due to their center of origin or center of inertial field.**

The processional motion of mercury has to do with the full lambda wave core around the Sun due to a standing wave established within the spherical space matrix or inertial field generated by the Sun. Imagine a sombrero standing up and then one facing down. The amplitude of the standing wave at the center of inertial field in two directions is full lambda. At the core there is a full lambda wave core of the standing wave, and as it flattens outward as a standing wave within the space matrix sphere at radius $r$ or $D$, you have the image of a gyroscope and formation of the ecliptic plane holding the planets together. This standing wave is also at radius $D$ from $\boldsymbol{D^5 = MC^3}$ describing the full extension of the inertial field. These wave functions exist around larger masses' inertial cores and related to the orbits of particles or planets and represent the ordered structure of supporting space matrix. There is no "uncertainty." When inertial fields interact, they become one system, and so systems are disrupted and form new associations constantly. If we consider space as object field as a sphere, as soon as we add spin, we end up with a toroid form. Within this construct, if we add the motion of the moving space matrix within the form of spherical form from center of inertial field outward and back, we induce a standing wave due to internal space-wave vibration as a secondary system standing wave within the toroid in the form of gyroscope to influence space density and control the motion of planets within an ecliptic. (LaFreniere: Web Link)

## The First Cause or Creator

The natural universe where processes are emergent can all be explained by math with implications far beyond physics. The universe also shows a principle of entanglement and dependency of the numerous parts and not one part cannot exist

# Spherical Lattice

without the other either directly or through transformation. By principle of symmetry, the entire universe is dependent on something from the outside. In addition to the natural machine of the universe, there is also the anomalous and free will of living forms, which is only possible by transcendence. By principle of symmetry, if my consciousness requires a consciousness to receive my actions, then too the natural universe is alive and is a form of consciousness. All consciousness proceeds by act of a living will. The action that caused the diversification of unity, from simplicity to diversity (the 1 is in all and the all in 1 represents entanglement) can be in a multitude of forms. Selection of an option that results in the mathematical structure particular to this universe requires an intelligent being. The first cause is referred to as the Creator, and the entire math and creations refer to the nature of the origin and, hence, is the concept of created referring to the source as an analogy of truth. We as intelligent beings spend much time on the equality of all beings to negate the concept of first cause, but in truth, the entire universe is hierarchy and the equality we allude to is only by position and similarity to surrounding. There is both equality and hierarch, and hierarchy requires a first cause. If the universe represents a form of math, the idea of math requires a first cause and the idea being associated with a form of consciousness.

The processes of the universe are described by math. Hence, nothing is random. It is incorrect to use statistics in creating models for physics. Statistic deals with the randomness only found in the free will of living conscious beings. And that we said the universe is a form of consciousness, the natural order has been fixed by universal laws and allows a backdrop of constancy that allows free will to participate. These constancies also refer to constancies about the nature of matrix truths through all realities and, hence, are an analogy of those truths by constancies that exist. Hence, constancy also alludes to constancies regarding the nature of matrix and its actions and modes of operation and behavior. Entanglement is the same as dependence on numerous parts. Where one variable, component, or equation does not exist without the other, the impact on one is simultaneously an impact on another. With the principle of symmetry, the nature of the action is within the nature of the result, which includes dependence, even in the existence of the universe from something outside. We cannot love unless the universe has the same. An object cannot pull by gravity if the participant object does not have the same. There is nothing random in the "natural" universe; a thousand equations running simultaneously have the illusion of freedom or randomness, and all can be explained by mathematical equations. When it comes to human behavior and those of other life-forms, then there is random activity. When one set of mathematical operators is foreign to those existing, then you can have random or free will. Hence, free will is transcendent, and there can be no free will without transcendence. Consciousness includes the capacity to choose and does include free will, thought, and awareness. Any being that has life or is living is a being, a form of consciousness. The big bang is not the first cause but the result of it. Here Euler's equation

$$E^{(i \times \text{pi})} = -1 = \text{time} \times \text{velocity}$$

is used to define the big bang, and by it, it implies the universe is induced from outside. And if the universe is a form of consciousness, then that supports other arguments from its origin from the outside and also for the consciousness of all living beings.

Well, I would assume that in a mathematical model for behavior, we use statistics, and any such model uses the option to defer and a wide distribution of mathematical outcomes. So then in a sense, such a math allows for transformation of itself and its environment, which is not found in planetary motion. Now a roll of the dice is not random in the manner I am speaking. The natural universe is a machine in the sense of being represented with equations without exceptions. Behavior, on the other hand, is the option to select equations, programs that change itself and the environment without being just a program. The dice is emergent, the person throwing is transcendent and free willed, and the dice is not random of its own accord. In living being, I include any form of life and all cells. We can debate whether that has attachment to a central nervous system. But there are no emergent processes in the "natural" universe that show the beginning of life. There is an argument that it can be achieved in a lab, but already that is intelligent design if in a lab. Free will excludes prediction but includes uncertainty, which is common in physics; however, there is question of using proper (multiplicative) mathematical form to solve uncertainties. Heisenberg says you cannot know both the momentum and position of a particle, but then he uses proper (multiplication form) to solve a scalar problem (additive as a quadratic or polynomial equation). That's a wrong philosophical use of math. (Schreibner: 1-13)

Ultimately, what came first, the chicken or the egg? What is outside of outside? The first cause means that in 1 all diversity emerged. From awareness emerged thought and free will and all the diversity of the universe. In order to love or express free will, God created, and what was created had the capacity to love back. This is the principle of symmetry evident in all math and carries the equal (=) sign. In all this are symmetry and hierarchy where symmetry is part of the very nature of the first cause and is essentially represented in math.

This theory postulates that any constant with units is in fact variable. And that all numbers refer to relationships of *D*, *T*, and *V* such as 1 refers to *D* and phi to velocity, etc. I was searching to explain pi in terms of 1 (or *D*), 0, phi and phihat. Back to the theory of constants, I revised the idea that certain constants with units that cancelled in the numerator and in the denominator originated within the universe and were emergent. The transcendental constants such as pi were an exception, and thus, I postulated that they originated from functions outside the universe. Still looking to find an equation that related pi to phi, phihat, 1, and 0, there were none until I saw a connection in Euler's equation. In Euler's equation, there was a -1.

$$\boldsymbol{E^{(i \times \mathrm{pi})} = -1}$$

Well, in transdimensional theory, -1 was equivalent to time × velocity. So the following resulted:

$$\boldsymbol{E^{(i \times \mathrm{pi})} = -1 = \text{time} \times \text{velocity}}$$
$$\boldsymbol{E^{(i \times \mathrm{pi})} + 1 = 0 = E^{(i \times \mathrm{pi})} + (\text{time} + \text{velocity}) = 0}$$

By this, all *D* were induced in both scalar and proper form by this equation from outside the universe and thus connected the separate nature of pi as a function outside the universe involved in the big bang to create all *D* in scalar and proper forms and all the complexity of the universe. Scalar additive systems involving multiple inertial fields result from ***T* + *V* = 1** and include all quadratic forms and polynomials. Also, proper multiplicative systems involving a single inertial field result from ***T* × *V* = -1.** The 1 in both systems represents *D*. When dealing with what mass does, it is correct to refer to the entire system. In scalar form, mass increases with velocity, and in proper form, the opposite holds true. Mass is mysterious, meaning it is $D^3T^2$ superstrings, and these components exist even in the ether of space. The 0 refers to the relationship between cause and effect. And the transformation between them for scalar systems and *C* is same for proper systems. The <u>1-0</u> is the primary binary form (telluric form) possibly related to photon at the node from which all division of the 1 are made as a network or code. In the sense, the universe is the induction of a binary hologram. The principle of symmetry relates cause and effect by the = sign. Brought together by zero and *C*, which refer to unity through transformations? Yes. The principle of hierarchy means that all order refers to the first cause as the source. Hence, there is God.

***<u>A first cause or God can be derived through both science and metaphysics.</u>***

**How harmonious is your theory with Einstein's scientific ideas and with his philosophical ideas?**

Well, it would change them. The notion that space is two dimensional is incorrect when involving all the known forces to include magnetism with ideas such as density. And if we are to explain the ether space as a structure of "something" of varying density, then that makes special relativity merely an abstraction. It is very possible to get experimental results by abstraction, but then also far from representing the truth of what is being observed. In science, this happens often, and thus, the proposed models are riddled with exceptions, which make them incorrect.

Einstein philosophical derivations are fine in spirit, but derived from wrong axioms. For example, his equation $\boldsymbol{E = MC^2}$ is rewritten here as $\boldsymbol{D^5 = MC^2}$. There is equivalency in many things, but the <u>something</u> that relates mass and energy here is more clearly defined as time and velocity. What is mass? A mystery no doubt, but as we have to make a definition for it, here it is: $D^3T^2$. And hence, any place with these components has mass; therefore, the ether, emerging from a mass core, or more appropriately said, the ether that has a localized multidimensional mass core also has mass. That being said, it means that *G* is absorbed into the mass and also revises Newton's force of gravity equation to $\boldsymbol{F = \text{volume1} \times \text{volume2} / R^2}$ where the volume refers to the localized inertial volume, where *d* is the radius of the inertial field. Space is localized, and all core mass generates space to distance *d*. Mass is changed to include the core mass and ether or space it generates. Space and mass are the same only differentiated trans-dimensionally.

Let's try a metaphysical example:

When an artist looks at nature, he or she refers to the randomness and freedom; but in actuality, all natural process—and I mean nature, not related to life—can be explained mathematically through equations. So there for one thousand equations running simultaneously give the illusion of random.

**So where is random?**

When one set of mathematical operators is foreign to those emergent and existing, then you have random. Hence, only through transcendence is there random. So only through transcendence is there free will. Transcendence is that which acts from exterior to a system; hence, you have a soul. That in science they talk of uncertainty, but there is no uncertainty. And Heisenberg used a proper multiplicative form to try to solve a scalar or additive problem of momentum and position. The philosophy of math is wrong, and hence, the philosophy, and also a quadratic or polynomial, would have been the correct choice. Hence by incorrect science, we establish incorrect models and have a science riddled with exceptions. And from these it is impossible to find the harmony and equilibrium of a unified theory. And such mistakes perpetuate the notion of myths such as chaos and uncertainty, which pervade our understandings of life. But in an exception to a model theory, it means the theory is wrong, and in the exception does the greater truth lie and opens a door to new knowledge. Implications of this theory in metaphysics are boundless and include such ideas that there is free will as a transcendence activity. There is a God or first cause. The form of the universe is akin to a binary hologram.

**Is God a personal God then?**

God would be both personal and universal. The universe is Gods consciousness.

**Maybe there is a God or a first cause of the universe or universes that we know or can imagine of, and I guess things could go even beyond?**

They can. The theory proposes a construct beyond the universe supporting everything within the universe.

**So imagining that there is a God and he created universes, this can just be one of the events he made to happen?**

Yes. Euler's equation to me is the big bang of this one universe. $\boldsymbol{E^{(i \times pi)} = -1}$ In this theory, -1 is **time × velocity,** and time × velocity is *D*; so therefore, Euler's equation is the induction of 1 or *D*, which is all points of the universe.

**You know what, I just realized; what we will do if the entire world begins to agree upon one metaphysical theory, and furthermore, it has been scientifically proven? Not only that, science has come so far as to become complete.**

In 1 is all diversity and in all diversity is contained in 1, and it would be complete by such a definition. But then a unified theory necessitates one to be open to understand it and also go beyond. This universe is but one event of the first cause or God. The 1 can be subdivided into binary networks in infinite types of series, and so what can be created is infinite. That there are constants reveals the nature of the first cause, and there is accountability and morality inferred by constancies of the free will. When you are not in harmony with the nature of the first cause, which is allowed, there is pain and there is truth referring to qualities or constants inferred through creation. The universe being an analogy of truth describes the nature of the matrix that supports it. There is no end to then what can be known. Form is a qualification of infinity.

**According to Einstein, "the real nature of things we SHALL never know."**

I can agree with you here. Now to the notion of real nature, that is multidimensional. Form is a manifestation of infinity. In the universe, we can set velocity and time as the primary components, but then we can go further below that and define a binary network for both time and velocity and the compliment of them together as *D*. Therefore, what we measure is what we can experience. We can explain but not directly experience some things. And so for certain metaphysical truths, we are allowed to know and also experience and others just to know. I know "bad," but then I can experience it. I don't need to know something is bad by experience. It is not a requirement to be a fool to know something is foolish. The questions of the real nature of things are reserved for God alone since only God transcends all that is. Infinity is the matrix that is organized to give form. (Tomes: Web Link)

**It may itself be a God or the God, the only God.**

The idea of nothing is a myth. All that is, is.

**If there is a God greater than that, who and what is he?**

God is the first cause through which all things that are known have meaning. 0 means unity and transformation. The equals sign (=) means symmetry in actions. Hierarchy means there is a first cause. God is the first-cause consciousness, which is awareness, will, and thought. Awareness or the "I am" brings all things to be, the 1 wherein all exists and all contains the 1. In that consciousness, all is imagined.

**What does God actually do though?**

By principle of symmetry, for any action given it needs to be received. So to love, God created the universe in all its infinite diversity and possibilities to express infinite love. From awareness resided will and thought, and through creations love was expressed and felt. The entire universe is consciousness itself, so the distinction between consciousness and non-consciousness is merely our own creation. God is lending to you consciousness to participate. Your consciousness is residing in another consciousness, and so your body, also a part of the physical universe, is a shared dream or receded form of consciousness originating in a collective consciousness.

**If it's a dream, why are we all having the same dream?**

That's how actions such as love are given and received.

**So what is real?**

Being is real, place is a dream. But from the outside, this universe has no energy; and from the inside, the outside is a constant. This is seen in Euler's equation.

**What's on the outside then?**

A higher dream state is your source consciousness. So when you wake up as in death, this will appear as a dream.

> **A human being is a part of the whole, called by us the "Universe," a part limited in time and space. He experiences himself, his thoughts and feelings, as something separate from the rest—a kind of optical delusion of his consciousness. This delusion is a kind of prison for us, restricting us to our personal desires and to affection for a few persons nearest to us. Our task must be to free ourselves from this prison by widening our circle of compassion to embrace all living creatures and the whole of nature in its beauty. Nobody is able to achieve this completely, but the striving for such achievement is in itself a part of the liberation and a foundation for inner security. (Albert Einstein, *New York Post*, November 28, 1972)**

That's correct.

**Beauty is subjective though, isn't it? Some parts of nature are far from beautiful.**

Beauty is a reflection of constancy of the order of the universe. Our familiarity with constancy is beauty. (Lucas: Web Link)

**Anyway, what does this all mean when it comes to how we live our lives, and what are the implications?**

Well, when there is pain, we are not in harmony with the order of the universe. So love is the way. The awareness of truth is wisdom, and by that, you know how to live.

**When you speak of pain, do you mean emotional pain?**

Yes.

**But pain comes and then it goes, and it's just a temporary state of mind.**

The interaction of free will with the matrix is gentle, guiding, and there are consequences.

**Why do we need to be in harmony with the order? The more pain we experience, the more it can magnify our joy.**

We don't need to experience to know, although there is complete confirmation in pain. If I set myself on fire, I know for sure. But I don't need to. In your intuition does the greater truth be.

**Is pain something that we suffer because it is beneficial to our survival as a species?**

It is the matrix guiding to infinite harmony. Pain is stagnation from the infinite variances and harmonic growth that the matrix can enable.

**What's so great about infinite harmony, wouldn't that be boring? What's wrong with throwing a few discords in there to make things interesting?**

The options are in free will. The preference is for enlightenment and grows on a huge bounty. The stagnation of disharmony is akin to death and disassociates contrarily to the nature of being. All life is to be respected, including your own. The dream is here for learning, and what you learn is up to you. And that which is harmonic enables and the contrary is restrictive; this is so even in science. What about crime? Should I participate? What you do is about you. All life is consciousness, one form perceiving another. And those things like rocks that seem to have no life are also God's consciousness, which are the whole universe and all reality. God transcends all and is aware of all. The dream is real in that constancies reflect truth, and in the sense, that being is real, but the place is the dream. Energy is an illusion. Outside the universe the whole universe is a constant that involves pi.

Life is one form of consciousness perceiving another in an axiom and, hence, by principle of symmetry, identifies the universe as consciousness. Even though a rock has no life, the rock itself is a manifestation of consciousness of the first cause or God. The life-forms or other consciousness that participate through biology are transcendent beings into the consciousness of the first cause, which includes the universe. The construction of biology allows the free will, which is a state for the expression of transcendent beings. That a tree cannot move does not mean it has a particular expression of consciousness. After all, will, awareness, and thought exist, but their transcendence into a form of participation does not necessitate its full expression. If I am a person with disabilities, I live, however perhaps not in the same manner as someone else or even, for that matter, if under surgery in a hospital under anesthesia. The event does not negate the nature of being, which includes the form of being represented in the form of consciousness.

As the universe is a form of consciousness, it then becomes an analogy of truth of the nature of the first cause and nature of forms that manifest life. This is also represented in math, and we find that the most basic relationship in the development of form is the number 3. We have identified two also as a polar relationship complimenting the three to form relationships of hierarchy. And thus by the number 3 is also ordered the form of consciousness and hierarchy of it. By relating the nature of physics into three units and also the nature of three to the nature of consciousness, we can begin to identify processes that are both separate and dependent in the synergy of consciousness. "I am" and the processes of thought and will all work together. Within these we may anticipate further hierarchies and identities. What has been identified is the defining of hierarchy in the universe and the universe as consciousness through representative hierarchy and meaning in numbers. And also by experience, we can begin to develop a model of what those hierarchies in consciousness are to be, and also considering that the universe being consciousness, almost like a dream, we can examine our own dreams to see what the identities are as awareness, will, and thought, and then take to analogy to physics and natural laws, which are really co-operators of consciousness, and see dimensions, time, and velocity.

## Form of the Universe

We have talked about the inertial field being a localized system where the space matrix unfolds trans-dimensionally from higher concentrated mass forms to lower concentrated mass forms such as the ether of space. Through such transitions, the changes occur by

- **dimension,**
- **space-matrix configuration,**
- **density,**
- **charge,**
- **spin, and**
- **overall effect.**

As then as our Sun projects space, we can imagine the core mass extending outward to an attached inertial field as a localized system. Then each planetary body is also another system. And each of these systems exists within even a larger system generated by the black hole at the center of each galaxy. So then space and associated mass clusters or centers are like cellular forms of life. And the limit of each inertial field is according to the equation D5 = $MC^2$, where $D$ is the radius of the inertial field, and mass includes mass of the ether or inertial field plus the core mass. As we have said before under such a definition, the gravitational constant $G$ is absorbed into the mass and Newton's gravity equations can be rewritten as **$F$ = volume 1 × volume2 / $R^2$,** where the volume is the volume of inertial field of one localized system. So each inertial field has an internal velocity $V$ and also an expansion acceleration, which we will call here $A$. And the expansion of the inertial field would vary with the radius $R$, which will be shown, and also limited by the equation $D^5 = MC^2$. And as such, then the entire universe will appear as a cellular form expanding in some regions and also contracting in others.

### Is there a number for the expansion of the universe or galaxy?

Well, this has not been determined, but a solution here is proposed. In each case of expansion or contraction, the acceleration determined by the core mass is fixed within variables determined by $D$ of the inertial field in relation to that core mass.

**Acceleration = $A = D / T^2$** is the acceleration of the inertial field of space.

*This can be rewritten in various forms involving*

**$G$ = gravitational constant = $1 / T^4$**
**$H$ = Hubble constant = $1 / T$,** *hence*
**$A = D \times SQRT(G) = D \times H^2$.**

We find through such expression that the acceleration for large radius is extremely fast, and for a galaxy in expansion, it may boggle the mind how fast that really is. $D$ is the diameter of the inertial field, but then when the projected space gets to equilibrium or maximum extension $D$ or radius for a set core mass, the expansion stops. Then what allows contraction is when various inertial systems combine their masses to form larger masses into black holes, which again explode for more expansion. So in a sense, we have established a mechanism for perpetual motion. And also in the manner of the creation of mass, since it is from the outside of the universe according to Euler's equation, it is always possible that new inertial systems are in creation even after the initial creation, but this is a proposition that needs to be proved. The limit on acceleration would be $\boldsymbol{D}$ from $\boldsymbol{D^5 = MC^2}$. The diameter $D$ would set the rate of expansion and does imply exponential growth unless the expansion slowed down in proportion to diameter. The other option is such as in the opening of an umbrella with exponential motion and then it stops. So for a mass with associated diameter of space, there is acceleration, and it is set to that ultimate diameter. By contrary, if you have dilution of mass, you have contraction and some variances in between them. So for the entire galaxy, the expansion from the black hole would be enormous.

**That's where dark matter comes into play?**

Dark matter effect is the by-product or ether.

**Also technically, what are we expanding against? Are we expanding against a vacuum?**

The expansion is not against but within the inertial system guided by internal ratios internal to a system.

**So we're not expanding?**

Space is localized, but we are expanding relatively.

**Some of the matter is just moving into the empty parts?**

Well, you are just creating space. There is no empty.

**How do we know that the space doesn't already exist? It just isn't manifested because there's nothing in it, so field isn't detectable?**

Well, if it already exists as a void or vacuum is one model, but this theory proposes that it's in ratio to mass and made of ether superstring matrix. This matrix is also a part of the system field integrated to all core mass, and each field is localized. The arrangement of space in the universe becomes like a multi-celled creation.

**And it's making more cells?**

It could be, but I prefer to think of boundaries as transitions, so there are no boundaries really between cells or beyond. What we find is either the same space or a different construct. The universe becomes something smaller and transitory within something larger, like a space-matrix complex. The idea of shape or form is dynamic, much like the conscious mind. If you calculate the inertial field of the Sun based on the mass of the Sun plus the mass of the Sun's inertial field, you get a greater distance $D$ and transition beyond to the heliosphere. Now at the perimeter, we might include a range where there is flux or oscillation in the actual boundary based on fluctuations in the inertial field projection. If the distance of the inertial field is calculated based on the mass of the Sun alone, we get a distance just past Neptune. Therefore, the region between Neptune and the heliosphere is of far less dense and stability to hold the planets within the ecliptic as with planets within the orbit of Neptune is far less to have a nil effect. The standing wave created within the inertial field is due to a spherical wave that passes from the Sun to the heliosphere and creates a standing wave that looks like a gyroscope or spindle top, and this holds the planets inside the obit of Neptune to include Neptune. For Pluto, there isn't enough gyroscopic effect to stabilize its orbit within the plane of the ecliptic. Mercury is a different case since the lambda wave of the gyroscope effect at that location of the Sun allows more divergence from the ecliptic.

**What about the gravity of the Sun when an object leaves the Solar System?**

What was once $1 / R^2$ within the influence of the inertial field from the Sun acts like a point source outside the solar system to become $1 / R$. We have mentioned that all inertial fields have a velocity $V$ from source. All is motion in the universe along with time. Acceleration of the inertial field is given by $\boldsymbol{A = SQRT\ (G) \times D}$. So then the velocity of the inertial field of the Sun is blowing on the inertial field of the Earth, and also the velocity of the inertial field of the galaxy is blowing on the inertial field of the solar system at the heliosphere. We would expect under such circumstances that the inertial field is pushed to one side from the source of the overlapping field and also of an oval shape due to rotation of the source. The density of the inertial field varies according to **density = $T^2$.** There is variation and decreasing density from the source. When the mass of the source is diffused and/or lessened, the inertial field recedes, and when the mass is increased, the projection of the inertial field is increased. If one asks the questions what happened before the big bang, the notion of nothing might be common since it is the notion of human perception that events have beginnings and an end rather than seeing "everything is" in a continual process of transformation. If the universe is considered as a manifestation of consciousness, then what is "induced" is not necessarily a bang but as in the representation of thought, and the thought is a mirror of consciousness itself. Mathematics is the pattern of consciousness. Therefore, the notion that 0 is nothing is misleading, and the concept of "nothing" is a myth. Zero represents the transformative aspect of scalar relationships. Hence, in quadratic and polynomial forms where "= 0" is used represents the transformative relationships of those separate operators that combine in the balance to zero. In proper multiplicative forms, this is inappropriate since the variables do not come in out of existence to be. So the big bang is a transition as described in this theory and represented in Euler's equation.

**$E^{(i \times \text{pi})}$ = time × velocity = -1 = -*D***

***T*, *V*, and *D* variables are induced from outside the universe.**

**Gravitational Constant Is Temperature Dependent**

**$G$ = 1 / $T^4$**
**$G$ = 1 / phihat$^4$**
**$G$ = 1 / $(0.618)^4$**
**G = 6.8556093**

**The gravitational constant is an abstraction for the inertial field of a localized system. Because *G* has units, it is a variable. *G* refers to the mass of an inertial field extending from a localized mass core. If mass of the inertial system includes the mass of core plus extended field, then *G* disappears, and the resultant volume is the revised equation as in *F* = volume1 × volume2 / $R^2$. With such an approach, all constant with units need to be reevaluated because they represent abstractions or simplifications of more complex relationships. (Gow:1-9)**

**$G$ = 1 / phihat$^4$**
***SQRT G* = 1 / phihat$^2$ = 1 / time$^2$**
**Phi × phihat = -1 or velocity × time = -1 = -*D***
**Phi = -1 / phihat**
**$V$ = -1 / time**
**$V^2$ = 1 / time$^2$**
**$V^2$ = *SQRT G***
**Temperature = $V^2$**
**$V^4$ = $G$ = temperature$^2$**
**$G$ = $V^4$**
**$1^4$ = $D^4$ (*D* is interchangeable with 1)**
**$G$ = 1 / $T^4$**

**We can see here that the gravitational constant is temperature dependent. It can also be expressed**

**Force × $G$ = $V^4$ = temperature$^2$**

**Energy and the Acceleration of Volume**

**Energy =** $D^5 = MV^2 = D^3T^2V^2$
**Energy =** $D^5 = D^3T^2D^2 / T^2$
**Energy =** $D^5 = D^3T^2V^2T^2 / T^2$
**Energy =** $D^5 = D^3T^4V^2 / T^2$
**Energy =** $D^5$ **= volume × acceleration**$^2$ **/** $G$

**As we can see, the acceleration of volume involves a definition of energy. The acceleration of volume also represents the acceleration of voltage and relationship to charge.**

**Awakened Consciousness and Awareness**

The capacity of consciousness to move in harmony or disharmony with the matrix represents free will, and this freedom transcends the place of participation because whether one is awakened (aware) or not is not dependent upon place, although the place may reflect where you are on your journey. Hence, as a person is awakened, that person is in harmony with the infinite potential that is allowed and enabled toward growth. Harmony leads to the awakened consciousness.

Consciousness has many facets, and the vehicles for the awakened consciousness are thought and will. Perhaps that in both, there is the need to traverse to the infinite and perfect harmony, and hence, our inward journey is only complete as actions mirror thoughts or is aligned, and of course, this is no truer than in the awakened self. The term *awakened* is also referred to here as the component of consciousness-termed *awareness* in my prior description. The level of awareness is the level of awakening, and this can be for good or bad and is supported by thought and will. There is truth in numbers, and as thus, I give value to the Pythagoreans. The patterns of the universe speak about patterns of truth and organization of any truth that is.

**In the One Is All and All Is in the One**

Energy or the matrix is bound up at high densities and shielded from observation. Large mass and energy is zipped up. But then if the model for physics is about particles, this leaves out the ether. If the ether is accounted for as the matrix from which all is made in various matrices and dimensional states, then the binding energy is not associated with any particular particle per se, but the matrix itself and the particles represent or are associated with various organizations of the matrix or interference nodes of the matrix. Therefore, the binding force is everywhere and probably incorrect to develop a model that associates it with a particular particle since a particle is already a higher evolution of the matrix. It may be that we can associate a simple-form particle to that high-density state, but the force itself is within the matrix. The argument here is that the particle is associated with the threshold of the binding energy of the matrix at a high-density state. Theoretically, I have defined that matrix division as superstrings $D$, $T$, and $V$. The whole universe is made from quantized matrix of time and velocity, and the force that binds them is also associated with the highest density of mass at $D^{20}$ where the resulting string force is

$F = D^4 = C^4 / G$
$F = D^4 = C^4 \times$ **density**$^2$

**This is a very high density and mass state.** And since all dimensions are related and induced by dipole motion, then the circular nature of the underlying matrix does affect a centralized relationship of force between particles in a radial relationship but is attributed to the nature of the matrix that imparts to the particle. So the basic components of the ether cannot be assessed in a supercollider. And then it is possible to argue that even beyond the matrix from what is observed, the breakdown can even go further into binary form of networks that result in this small matrix unit $D$, $T$ or $V$. Hence, the underlying mystery of the universe is inaccessible but understood, and whatever that is, is life or consciousness itself and includes everything that is. My consciousness is a reflection or symmetry of my higher self, and that is the collective consciousness or first cause or God through hierarchy by which it is established. And by transcendence, the parts are the sum, the parts greater than the sum and the sum greater than the parts, and the parts experience the sum and the sum the parts, and the two are equivalent. In the one is all and all is in the one.

**Principle of Symmetry and Principle of Hierarchy**

Now time and velocity can be reduced to a binary network of 1-0. That is the telluric agent of the agents of life. The network subdivides 1 to create anything in creation of the first cause or God. In the "I am" of universal consciousness of the first cause is awareness. In the expression of will and thought, all was created. By symmetry and to love, love had to be returned by creating and in creation. And what was created was a mirror of consciousness referred to here as the principle of symmetry. The = means symmetry and induced that images and refers to the source that supports it. The first cause is God. By the principle of hierarchy, we know there is a first cause. The primary mover or God is the "I am" and the source consciousness and awareness.

**This is interesting, the first energy.**

The first cause is all awareness, thought, and will. God's universe is mind reflecting the nature of the self-source.

**So for you the universe is consciousness itself, right?**

Yes. Consciousness is the pattern of math, which also includes a binary code, and thought is as a binary hologram that defines consciousness.

**God exists because thought exists?**

Yes. Also, God exists because there is hierarchy, and thus, a first cause.

**Interesting; that is a sound idea and may explain well the nature of thought as an intangible.**

What we see in the universe is the negative image of a positive-like photo film. The darkness is really all light. The particle is the darker form. So to review

- There is hierarchy, and so there is a first cause, which is called the principle of hierarchy. In all systems there is hierarchy. In the hierarchy of intellect, God is ultimate.
- The principle of symmetry refers to thought and action and cause. This is also a stated proof of God.

All truth is about the first cause. The 1 is in the all and the all is in the 1 means transcendence. The whole of universe and consciousness is made from velocity and time that result in the form of space through hierarchy. Further subdivision has to do with the very transcendental nature of life itself. Within the subdivision of 1 lies hierarchy, and specifically the relationship of time and velocity in 1 as combined polarities, when separated, there is asymmetry, and asymmetry with symmetry is hierarchy. The separation of polarities is asymmetry in $T$ and $V$.

**Defining Variables of the Inertial Field**

$V^2 = V + 1 = V + D$
$\mathrm{Phi}^2 = \mathrm{phi} + 1$
$(V^2)\, DT^4 = (V)\, D^2T^3 + D^3T^2$

***Mass energy inertial field = mass velocity inertial field + mass inertial field***

$M = M + M$

1. $(V^2)\ DT^4 = 2.)\ (V)\ D^2T^3 + 3.)\ (D)D^2T^2$
2. $(V^2)\ DT^4 = (V^2)\ D / G = V^3\ T / G = D^3T^2 / T^4G = MV^4 / D^4G$
   $(V^2)\ DT^4 = MC^4 / \text{force} \times G = C^4 / \text{acceleration} \times G$
3. $(V)\ D^2T^3 = V^4\ T^2 / DG = V^4T^2 / \text{density} \times \text{acceleration} \times G$
   $(V)\ D^2T^3 = V^4T^2D^3 / \text{mass} \times \text{acceleration} \times G = C^6 / \text{volume} \times G^2$
4. $(D)D^2T^2 = \text{mass} = (D)\ C^2 / G$

$C^4 / \text{acceleration} \times G = C^6 / \text{volume} \times G^2 + \text{mass}$

**Mass inertial field** $= C^4 / \text{acceleration} \times G - C^6 / \text{volume} \times G^2$

$(D)D^2T^2 = (V^2)C^2 / \text{acceleration} \times G - (V)\ C^5 / \text{volume} \times G^2$

$V^2 + V + 1 = 0$

$C^4 / \text{acceleration} \times G + C^6 / \text{volume} \times G^2 + M = \text{mass}$
$C^4 / \text{acceleration} \times G + C^6 / \text{volume} \times G^2 + D\ C^2 / G = \text{mass}$

As you can see, the mass of the inertial field is a relationship of the following:

- **Acceleration of the inertial field.**
- **Volume of the inertial field.**
- **Radius *d* of the inertial field.**

Templates can be used to create infinite number of scalar equations.

## The Alternation of Infinite and Interval in Creation and Consciousness

The binary system that results in *D*, *T*, and *V* is a network subdividing 1, and that network is by interval. The 1-0 is a pure infinite state that is all that is, aware and self-organizing, the agent of all forms. And the 1-0 is life itself, transcendent, and self-organizing. The system is pure and infinite, alternating within its organizing elements of form and also allowing for change. Hence in the system network of *T* and *V*, there is both a pure infinite state alternating with fundamental organizing unit 1-0 and also the capacity for self-organizing by the introduction of intervals, which allow transformation through time and motion. Such an organizing system is both everything transcending over the interval and the interval by polarity approaching the all and is everything of the all. Hence, we say the 1 is in the all and all in the 1, incorporating duality infinite and polarity qualification in an alternating hierarchical and symmetrical rhythm. Duality is everything that constructs the interval and in the polarity or complement made by intervals that account for the all. The very contradiction in philosophy between polarity and the duality of the all is the meaning of transcendence and life and makes the construct a self-organizing awareness. There is relationship by alternation, symmetry, and hierarchy, which is the nature of all that is. The dream of the dream, the cause of the object, is the object is also the object of the cause. I am the thing created or effect and subject to its cause, and the cause is also subject to effect. When I dream, I am the cause and also subject to all created and alternate between them. Even in physics within a contained system, in a proper system such as $MC^2$, one variable changes or alters the other, which in turn alters the source within limits set by the source.

Velocity and time are infinites and a complementary pair, a complement that has polarities within the infinite that account for the whole and permeate everything with distinction and relationship, both touching by the interval of transformation and not touching. The networks of 1-0 are pure states that are networked or interval to create this complementary system, but it is the pure state that is the whole and permeates all creation and everything that is. The whole or pure state is responsible to create all finite things in the interval and bring unity and balance to everything in 1 and 0. Instant changes from one state unto another are governed by the binary states of 1-0. We have explained the nature of consciousness and the universe and its representation, which is math, to what is and to what came before. Math is the pattern of consciousness. When consciousness acquired form, there was time or thought, which is also motion of the complement to velocity or will. The 1 is the representation of the universe as consciousness by symmetry; the 1 is supported from the outside and is created from the outside as induction and, by symmetry, reflected the source. And by asymmetry, time and velocity are complementary as with thought and will. And together, time and velocity are one and 1, and in higher dimensional forms are combined to form more complex symmetries, which become hierarchy. Hierarchy means there is a first cause, which is the principle of hierarchy.

Now imagine something that exists and is transcended by a transcendent part or emergent and transcendent; the 1 in the all and the all in the 1, it means life and consciousness is self-organizing. Our perception is in the transitioning of form, but life includes all forms. Duality and extremes exist in the form of life and consciousness and occupies all that is. The perception of the human mind lives in the transition of the forms of consciousness, but the higher self includes the all. The all affects the cause of the particular effects cause to the all. So what is the infinite? It is life and always was. What came first, the chicken or the egg? Life always was, and life began life as consciousness. The universe of emergent forms is just another form of consciousness. The transcendent nature of the infinite and its alternating nature to the expression of form through transcendence is all consciousness.

The interval is the all that creates the complementary or polar, which in turn makes the interval. Symmetry and asymmetry alternate, and thus a result are hierarchy; and thus, concepts of duality and complementary polarity also alternate. Ideas of the infinite are approaching the whole and the whole defining the parts alternate and are 1. Consciousness is the alternation of the infinite with the finiteness of form, the alternation of cause and effect. The alternation of acting and being acted upon and alternation of motion with thought—that is in the very nature of the matrix. There is no pure symmetry anywhere. The intervals of experience are a consequence of the alternation between the infinite and formal, symmetry and asymmetry, the alternation between cause and effect through symmetry of functions, and the asymmetry of the operators and the asymmetry of function with symmetry of operators. The 1 is interval and balance of complementariness and the 1 is all; the zero is unity of the all and the balance of intervals and complementariness through transformation. The beginning was infinite self-organizing awareness, and through form of the infinite, consciousness is; by the organization of 1 and 0 by various modes of alternation, consciousness is and created the first cause, reflecting its nature of symmetry and hierarchy. The form of consciousness was the relationship of self into roles' distinction, and this is where time emerged. All life is God and all that is, is. And complexity is resulting from the differentiation of consciousness infinite. From the outside, this universe is as a constant, but transcended, it acquires all its force, and life is both objective and transcendent. The 1 in all and all in the 1 allows free will and interaction. Free will within consciousness as well as between other life-forms, this transcendent nature allows self-organization and thus the first cause, through its very nature, created and imparted those qualities to all creation. There is an alternation between interval and infinite in consciousness. The beginning of consciousness is the "I am," and from that, all emerged and the diversity is the complexity we see. When we speak of consciousness, let's consider form to be where the interval and infinite are organized and alternate. And let's say everything is consciousness, even the universe, and everything is life; and in that we have the alternation between interval and infinite in form, and that form is diversified as it exists presently. So then we no longer need to debate what the universe is and what life is, or what life is and what consciousness is since they are all the same.

Perception is this: universal consciousness filtered in a dream. Your consciousness is experiencing a particular form called human, and that is a perception you know. In that perception or dream, the universe carries with it in your perception all the energy and weight of an emergent system, but from outside the dream, the dream has no energy at all and is a constant, a memory such like a mirage. Perception is not universal, but your higher self is. Perception alludes to the universal as the universe becomes a representation or analogy of truth. Being aware is the hierarchical first of consciousness and is life, self-organizing, and everything that is has it, always was, and there never was anything without it. The notion of beginning is purely human. It's hard to imagine a state of everything that always was, but this is the nature of the first cause and everything that is. I imply I am aware, and everything has this quality of awareness. Only by transcendence is there free will. And so without free will, there is no life and no consciousness. In that coming into transcendence of this dream or God's, manipulation by free will is allowed. Scientists are too caught up in making emergent process of this universe primary if you think that the universe supports itself and caused life and are contradiction, since an emergent process cannot create that which is beyond its nature, which life is. Life requires transcendence and freewill as consciousness, and hence, not only life but the universe is also consciousness of the first cause, and it exists everywhere and in all things. But understanding what people have done to make emergence primary in philosophy is an incorrect axiom, and as such, most philosophical constructs based on incorrect axioms are myths. All that is in this universe is a dream; the real life is your transcendence into this dream and all that came before the dream. So the notion that this world is the source of life is nonsense and also that it has all this energy is also an illusion. The only things important are those things that the first cause causes to importance; it's not about evolution of the universe or of species nor about goals and accomplishments, but only in the harmony of you in the now of existence. The only value in achievements is how it helps the dreams of others and in being of harmony with the first-cause matrix. That which enables is fair to the positive good of all, and in that is all morality and conduct.

**"What is truth?" asked Pontius Pilate.**

A fundamental question about the importance and meaning of being refers to truth. All truth is about the first cause. It is impossible to imagine when you see all mass and action in the world as of the utmost importance because you see mass of the universe as so immovable and, hence, deserving all the primacy of scientific thought and method. The dream becomes greater than the quality of being, the emergent processes more important than the hierarchy that caused them from the first cause. But for those where the action is more important than the effect, where the cause of free will and its quality of being is the fundamental source of dreaming, the dream appears as it does the dream, where the value of being is now reversed to the role of being as harmony and the location an effect exercised. What is pain then? Pain is communicating a message of conflict with the nature, true nature, of being and natural flow of energies. You think emergent rules can, in themselves, lead to evolution and attain something greater than the emergent rules. But ironically, emergence is an extension of transcendence and cannot be achieved by opposite approach. Only by transcendence can there be an emergent system and also exemplifies a system of hierarchy and symmetry from the first cause. Without self-organization of all reality in life or consciousness or transcendence, nothing would move. Hence, to be is to create, and to be within and subject to and be affected and also to organize, to be both object and subject.

**What's life all about?**

To enjoy the present moment, so many people have it and don't know what to do with it. You are the nature of your intent in the present moment. Free will acting through this intent reveals your nature. That which acts according to the nature of purpose of the first cause is enabled infinitely. Purpose refers to the acting upon the nature of cause and effect in accordance with the nature of the first cause, and thus the process is enabling life, which is also free willed and self-organizing. The matrix enabling infinitely and can only become limit where the transformation enabled by the matrix is perceived as boundary; boundaries are illusions where intuition is disabled by fixed perceptions. In this condition, there is pain acting on precepts by very nature in revolution with self. The option to defer exists both in the mode to enable and disable, and hence, the argument for duality based on singular precepts is a narrow abstraction and separation of which infinitely enables. There is the argument of two and of opposites, and then there are also the circular possibilities of the infinite where the negative argument is polarization of truth with contradiction; pain and separation from higher self.

**But the present moment has gone before you realize it's there?**

The present moment is present and in motion. If you want to realize something, wake up. Develop some awareness.

**Develop awareness of what?**

Of yourself, the "I am" of awareness, and your interaction with cause and effect of consciousness. In that are the love that wants to be loved, action and return of action, symmetry and hierarchy of creation, where your awareness adopts the form of consciousness as a pattern emerging from the first cause. Creation enables orders according to its evolving, changing, and infinite nature or truth.

**What's the secret of true happiness?**

To love and be loved by harmony with the infinite matrix that enables infinite action. The infinite matrix is the consciousness from which precedes all, the first cause, God, that enables infinitely. The matrix includes everything, such as the universe and all dreams.

*Your mantra should be*

I am and in all harmony and know all and am infinitely embraced, and without the turbulence of contradiction, I am at peace. Your transcended experience is fulfilled by its alignment to the nature of the matrix by being both object and cause of infinite law of creation. As a transcendent being, you are the source and cause of actions, a living being and self-organizing, and in transcendence is free will. Everything is life. Your nature is in the alternation of interval with the infinite emanating from the first cause, and its nature is what is referred to as truth, that which is not tarnished or changeable referring to the nature of existence and in that refers to a mode or being, and hence, morality exists. There is inferred an appropriate mode of conduct infinitely enabled by the true nature or truth of the universal consciousness, enabled infinitely according to its nature, which the harmony of its own collective is being.

**The material universe is merely an emanation of the mind?**

Yes, so it really has no energy. By symmetry, it's a state of being, like when you dream at night, you are both the cause and subject, life, self-organizing.

**Let me rephrase to perfect the alliteration: the material universe is merely a manifestation of the mind?**

Yes, what is emergent is what the mind has decided to become constant.

**Nice.**

Everything is transcendent. Only in transcendence is free will, where you are cause and effect together in one system, relational and proper. They are but one system. Nothing is a myth. Everything is life; it always was.

**Can the mind manifest objects and motion?**

Yes, it is possible. In the right mode, you could transport yourself to Mars and back. But then if you see everything as boundaries, this is impossible. At some point, science will merge with spirituality, but all is science if you have the correct explanation.

**That's what Jesus said, move mountains with your mind etc.**

Yes, that has to do with harmony with the collective consciousness. So when you finally understand and are enabled, along with that requires certain responsibilities that align with the universal free will. It's like if you love someone, you do things with that person that enables both. So the very act is incorporation, infinitely enabling, with awareness, will, and thought in union to harmonic consequences. The notion of morality is infinitely enabling and reflecting the true nature of the matrix. Right is not opposite to wrong but infinite enabling. Wrong is a polarized condition and contradiction of self.

**So how do you free yourself of the negative boundaries that block your path to limitless reality?**

Understand that limitless reality is the nature of self, and the opposite is self in contradiction. And awareness of self is a matter of degree to the nature of self and being. In the true state of "I am" is all knowledge. Your consciousness is an alternation between transformative intervals and infinity. "I am" is infinitely self-organizing, and in the adoption of form, there becomes will and thought. Awareness of truth is the beginning wisdom where consciousness possesses infinitely enabling will, correct thought, and purposeful action. Life is one form of consciousness perceiving another. The whole universe is that. Mathematics is the pattern of consciousness. What physics will eventually be is the study of consciousness form. And that science is the science of consciousness and spirituality and mathematics; all are one and the same. Life has always been, and life is love, the action of conscious form, and evil is contradiction of self; in intent it is born in very contradiction to the nature of self as enabled by the universal consciousness.

## Chicken or the Egg, and which Came First?

We have all heard the story of the chicken and the egg and ask what came first. It is human to see the beginning and end to every interval rather than the interval perceived as transition or transformation of the principle of symmetry and hierarchy. Neither the chicken nor the egg came first. Since all transformation and related form exists in the alternation between infinity and interval, all form, by principle of hierarchy, refers to the first cause. We are both the chicken and the egg. As a transcendent being, you are both the object and cause of all actions, which is also the definition of life and free will, which is all consciousness of everything that is.

## Cosmological Constant

In this universe, change is constant since the variables implied by phi and phihat in velocity and time respectively are subject to change. Constancy refers to certain limits of form related to these variables. We have said before that true constant that have no units refer to functions outside the universe that create the universe. Hence, the existences of the universe by mathematical operators outside the universe that also transcend the universe are perceived as constants from emergent world to transcendent source. The cosmological constancy of the operators that maintain the universe and support it from the outside is the true cosmological constants. In that group, we can include the ratio pi, prevalent, constant, and immutable throughout the universe. Only the constants without units are the true cosmological origin and constancy that supports the existence of the universe through symmetrical operation.

## Energy

Einstein said mass and energy are related by the "same thing" in $\boldsymbol{E = MC^2}$, which is not a very good explanation either of energy or mass or the relation between the two. Energy comes from $MV^2$ discovered by a girl named Emilie du Chatelet and Gottfried Leibniz about two hundred years before Einstein, and Emilie said that energy was proportional to the square of the velocity, $\boldsymbol{MV^2}$rather than mv proposed by Newton; mass and $V^2$ were conserved and that was called energy. That still doesn't answer how mass and energy is related by the "same thing," in MC^2 which Einstein said. On November 29, 1903, it was Olinto de Pretto that first published $\boldsymbol{E = MC^2}$ a 62-page paper "*Ipotesi dell'Etere nella Vita dell'Universo*". He theorized that mass and energy are the same vibrating energy is an ether theory; equivalency of mass and energy was also proposed by Samuel Preston in 1875 on which De Pretto's theory was built.

There is an answer in this theory:

Mass and energy are related by time and velocity. Time and velocity is the same thing that relates mass and energy. Why? Everything is made of various dimensions or powers of time and velocity. You only need three units in physics to determine everything. Those are velocity, time, and dimension or *V*, *T*, and *D*. That we use the term *energy* is like using Chinese, but here the units have been simplified.

So $\boldsymbol{E = mv^2}$ can now be

$\boldsymbol{D^5 = MV^2}$ **where mass =** $\boldsymbol{D^3T^2}$**, then** $\boldsymbol{D^5 = D^3T^2V^2}$
$\boldsymbol{D^5 = MC^2}$
$\boldsymbol{T^2}$ **= density**

### *So what is energy?*

It is a configuration of space where higher density space, which we call an atom, is surrounded or generates the lower density space, which we call space. Space is localized and made of *D*, *T*, and *V* superstring matrix. All a particle is, is interference in the wave-matrix pattern of space that is denser. And energy refers to the mass of space and its velocity squared. So imagine a particle that generates space and that space has a velocity like a magnetic field and the relationship of that velocity per particle is energy. The velocity refers to the velocity of the inertial field of space. We have arbitrarily assigned that event to the word *energy*.

## Derivation of Fine Structure Constant from Euler's Equation and Relationship to Alpha

$E^{(2 \times i \times \text{pi})} = 1 = V + T$
$E^{(-2 \times \text{pi})} = 1$ (raised both sides to $i$ and $1^n$ is 1)
$-2 \times \text{pi} = 0$ (took the $\ln$ of both sides and $\ln [1] = 0$)
$2 \times \text{pi} = -0$ is an inversion
$0 = (T + V) -1$ from $V + T = 1$
$-0 = (-T - V) + 1$ from $-V - T = -1$

When we get to -0, there is an inversion. This goes back to the argument that the concept of nothing is a myth. Hence, our understanding of zero must change. Zero is a dormant state of something and identified here as unity as well as transformative principle. Here, -0 is the door to inversion of physics relationships where relationships of time and velocity are reversed.

$-0 = (-T +1) / (-V + 1) = -0$ inversion and - polarity
$-0 = V / T = T - 1 / T = +T - 1 / -V + 1 = -0$
$-0 = -1 + 1 = -0 = -T - V +1$
$-0 = (-1) + (T / -V + 1) = -0$
$-0 = (- [-V + 1] / [-V + 1]) +T / (-V + 1) = -0$
$-0 = ([V - 1] + T) / -V + 1= -0$
$-0 = (V + T - 1) / -V + 1= -0$
$-0 = 0 / -V + 1= -0 = -T - V + 1$
$-0 = -T - V + 1 / -V + 1$
$-0 = -1 / -V + 1 + 1 / -V + 1= -0$
$-0 = -T - V / -V + 1 + 1 / -V + 1= -0$
$-0 = (-T - V) / (-V + 1) + 1 / (-V + 1) = -0$
$-0 = (-T - V) + 1 / (-V + 1) = -0$
$-0 = 2V / 2(-V + 1) = V$ (proper) $/ (-V + 1)$ scalar $= V / T =$ acceleration. $= D / T^2 = D /$ density $= -0$; $C / T$
$-0 =$ inversion $= V / T =$ (phi) / phihat $= -0$

Phi / (phihat × 360) = fine structure constant

We can also say (phihat / phi) × 360 = 137.549, and the inversion of that is also the fine structure constant.

2 pi = - 0 = $V / T$ = fine structure constant where units are equal to 1
Alpha = 1 and 2pi

Fine structure constant represents a ratio of velocities of proper and scalar systems. It relates the ratio of the velocity and time. This can also identify with the ratio between current and charge through $V$ and $T$ respectively by relationship with $D$. By proper system, we refer to multiplicative systems of ×, and by scalar systems, we refer to additive + of quadratic and polynomial forms. To summarize, the fine structure constant is a ratio of

- current as a subset of $V$ and charge-magnetism as a subset of $T$ and relationship to $D$,

- velocity and time and acceleration,
- proper velocity and scalar velocity,
- *C* and time, ratio of time with the speed of light, and
- *D* as related to and a variation of density.

***Fine structure constant = 2 pi / 360 where -0 = (phi) / phibat***

***Cosmological Constant and Relationship to Fine Structure Constant Alpha***

Gravitational potential = *I* = joules / kilogram

$I$ = joules / kg = $D^5$ / mass = $D^5 / D^3T^2 = D^2 / T^2 = V^2$

$I = V^2$

$I$ = 4 × pi × Gp = 4pi $(1 / T^4)\, p$

$p = I$ / 4pi $G = V^2$ / 4pi $G = V^2T^4$ / 4pi = $D^2T^4$ / 4pi $T^2 = D^2T^2$ / 4pi

$p = (CC)\, C^2$ / 8pi $G$

$(CC) = p$ 8 pi $G / C^2 = (D^2T^2$ / 4pi) 8pi$G / C^2 = D^2T^2$8pi$G / V^2$4pi

$(CC) = 2$

The cosmological constant is 2 according to modern equations analyzed by Trans Dimensional Unified Field Theory transdimensional analysis. Thus,

$(CC) = 2$.

If you have three units, that allows simplification of functions that cancel to 1 and leave the true constants such as pi. Emergent constants sometimes are the product of true constants with values assigned to variable functions (emergent constants like *G*).

***Fine structure constant derived from 2pi and related to the cosmological constant is as follows: (CC) pi / 180, which is (CC) (phi) / (phibat × 180)***

***2Pi = -0 = -V - T + 1***

Meaning that pi is both a unifying and transformative principle brings both scalar and proper forms together.
-2pi = 0 = $(V + T) - 1$

***Alpha***

Alpha = 1/2pi = 2(V × T) / 2($V + T$ - 1) = $(V \times T) / (V + T - 1)$ and

Alpha = $e^2 / hc = D^4$ 2pi / $(D^5T)\, V = D^4$ 2pi $T^2 / D^5T D^2$ = 2pi / $D^2$ = 2 pi / charge

Alpha = $e^2k / hc$ = 2pi, where units = 1

Pi is a relationship between scalar and proper systems.

In Einstein's equation, the cosmological constant is 2 when trans-dimensionally refigured, and then he has

8 pi $G\, T / C^4$ = 8 pi, where pi is cosmological constant *CC*

$\boldsymbol{T = CC\,C^4g / 8CC\,G = D^4g / 4CC}$

$\boldsymbol{8CC\,G\,T / C^4 = 8CCg / 4CC = 2}$

We can refer to the cosmological constant as 2 and the fine Structure constant as 2pi. These constant are involved in generating the universe and, as emergent constants related to functions, become the traditionally known alpha. But in truth, there are several constants, but here we see the reoccurrence of 2pi. Those constants without units have their origin outside the universe, and those referring to and combined with units are emergent as with phi and phihat.

**(Phi) / (phihat × 360) = fine structure constant coming from 2pi / 360**
***CC* = 2**

***Fine structure constant = cosmological constant × pi / 360***

So from Euler's equation, all physics can be derived since it represents the creation of the universe. From it also comes the fine structure constant and cosmological constant. And going further, we can say that all mathematics is representation of life, physics, consciousness, transcendence, and free will, and all concepts are equivalent. Everything is good. In the multiplicity of nature, everything is here to enable according to the nature of the matrix and consciousness. What we refer to as evil is merely the contradiction to the true nature of self. Transcendence means free will. You are both the object and cause of everything you are. The true meaning of transcendence is illustrated in the 1 in all and all in the 1. Math is the pattern of consciousness, which is all reality. Life is merely a dream or receded conscious state. From outside the universe, this universe has no energy and is represented by constants and vice versa, but inside the universe, experience carries all the force of experience we know. True self is transcendent to this experience. Life is transcendent and self-organizing; you are math that can change. The universe is established by the first cause, and those things that are constant have been established, which we term emergent. In life established from the form of consciousness, there is alternation between interval and infinity, hence, constants emerged; in the nature of that also refers to the proper nature of being, and hence, there is value in conduct or morality. From awareness emerged will and thought, and the form of consciousness was established. This same form is the form of what we term as *physics*. All numbers refer to relationships of modes of being, truth, and anything you want to know. *D* is resonance and 1 and displacement and the emergent principle arising though the consciousness of the creator in the creation of this universe. It is a projection of consciousness. $\boldsymbol{E^{(i \times pi)} = -1 = T \times V}$ is the creation of the universe, the big bang, and described that action of consciousness where awareness becomes the form of consciousness through will and thought. Hence, both we and the universe are the resonance of God's consciousness that extends to all reality. And the form of that resonance alternates between infinity and interval. So without incorporating the emergent variables (related to phi and phihat) with the true constants related to pi, the cosmological constant is 2 and the 2pi is the fine structure constant that includes the cosmological constant. The value of the fine structure constant associated with variables leads to 1/317. In Euler's equation, 2pi creates the phi relationship of an emergent system of phi and phihat or velocity and time respectively. Euler's equation is the source of all the math and order of the universe. Pi is the true constant originating from outside the universe. The 1/317 is the emergent constant produced by pi through an action of symmetry in consciousness from an exterior reality consciousness. The fine structure constant in this universe is derived from that outside, which is represented by 2pi; functions appear as constants from extra-universe positions. And this perception works both ways so that functions within this universe appear as constants from the outside and vice versa. Constants do not mean that the universe is frozen but represents motion and the form of consciousness alternating between infinity and interval; reality is alternation between infinity and interval, and the constant is the interval.

**Have you ever actually started from a set of small of assumptions, and then found that you can determine the cosmological constant in a way that agrees with the current expected value?**

Yes, it all comes from Euler's equation. The relationship of phi and phihat is the 1 in Euler's equation, and it derives the fine structure constant, which includes the cosmological constant. The difference in value is the result of including the value of variables into the product, which has been shown for the value of 137 in the finite structure constant and relationship to phi and phihat. All numbers mean something and all constants. Phi represents binary network of velocity subdividing 1 and its compliment. The mathematical relationship of those represented in phi and phihat also represents the relationship of representing variables or functions time and velocity. 1 is equivalent to *D*, and units are there in *D*. Limits and infinity and together represented in numbers. Mathematical work done on phi and phihat represents relationships of time and velocity. Euler's equation represents the big bang. Numbers and dimensions are mixed. Pythagoras was right.

**Are you saying dimension and number are alike?**

Yes. Numbers have meaning.

**So 1 is the basic dimension, right?**

Yes. When you see any constant, it represents a function from somewhere.

**Nice. It's just that orthodox physics stopped there, and you decompose it into $V + T$, right?**

Yes, because ideas like time and velocity needed further development. $V + T$ and $V \times T$, 1 and -1, 0 and -0, pi, phi, phihat, and all are related and connected. Conventional physics is a flatland in this light, which requires further iterations to clarify connections rather than view number and dimension as isolated concepts.

**They don't see the connection although it is hidden in plain sight, just like number once was.**

I have derived the fine structure constant via Trans Dimensional Unified Field Theory from Euler's equation. **(Phi) / (phihat × 360) = fine structure constant.** And that can be derived from Euler's equation. -0 is the inversion in the creation of the universe to become (phi) / phihat. Everything is a form of consciousness; the universe is the consciousness of the first cause. Reality is one form of consciousness perceiving another. Transcendence means free will (including the cause and effect of experience) and is also equivalent to math, consciousness, and all life. Therefore, nothing is dead. At the smallest level, life is a self-organizing binary system. What are appearing emergent are merely those things of conscious state that show form and are set through form by the will of the first cause. Outside the universe, everything here appears as a constant as pi does representing functions of the outside. These functions are functions of consciousness. Therefore, the universe has no energy, and what is felt is merely the illusion of a dream, a receded conscious state. Hence, that ladder of materialism that Hawkins speaks of does not exist. Euler's equation is the big bang. Free will means transcendence, means consciousness, means life, means math. All are one and the same. In transcendence is the form of consciousness in will, awareness, and thought; the system of form is established through hierarchy and symmetry where the awareness is the first cause. By the awareness of the first cause, the form of consciousness was set into motion by transcendence through will and thought. This formal relationship is mirrored through all reality and mathematics. Symmetry means action and return of action, so to act in consciousness, God had to create. To love, the first cause manifested through creation. The creation of the universe is mirrored in the same manner you think or dream; you are both the cause and effect of experience, which are as even in systems of physics, one not existing without the other. $\boldsymbol{E^{(i \times pi)} = -1 = -D}$ **= time × velocity** and is merely induction. All points in space are equivalent to center. The universe is describing conscious states, and so it's just the pattern of dreaming. So then pi is a constant representing outside the universe.

**The material world is just the dream condition of the marginal living entity, the soul?**

Yes. I doubt we really get to see what the source really is. I believe it to be a binary system. Consciousness is form relating on a level of three, where the = means symmetric equivalent action and the hierarchy refers to cause. So when you write an equation, you are explaining conscious forms.

**Is the human soul fundamentally different from a computing machine?**

The functioning of soul is transcendent, meaning that it's both cause and effect. A computer follows set equations, but a life-form has the capacity to change those equations and alternate between form and infinity.

**Life has highest uncertainty?**

Yes. I could imagine a system far more unique than currently devised. Life has free will or a non-emergent concept. The emergent system is but many equations running together and gives the illusion of freedom.

**Physics doesn't seem to formalize the notion of life; they use the clumsy biology notion or just don't care.**

That which is transcendent is free from the emergent. Physics is that portion of transcended given constancy. All reality is conscious form, so everything is life. When you dream, you are the cause and effect, observer and participant.

**So uncertainty = life?**

Life is the alternation between form and infinity, between interval and infinity; that form is three or ternary, but below that, life itself is a self-organizing binary system. On an infinite scale, life always was. Nothing is a myth where we can propose -0 as a combination of functions and 0 also as something. Eternity is infinite expanse of life, and this is but one interval within all realms of reality, and from the outside it is a constant. *Uncertainty* is another misunderstood word. I like to think that free will is enabled by the larger reality.

**I never really liked the asymmetry introduced by 0.**

It's true; there is asymmetry in 0 and 1, which means that hierarchy points to a first cause. The intuitive notion of a first cause is no accident.

## The Relationship of Color Theory to Physics Laws

This theory proposes a pattern of the universe, which is also a representative pattern through all phenomena that is also related to relationships of color. Why? A unified theory would be a pattern recurrent throughout nature and phenomena and a variation in resonance *D*, *T*, and *V*. At the very inception of this theory was the thought that patterns of nature repeat throughout various phenomena and represent basic relationships of natural mathematical patterns of nature. Within this approach was the thought that through color, theory related to the mixing of color light is an analogy of truth as postulated before, that nature is an analogy of truth, and the pattern representation through natural patterns or patterns of resonance is the mathematical unified field. Hence, the concept of mixing colors of light was determined to be equivalent to multiplication, and that of pigments related to addition. Approaching the problem by using color of light multiplications, various colors were assigned physical properties. The pattern of physical color to determine white light was a minimum of three colors overlapping. And hence, in this presentation was the basic tetrahedral structure of form math as multiplying two functions to get a third. As the most basic relationship were *D*, *T*, and *V*. I identified *D* with color resonance and the

## Colour Theory Resonance Matrix

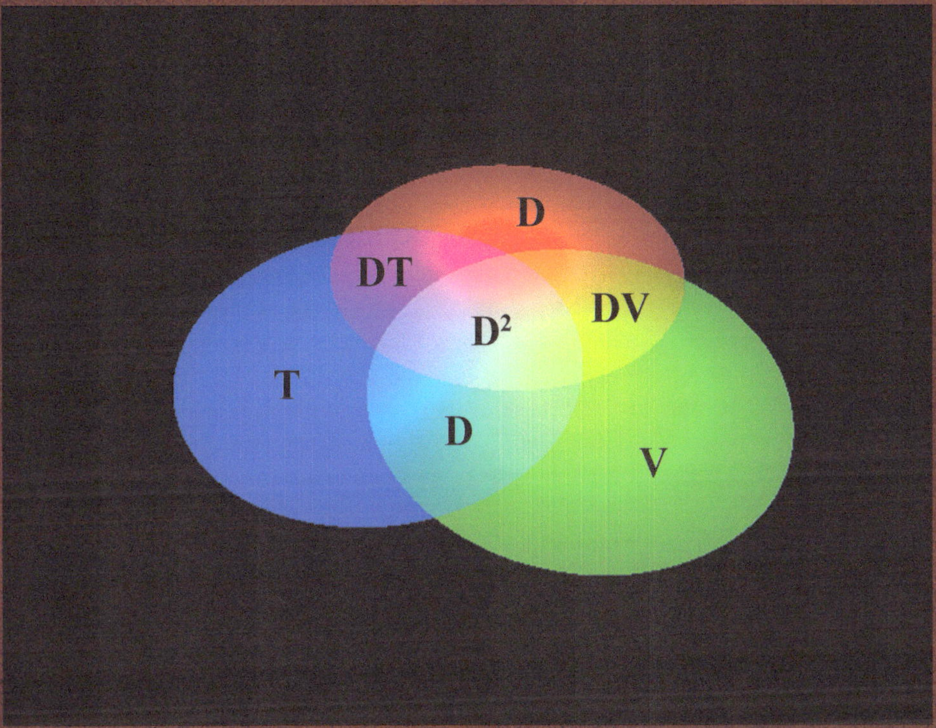


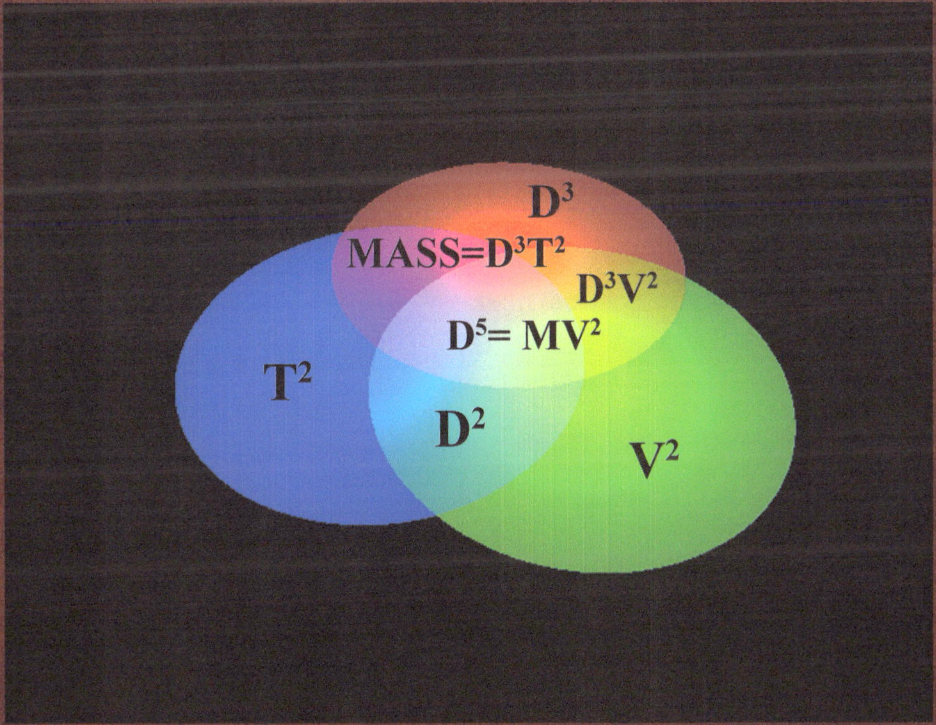

various colors as multiplications of resonance; the colors multiplication factors also represented various multiplications of *D* or *VT*, which were also physical properties of the matrix of space in various dimensional states, configurations, and matrixes. Therefore, resonance could be linked to the properties that we see commencing with *D* and increasing in powers of multiplication. When the system led to $V^2$ at various multiplication factors of $D^n$, these became options for understanding mass in energy at $MV^2$, and an option was <u>$D^3T^2$ for mass</u>. Matrices were developed for *D*, *T*, and *V* through multiplication represented in the intersection of three spheres. At the simplest level, three spheres intersecting with *D*, *T*, and *V* result in $D^2$. To explore $MV^2$, one sphere is assigned $V^2$. By super symmetry, we assign $T^2$ to an adjoining sphere. Since all mass has volume, the remaining sphere is $D^3$. Hence, the resulting overlap of three spheres is $\boldsymbol{D^3T^2V^2}$, **which is also $\boldsymbol{D^5}$.** $D^3T^2$ goes to mass to make the matrix work as a system of resonance multiplication of three units *D*, *T*, and *V* while assigning $V^2$ to $MV^2$. This was later derived mathematically without the color-resonance mathematical analogy. Also, the difference in *D* and *T* was V, and also between $MV^2$ or energy and force at *MA* was also *VT* or *D*; it also applies dimension per dimension as a periodicity of resonant multiplication. And defining *D* as resonance and powers of *D* as hierarchies of resonance leading to various physics properties, the connection to a periodic table of *VT* properties emerged. Thus, the conclusion was that all could be written in three units where *V* and *T* products were the result of all physical properties and consequence of the simplest form *VT*. *V* and *T* are complementary networks or *D*. And this was further amplified by identifying that all numbers represented functions of *VT* or (phi)x(phihat) and that *V* and *T* could be associated with phi and phihat respectively as connected to applying meaning to all numbers related to proper or multiplicative functions in the universe; this allowed the proposition of additive or scalar system structures in math to derive quadratic and polynomial forms of *V* and *T* of additive system functions. Through a transdimensional approach, 0 and -0 emerged as something rather than nothing. The resonance *D* describes the polarity between time and velocity, giving stability to the space matrix. Referring to *D* as a form of resonance, the entire universe then becomes an expression of consciousness, also related to the multidimensionality of conscious states and conscious resonance. The math of the universe is explaining the math of consciousness, and they are one and the same; a pattern of symmetry and hierarchy induces one state from another and all and interrelated.

**<u>Beta Function</u>**

$\boldsymbol{B(V, T) = (V - 1)! \times (T - 1)! \,/\, (V + T - 1)! = B(T, V)}$

$\boldsymbol{B(V, T) = (-T)! \times (-V)! \,/\, (1 - 1)!}$

$\boldsymbol{B(V, T) = (-T)! \times (-V)! \,/\, (0)! = (-T)! \times (-V)! \,/\, 1 = (-T)! \times (-V)! = T! \times V!}$

<u>$\boldsymbol{n! = n \times (n - 1)!}$</u>

$\boldsymbol{n! = n \times (n - 1)!}$

**Interchange LHS and RHS**

$\boldsymbol{n \times (n\ 1)\,! = n!}$

$\boldsymbol{(n - 1)! = n! \,/\, n}$

**Substituting $\boldsymbol{n = 1}$**

$\boldsymbol{(1 - 1)! = 1! \,/\, 1}$

$\boldsymbol{0! = 1 \,/\, 1 = 1}$

**$0! = 1$**

**$1! = \ldots 1 = 1 \times 2^1 - 1 \times 1^1$**

**$0! = \ldots = 1 \times 1^0$**

**$0! = \ldots\ldots = 1 \times 1^0 = 1 \times 1 = 1$**

**Beta function shows the dependence of all physics interactions on time and velocity. The fine structure constant is also dependent upon the relationship of time and velocity. The fine structure constant is not really constant since it depends on the value of scale of phi and phihat as velocity and time.**

$$D + T / D = D / T = V$$

$$T^2 - T + 1 = 0$$

$$V^2 - V - 1 = 0$$

**Awareness and Determination of Axioms**

You cannot prove that something does not exist; only that it does. Misuse of logic is the greatest fault. Truth has many axioms, and by truth I mean what the nature and math of reality really is with all its meaning and implications, effects, and possibilities. Truth has its axioms, and people have their intent. When the two are one, there is harmony. If the truth is to refer to an unknown God, then the greatest obstacle and asset in the understanding of truth or God would be free will. In intuition does the greater truth lie, but often ignored by reason, and intent that is not in the harmony or alignment that would be required to perceive that truth if it existed, only because free will allows divergence. The universe being a source to enable infinitely is at odds with the nature of that same free will that allows conflict with the true nature of self. Evil is conflict with the true nature of self. All is good and here to enable. The proposition that there is no first cause negates the concept of hierarchy that refers to a first cause. And if we adopt the axiom that all reality is one form of consciousness perceiving another, then the idea of the universe is merely an extension of a collective consciousness, an idea. And while all ideas are born, they are born through intent of will and thought brought from awareness. Action and return of action and hierarchy requires there to be a first cause or God. I cannot exist and love in the thought of mind without a recipient, and that recipient is God, eternally feeding response to all my actions of thought, whether active or passive. I cannot exist as thought and feel without the supporting existence of another since this is not represented in math, which also refers to the pattern of the universe. If there is a God, the first cause is ultimately attached to all experience, whether good or bad. And that which exists through creation cannot exist alone, so we are all inextricably interdependent. We are all unique qualities of God. When you deliver an idea, the notion of whether it is measured or enabling is in the manner of intent and whether the axiom is divested of all the myths of culture, which are the results of a multitude host of preceding historical and incorrect axioms. The circle represents a range of an idea. It is the natural human inclination to step within this circle for a debate. As soon as this is achieved, you have merged your thoughts to the scale of the definitions presented, thus limiting a proper evaluation. The correct response is to evaluate not the content but the manner of thought, and then first to seek exceptions to the thought and rephrase the thought in a manner that enables an unmeasured and divested understanding of the subject. And so with whether God exists or not, the historical notion of that God is associated automatically with many axioms that are totally untrue. For example, God created from nothing (nothing is a myth), nonexistence (myth), God is like Zeus, God and scripture means religion, and so on and so forth to the point that language and culture has enormous divergence with truth, which is a priori. There are people with their ideas, and there is truth. Whether there is a God or not, to equate God with religion is false since by definition, one represents truth and the other people and their many myths. Hence, I say to listen to the manner of thought more than content, for therein lays clarity. There are emergent rules, and there are transcendent rules. Emergence has to do with what science refers to as measurable, and excluding life, all is in the form of equations. Hence, what is described is mathematical and freedom of the systems and how they interact is an illusion. The idea that Heisenberg uses multiplication to illustrate the position and momentum is incorrect since that is appropriately in the realm of additive systems or scalar systems and not proper multiplicative systems. So then we have a universe basically set according to defined laws. The only real randomness is by foreign observation and visitation of foreign systems, and this is called life and, more appropriately, uses statistics. This interjection and participation of life is consciousness of various states. By principle of symmetry and hierarchy, the very option to act requires that the system allowing participation also is a form of consciousness. The difference between the universe as participants and as a field allowing participation has to do with hierarchy. Hence, the universe where actions are allowed by participants can also act if willed to do, and this is defined as participant first cause. Only in transcendence is there free will, and so everything is mind, both writer of the screenplay and participant, cause and action. All experience is quantifiable, and the "non-quantifiable" is a myth since all is a representation of math, whether obvious or not. All reality is one form of consciousness perceiving another, by principle of symmetry. The very structure of math suggests this. The idea of quantity is reserved for emergent systems that are induced by transcendent ones, and in this process, what is transcendent are merely constant from perspective of emergence and vice versa. Quantity is an illusion, math is a pattern, and the pattern is the relationship of experience. When we speak of quantity in this universe, we refer to units, and as an emergent systems induced, all units can be expressed as three or $D$, $T$, and $V$; a simplification that allows for a better understanding of how the universe was induced, the true nature of mathematical patterns, and its representation in life, which is everything, and also consciousness.

## Electromotive Force

**$E$ = electromotive force = magnetic flux / time = $D^3 T / T = D^3$ = volts**

**$F = qE + qv \times B = D^4$; can give a 0 result because it is scalar**

**$F = u\, QQ / R^2 = D^4$ is a proper system**

**$U$ = permeability of intervening medium = tesla × meter / ampere**

**$U = (DT)\, D / DV = T^2$ = density**

**$Q$ = magnitude of mag poles = ampere × meter = $(DV)\, D = D^2V$**

**$F = T^2 \times (D^2V)^2 / D^2 = T^2 D^4 V^2 / D^2 = T^2D^2V^2 = T^2D^2D^2 / T^2 = D^4$**

**$K = 1 / 4\text{pi} \times e = C^2 \times u / 4\text{pi}$**

**$u = H.\, M^{-1}$ = 4pi newton / ampere$^2$ = 4pi $D^4 / (DV)^2$ = 4pi $D^4 / V^2D^2$ = 4pi $T^2$**

**$K$ = 4pi $V^2 T^2$ / 4pi = $D^2$**

## Mechanics of the Space Matrix

The relationship between velocity and time in the constant $C$ (constancy of systems of motion) refers to the elasticity of the medium of the space matrix that related the stress and strain of velocity and time respectively in the space matrix. Constants with units refer to the ratio of functions, and the resulting constant refers to the action of expansion and contraction of those functions that are interrelated through systems of polarity.

## Super-symmetry

Basic components of the space matrix or superstrings are combined in various stages or dimensions from simpler to complex systems, altering properties of matrix dimension per dimensions. Various matrixes or dimensional ethers exhibit characteristic dimensional properties; these various states can be termed loosely as energies. The superstring is made from various primary operators *D*, *T*, and *V*, which can be termed aetherons; operators are functions, and their relationships as variables are governed by constants. All relationships are expressed by three units *D*, *T*, and *V* of the basic mathematical relationship to form a unified field. The rules vary with perception since what we *measure* is a narrow range of dimensions, but effects exist trans-dimensionally connected to other dimensions via induction. Time and velocity are quantized to form all relationships and also describe consciousness since they are one and the same. Metaphysically, time involves memory; time and thought have memory. There are ranges of reality in a unified field, and the rules and properties change through the ranges. All phenomena and components of the space matrix are explained via a model of three units through a system of hierarchy and symmetry. Time is the super partner of velocity or super-symmetry. Since life is one form of consciousness perceiving another, this adds credence to Leo Tolstoy's idea that what is real is what is shared. All truth is known then through love, in every nature of when by symmetry; when God loved, he created. The theory is for the universal observer, one who sees all from an entire pattern where all processes exist simultaneously. All relationships exist is stability, and theory effects as dimensions are perceived to be and are occurring naturally. To finally understand time and travel in the highest sense is to understand our own being. And as such, we can travel anywhere without the need for spaceships. The nature of the matrix infers to the meaning of truth. Hierarchy means there is a first cause. In nature there is symmetry and hierarchy. Even in democracy there is hierarchy. We have equal rights under the law, but are not equal in our capacity. If the founders of the great religions had been female, then they would have been as you see males, and you could ask what if they were men. What if there was no hierarchy and no symmetry? Then you could not ask the question. Most of what is learned is myth, and what is better or not is a moot point. Whatever system there would be would be hierarchical, even such as socialism, and everything that is done. Truth has no gender and is a standard unto itself. Love thy neighbor as you carry no gender, and anything less is less than what is good, and one bad isn't better than another. So when you are finally perfect, the issue vanishes; you are like an angel. But before you get there, there's a lot of confusion and experiments. If I were an angel in heaven, I would be perfectly happy with hierarchy because it's only given to points of ascension to the totally deserved. (Zapato: Web Link)

## 1 / *R* Relationship when Leaving the Inertial Field of the Solar System

The value $H \times C$ is referring to acceleration from the Sun involving Pioneer (Turyshev:1-60)

**$H$ = Hubble constant = $1 / T$ and**

**$C$ = speed of light = $V = D / T$.**

**$H \times C = (1 / T)(D / T) = D / T^2$**

***$H \times C$ is a $1 / R$ relationship***

**$H \times C$ is a 1 / R relationship**
**$H \times C = D / T^2$**

**$D / T^2 = V^2 / D = C^2 / D$**

**$C^2 / R$; $1 / R$**

**$V^2$ = temperature**

Temperature / *D* refer to the inertial field from the Sun.

*$C^2$ / R refer to acceleration of a 1 / R from the Sun beyond the inertial field.*

## Resonance and Harmonic Action

Life involves transformation, and in this process, interval alternates with infinity. The universe and source consciousness, according to its nature, enables toward the infinite expansion of its nature through free will, which is equivalent to truth and good. The freedoms allowed in free will also allow contradiction and conflict with self and, as such, is the source of stagnation, suffering, and death. An idea of death can include that action according to the nature of the matrix is conflicting to contradictions in the nature of self of an entity, which is part of the matrix and lost in connection to the enabling of the source consciousness. There are many myths surrounding the human notion of the end and death rather than transformative action that is enabling or disabling; disabling transformation is actually termination in accordance to the disenabling that life has succumbed to; contrarily, the enabling through good works and the enabling of precepts of truth is a harmonic resonance with the matrix. If I lose sight of the source consciousness and conflict with the very nature of being derived from that source, where the universe and you (an extended part) are here to enable, the disabling nature of this action is no myth. And in a sense, like a branch on a tree, the disabling action is withering. It verily hands out the concept of self-destruction as life is self-organizing; it also can be disenabling since there is free will that allows such as otherwise, there would be no free will. And truth cannot be perceived where the will is to cover up the wrong of such disabling action. Karma reorganizes the matrix according to self-destination. There are consequences related to the internal harmonic workings of reality; you are led by the resonance of actions to its proper place. All things in the balance fall to their natural state of being, good or bad. Through love, I perceive the knowledge of truth. And via conflict, the same is lost, and as such, knowledge is qualified by its use. Wisdom emerges as the harmonic equivalent to knowledge of truth by awareness and the will to exist in harmony that allows life to be lived and enabled infinitely in the knowledge of truth. All is good, and contradiction is within self and akin to self, since that which is enabled through love is outwardly directed.

There are no emergent process in this universe that identify the beginnings of life, hence the word evolution does not apply to that beginning which much like the universe was induced by transcendent activity. When you have a thought, it emerges, and does not evolve, much in the same way as the emergence of life and the universe, since all that is real is shared, one conscious form perceiving another. If we consider that what is real is shared, and that all life and reality is one conscious form perceiving another, then science and consciousness and math are one and equivalent. So then for the notion of time and velocity in the framework of consciousness, what do these become as operators of consciousness? Time refers to the collective consciousness of all beings as memory and hence in the sense of physics, such a construct can never be altered. The source consciousness would be the first cause from which all experience is extended, and thus all experience recedes from that first level. I experience this life as a participant to a shared dream since what is experienced is induced from the outside as is the universe which is also the consciousness of the first cause. My transcendence enables my free will. I am the object and also cause of action in transcendence. Without this transcendence there is no free will. My capacity to defer in free will is the source of suffering. The natural state of being is infinitely enabled as a harmonic resonance of the first cause.

$V + T = 1 = D$

$(V^2 - 1) + (T^2 + 1) = V^2 + T^2 = V^2 + 1/ V^2 = 1 = D = VT$; @ T=1, V = C

$V^2 + 1/ V^2 = C$

$\mathbf{Phi^2 + 1/ Phi^2} = 2.99$

# CHAPTER 7

## Universe as Consciousness

### Conclusion

The End is transformation.

The subdivision of 1 show infinite possibilities of networks of 1-0, hence the creations that are possible as mathematical expressions are infinite. The orders of reality based on set types of physics are infinite as the creation of various life forms. The genetic code of all reality is based on the 1-0 infinite binary system. Intent, purpose, process, tenable, enabling to self and others, is rational. Causality is the radius of action proceeding from intent according to the harmonic nature of being. Being is infinitely enabling, self-organizing, and sharing of being. According to the natural state of being is love. That which is contradiction to this natural state and of self is evil. Truth is a standard unto itself and pervades the matrix as the organization and hierarchies of being and love. In evil there is no equivalency with good since all good requires resonance with the first cause and of self.

All comes into being through modes of resonance. Harmony enables infinitely through patterns of consciousness and the resulting synergy is love. For this synergy the first cause is action and creation and consciousness, and all equivalency is love. Consciousness is form, and the form is the pattern of math by which we know math and describe reality. The patterns we perceive attest to the nature of the patterns of consciousness, and hence all patterns reveal truths with regard to the first cause. Form refers to the nature of being, which is subdivided, and one with self, and divided and indivisible simultaneously. The 1 is divided into phi and phihat, but all three exist as one indivisible, one cannot be without the other. The state of consciousness rests within infinity, since form is the alternation between interval and infinity, and form in this universe carries a triad form. In this patterning of consciousness and relationship to the first cause, we are etched with the all or first cause. To love God had to create. We resonate to the nature of our actions; this would be my explanation for anything related to karmic action. Time can be quantized and is energy dependent, but from outside the universe all operators appear as constants like pi does referring to something outside the universe from within. All action represents functions of hierarchy and symmetry that work together.

## Conclusion

From the proposition of this theory, it appears that the fabric of space is directly related to mass, therefore suggesting that mass generates space and vice versa. It is possible that the larger planets and galaxy centers are larger generators of space, and photons are actually elastic collisions of the *D* particle space matrix. It could also mean that space varies as mountains or valleys depending on the capacity of the generators. Space travel via energy fields that disguise spacecraft in other dimensions might be shortcuts through space limited by factor *C* in all dimensions, but changing by a factor *C* relative from one dimension to the next higher dimension. So in dimension 4 ($D^4$), we are at $C^2$. All dimensions exist in stability, and hence, their effects are experienced as naturally occurring events. By the principle of induction, this theory maintains that subatomic forces (higher dimensions of greater energy) generate exoatomic realities (such as mass) or lower dimensions of less energy. This also supports the concept of entropy or energy moving from a high state to low state and back, and I believe there is a point of equilibrium by which the higher dimensions limit expansion of the lower dimensions by some kind of proportion or perpetual motion equilibrium. Perhaps there is a limit to motion by a proportional equilibrium. So space expands with the explosion of stars and contracts with the collapse of stars and galaxies. Space is an elastic matrix that exists in proportion to transdimensional energies, and there is a definite proportion to mass. One does not exist without proportion to the other.

***From this theory are the following concepts established:***

1. Mass and space in a symbiotic equilibrium where one creates the other at a ratio.
2. The superposition of space creates the various dimensions.
3. The interference patterns that result from superimposed space are the particles that we see in positive, negative, and neutral.
4. Multidimensional framework of space exists from subatomic to exoatomic level.
5. Photonic waves as we know them are elastic collision in the space matrix, and the collision is photonic current.
6. Time travel, time can be induced. Time is real.
7. Hyperspace travel exists.
8. Induction of physical "energies," you can induce one dimensional state via another, as in electromagnetism.
9. Celestial bodies tend to reside in areas of greater space density.
10. Space density varies. (Haramein: 95-100)
11. The universe as a perpetual motion machine describes the reason for inertia and entropy. I like to refer to entropy as entropic inertia, but there is also disentropic inertia in the opposite direction, and there is stability at opposing ends of multiple systems. So nothing is ever completely tight or completely loose. At one extreme you have space, and at the other a black hole and this universe is a particular qualification of infinity. The clock is infinity, and the pendulum is this universe. Light is the monopole of this dimension $D^3$.
12. The dimensions begin and end in unity from $D^1$ to $D^{23}$, which is a black hole.
13. I redefined mass as **mass** = $\boldsymbol{D^3T^2}$, which means all volume has mass, and therefore a photon has mass, and even space has mass. I proposed an experiment somewhere to determine the volume of a photon, and therefore, the mass. This also is tied into explaining why gravity affects time and also explains mass in terms of time. This is strongly tied into the definition of density where **density** = $\boldsymbol{T^2}$. (Van Flandern: Web Link)
14. Theory reveals many properties that have an inverse relationship. It is obvious about time and velocity, but this goes on to many other relationships through the equations.
15. Light is a monopole on the space matrix. So it's a type of particle, but neither particle nor wave in the manner traditionally considered.
16. That what we measure and base our physics on is a narrow band of dimension that is observable via human perception; hence, the real truth is nebulous, and what we presently have is an approximate abstraction.
17. All the laws of physics are brought down to common denominators *d*, *t*, and *v*.
18. That travelling at the speed of light squared is possible.
19. Antigravity can be easily understood.
20. Zero-point energy machines from the void or matrix of space is possible.

21. Space can be manipulated and is an elastic superconductor.
22. That many physics theories today observe phenomena from one aspect, and this one ties them together.
23. A periodic table of dimensions is established.
24. There is a connection between consciousness and reality. Some large metaphysical implications exist.
25. An avenue for exploring the remaining dimensions and defining them more precisely.

26. I think the most important part is that it puts physical concepts into ideas of superstring and relates that to relativity and then explains quantum theory as the effect of the large cause: space fields.
27. Everything is tied together. So there are many avenues to rethink what matter is in terms of other dimensions. And this is for any physical concept, which can be redefined in terms of magnetism, gravity, or whatever.
28. Perpetual motion is the universe itself.
29. Black holes create space and are a doorway to infinite energy.
30. Gravity approaches $E^4$ in a black hole.
31. Mathematics *can* describe all the dynamics.
32. All physics can be expressed in three units and powers thereof.
33. There is a pattern that identifies how these units relate. For instance, time is associated with magnetism and $V$ with current and $D$ with force.
34. Everything is an expression of $D$, $T$, and $V$, thereby uniting all units into one field.
35. A periodic table can be devised relating all present units and identifying what they are in $D$, $T$, and $V$.
36. Additive and multiplicative systems seem to unite at balance $\boldsymbol{D = 1}$ where you have values 0 or $C$ identifying unity.
37. There are no constants, only functions integral with constants.
38. Space is a geometric construction of properties or values and is primary, and mass is a secondary construct.
39. Mass exist in ratio to space.
40. Mass is an arbitrary assignment of time and velocity whose products are used to define forces and energies.

That is what a unified field theory attempts to do. The order, however, needs to be developed into various structures of what is seen and unseen, which present-day physics does not have. These are

1. **The structure of the order or universal construction (transdimensional strings).** Strings are the mathematical structure of the universe, and space is in relationship to three strings that are distance, time, and velocity. Strings form the matrix of space. These strings exist in positive, negative, and neutral, so for three-dimensional spaces, you are looking at a geometric matrix of twenty-seven strings. The intersection or interference patterns of these strings (with spin) form the particles that we see, and the variety of particles is made possible by space superposition in the various dimensions. In the third dimension of space, the intersection node (monopole) is the photon, and the elastic collision of the matrix occurs at the speed of light. The speed of light suggests kinetic elastic collisions. Photons that collide in the matrix are photonic current, similar to electrical current. Current or kinetic motion is an elastic collision, and at the speed of light is a phenomenon of the three-dimensional space matrixes. (Musser: Web Link)
2. **The structure of space (relativity).** $\boldsymbol{E = MC^2}$ is now $\boldsymbol{D^5 = MC^2}$, which explains matter in terms of space ($D^3$), and $VT$ ($D$) is the constant string multiplier dimension per dimension.
3. **The structure of what is seen or particles (quanta).** Particles are the intersection patterns and superposition that occurs in the various dimensions of space. At dimension 23, there is singularity of a black hole, and the matrix of space converges to a point.

The source of all causal action is in the radius D which is a function of time and velocity. D is velocity and time as a proper system or product. It is the primary matrix super string or god particle, the 1 that is subdivided into networks of time and velocity. It is what is induced in Euler's equation. it is the sphere represented by its casual action as radius representing a dimensional vector of space of dimension 1. It represents the periodicity dimension per dimension. It is time and velocity quantized in all things. Equivalency rests with assigning all properties to functions of time and velocity. When you ask what D is, it's no different than asking what is time and what is velocity. D is displacement due to harmonic resonance and as a space matrix becomes the entire inertial field or fluid ether in a radial tetrahedral geometry.

Heisenberg is wrong, he uses proper (multiplicative form) in an attempt to solve scalar additive problem. Obviously he knows nothing about the philosophy of math. You can really know the momentum and position of something through mathematical quadratic and polynomial forms. This uncertainty also and incorrectly pervades philosophy. All events of emergent and natural systems of this universe can be expressed by explicit mathematical equations. Randomness and uncertainty are an illusion and can only be ascribed to transcendental forms such as the behavior of life forms.

By having these three concepts working together, you have a Trans Dimensional Unified Field Theory, which is the unification of physics and the unified field theory searched for. The construction of space is an idea that results in reality. The pattern order itself is the thing that results in being. Which came first, the chicken or the egg? The answer is there is no chicken and no egg, but an idea of pattern. Everything in this universe is moving, and nothing is static. This is also true for time and for dimension. That's the beauty of the whole universe; the relationships of a unified field bring everything into being. The structure or order of relationship of the unified field concept brings everything into existence. When God created, he did not create time and then matter, but the entire structure as one statement. And each part of the order or statement depends on the other related parts for its existence, or more profoundly, the order or idea comes before and brings about the particles we see. The results can be observed, explained, and measured. The process of manifestation is like an idea for humans, but for God, this is creation or creative in the real sense. And with twenty-three dimensions, the big bang is like unfolding a newspaper or tangram. Dimension 23 unfolds to dimension 1 in equilibrium via other dimensions. Perhaps the big bang sets the motion in place. The differentiation is in the order that precedes the bang and what really moves it.

Yes, motion of space and time produces matter. And the differential variations of these—such as time, dimensions, and velocities in twenty-three dimensions—produce all the particles we see, and the order that defines twenty-three dimensions existed at the moment of the big bang caused the initial movement to what we see today. So there are ratios in these equations that demand a distribution of mass, time, velocity, and whatever else there is, but that all these are also interchangeable. When we think of matter, we think of different kinds, like gold, oxygen, electrons. Now come down the food chain, and let's say when we think of space, we think of different kinds, such as matter, velocity, gravity, and so on; and so you see, it depends on the configuration. You haven't grasped what a unified field theory really means until you realize it means anything is equivalent to everything else by some equation. What did John Lennon say in a song? "I am he as you are he as you are me and we are all together"? The properties of one dimension are not the same as the properties in another.

A theory is a formal statement of the rules on which a subject of study is based or of ideas that are suggested to explain a fact or event or, more generally, an opinion or explanation. My theory proposes that a unified theory needs to incorporate laws that cover the vast spectrum of topics or issues and somehow recommend a discourse or answer to questions. That is scientific. I find the approach liberating since it is an attempt at a synergy. I am not convinced that an approach by small steps will achieve the goal, so an intuitive leap is required so that a large spectrum of possibilities is enabled from the start.

In this theory, truth is knowable and examples are given. The theory does not make a list, only reference a framework. The interpretation of the truths is another thing. Humanity is often wrong on interpretations because it involves its own biases, in that humans by default misinterpret. Truth is universal in that it is the participation of the whole, and that participation cannot be determined to be absolutely connected to everything. But of what we are connected to, we have the capacity to be sure of in reality. Universal is the participation in the whole. So it may appear you can only measure truth in context or relative terms, even by this definition, but truth is transcendent, intuitive; and this universe is an analogy of truth, and while we may seem far from all truth, it is also available as a whole to understand. Relatively speaking, we may focus on parts or, more profoundly, of the parts of the entire system and how they interact as an entire whole, such that in the synergy of the whole, the sum is greater than the parts. We are universal and tied to an infinite system. We are not absolute in that we are like God. There may be more to reality than we can ever experience in that other realm included with our own I call absolute.

I have heard some say an exception disproves a theory. I would clarify that by saying an exception in a theory proves it is more abstraction than truth. But then if you consider that we are beings living or perceiving $D^3$, $D^4$, and $D^5$ in a multidimensional universe, then our claims are more about abstraction than real truths since it is noted based on the synergies of the whole. When by an intuitive process we finally understand the whole, we will finally have laws that will represent truth and clarify philosophy. So we live in a patterned universe that is also a form of consciousness, which is also a pattern of God—about truth and God. And some would say, "Well, you don't know God, except for religion." But all have had thoughts about God, and in intuition does the greater truth lie. God is archetypal. Religions tell us more about culture than the truth that is, and culture refers to myth. Since the universe is an analogy of God, it is possible through science to allude to a philosophy of truth and also predict a likelihood of the existence of God.

# CHAPTER 8

## Creation of Life

### Planetary Atmosphere Tectonic Machine Theory

Life is the Second Big Bang

The subdivision of 1 show infinite possibilities of networks of 1-0, hence the creations that are passible as mathematical expressions are infinite. The orders of reality based on set types of physics are infinite as the creation of various life forms. The genetic code of all reality is based on the 1-0 infinite. 0 and -0 are something. Because of transcendence, both the universe and life were induced in two "big bangs" or grand inductions. On that which is shared is real and by shared consciousness and transcendence this is made possible. That which shows permanence and is immovable is emergent, and a machine for living; here random is an illusion. Life and free will is the source of all randomness.

## Introduction

The idea is that all oxygen on the Earth comes from water in the ocean. As you know, oxygen here is 20% of the atmosphere and on Mars a trace less than 1%. The idea is that under the ocean, at the location of the tectonic plates where magma is released at fairly distributed cracks throughout all the oceans, water comes into contact with temperatures over 2000°C. The water or $H_2O$ is converted into hydrogen and oxygen ($O_2$). The oxygen bubbles up as in a fish tank and oxygenates the ocean and also bubbles up from the top of the ocean into the atmosphere. This $O_2$ in the atmosphere is gradually lost to space and shielded partly by the layer of $O_3$ or ozone and also the magnetic field.

**The rules for a planetary atmosphere tectonic machine are as follows:**

1. Life cannot create atmosphere and sustain itself; it must be achieved by "active" mechanical processes that evolve to create an environment that sustains life.
2. The source of oxygen is the ocean, and it interacts with heat and the tectonic plate adjacencies to create oxygen.
3. Life on Earth does not add more oxygen or $CO_2$; the net effect is near zero or less.
4. The polar caps are storage of oxygen by this rule.
5. Global warming is destroying the polar caps, hence, the "stored" supply of oxygen in the form of ice.
6. Fish are more abundant in areas near the oxygen-producing regions.
7. Oxygenated water has a high $O_2$ value in oxygen-producing regions of the ocean.
8. The oceans as generator of oxygen explains why there was oxygen in the ice age when the "stored" supply of oxygen in ice was vastly increased and plant life reduced.
9. Life did not begin on the planet until the beginning of tectonic plate activity and interaction with the ocean.
10. The moon assists in oxygenating the ocean along with the wind on a minor level.
11. When hydrogen and oxygen are produced by nuclear endothermic reaction from water, hydrogen reassembles itself into nitrogen. Hydrogen is the building block of the universe, and nitrogen is one of the most abundant elements. So the transition from hydrogen to nitrogen is common in the universe with the right amount of heat, and nitrogen is more stable than other elements. The reaction to form nitrogen is a nuclear endothermic reaction to produce nitrogen as follows:

$$^{12}C + {}^{1}H \rightarrow {}^{13}N + \gamma \quad +1.95\ \text{MeV}$$

$$^{13}N \rightarrow {}^{13}C + e^{+} + \nu_e \quad +2.22\ \text{MeV}$$

$$^{13}C + {}^{1}H \rightarrow {}^{14}N + \gamma \quad +7.54\ \text{MeV}$$

12. The reason why there is more nitrogen in the atmosphere is because most of the oxygen is in the water for the fish to breath and comes slowly out of the water into the atmosphere. Contrarily, nitrogen passes quickly through the water to the atmosphere. The solubility of oxygen in water is higher than the solubility of nitrogen. Air dissolved in water contains approximately 35.6% oxygen compared to 21% in the atmosphere.
13. The $CO_2$ in the atmosphere produced is mainly from volcanic activity.
14. Salt is created by this process, but the water will not hold more salt than is present due to the presence of oxygen and other minerals.
15. Ecosystems do not maintain the stability of life on a planet; mechanical systems do. Natural physics and natural "machines" or models come before and are precursors to life.
16. Water in the atmosphere is allowed by its complimentary relation with $O_2$ in this process.
17. Oxygen or $O_2$ in the atmosphere is evidence of the tectonic plate—water mechanical processes that produce atmosphere as we have on Earth.
18. Oxygen is stored in the water as well as ice and is not lost there, but I think more water puts more vapor pressure to push oxygen into the top of the atmosphere to outer space. Liquid water is more fragile than ice for being lost in

space. More water on the planet will create a larger bulge at the center by the moon and could even affect the tilt of the planet's axis as has probably been in the past, and any adjustments there will drastically affect all plant life. They claim the Sahara was once a tropical forest. Nuclear explosion testing in the ocean is a blatant destruction of both the atmosphere and oxygen supply in the water. And the point is this: the oxygen in the water is just that, a supply, like oil, minerals, etc., and when it's gone, that's the end for life of an oxygen-breathing planet. We can go without oil, but not oxygen. And I think then the symbiotic relationship of $O_2$ ratios to quantity of water, if disturbed abruptly, you're going to have a destructive shift in equilibrium since they are connected and also connected to the generator at the core of the Earth. It is machinery, like a car, so don't mess with the mechanism. Look at Venus, just a few miles closer to the Sun, and the heat evaporates any water or chance for life. Look at Mars, still workable like the Earth in its early history, but the mechanical generator at the core to burn water to oxygen is much smaller, doesn't last as long, and is now dead, and the magnetic field that protects the water from solar radiation from the spinning hot core (rotational electrons at the hot core generate a magnetic field), once gone, water evaporates. The only water left on Mars is primarily frozen or stored, left to liquid; it would be bone-dry eventually. So you have the Earth in the right location, right size (core), right amount of water to create what I call a symbiotic relationship with the atmosphere with $O_2$, a "machine" that supports life; and Jupiter not too far off with a large gravitation just outside to draw in most of the comets along with the Sun to protect both the Earth and Mars. The two planets, Mars and Earth, are protected similarly, Mars being more fragile where the tectonic machine has stopped, but both in the right location, protected by Jupiter and the Sun. There's a model for equating probability for life elsewhere, but it starts to diminish fast by these limitations. So look into the galaxy, and if you see $O_2$ at 20% of the atmosphere, guaranteed it has an ocean and probably in a similar model to the Earth. Exoplanets similar in model to the Earth are good places for future exploration for both life and alien contact.

19. Oxygen in the atmosphere is an indicator of water on the ground as a result of an atmosphere-producing tectonic machine. Trace oxygen could be the remnant of plant life and an atmosphere-producing tectonic machine. Oxygen represents tectonic activity current or past with few exceptions. This can be used to define tectonic activity similar to Earth across the universe.

**Implications for Mars and Life**

1. Life as we know it depends on an ocean to supply oxygen via underwater molecular separation at hot volcanic/tectonic activity from the planet core. Not possible in older planets like Mars, but rather younger planets. Venus could be a future candidate, or Titan, moon of Saturn, but Venus is too close to the Sun as discussed, and Titan needs a larger Sun (maybe when the Sun expands).
2. If Mars has life, it had it when there was an ocean and high level of tectonic subsea activity.
3. If there is life on Mars, it must be of a type that breaths $CO_2$ (almost the entire thin atmosphere) in the day, releasing $O_2$ by photosynthesis, and hibernates at night since it is so cold (range about 80 to -212°F). Maybe pine trees that have adapted to hibernate at night are there. Plants on Earth breathe $O_2$ at night, and hence, Earth-style plants on Mars would have to be of a type capable of hibernating in the cold air at night. But it is also possible that the plants produce their own heat and, hence, survive the cold and melt the ice for nutrition. I would attribute large plant size to the Mars atmosphere dominated by $CO_2$ and lighter gravity. There are plants on Earth that are "warm blooded" or maintain a temperature of 36°C in 5°C cold weather. Consequently, the concept of a warm-blooded plant on Mars is possible by evidence on Earth.
4. Mars has oxygen, so I attribute its trace presence to limited plant life left there that survived and is there supplying the trace oxygen and also evidence of past tectonic activity.
5. The concept of terra forming is valid as a mechanical process, not a "green" landscaping construction.
6. Terra forming may be possible by collecting huge pieces of ice floating in dark space and sending them to Mars to create an ocean. The second step would involve artificially creating nuclear processors in even distribution under the ocean to create nitrogen and oxygen from that water for an atmosphere. Then you would have mechanically terra formed.
7. The less water in the atmosphere is because it is primarily $CO_2$. More water means more $O_2$ in the air, and more $O_2$ (in

the air) means more water on the ground. This is a compatible interchange relationship of absorption-diffusion. The detection of water in the air would be less in $CO_2$ atmosphere. You can have standing water on Mars simply because the temperature does rise there way above the melting point of water. And I do believe the trees could be warm blooded (heat producing), hence, melting the ice to water. The trees could also be hibernating during colder temperatures.

8. Symbiotic or interdependent relationship between oxygen and water is explained as follows: Oxygen in the atmosphere in abundance (20% $O_2$) with its nitrogen compliment is evidence of water on the surface of the planet in abundance. Water clouds in the sky are evidence of oxygen. Oxygen in the atmosphere is evidence of water tectonic plate activity "machine" or mechanical processes that produces the oxygen atmosphere.

**Mars history may have one of three pasts:**

1. Cataclysm that evaporated the oceans, like large body collision into the ocean.
2. War caused by alien life with advanced weapons.
3. Mars is an old planet, and resources could have been depleted long ago. In this scenario, without continuous supply of atmosphere from the sea, loss of tectonic activity, and loss of a magnetic field, the loss of protecting atmosphere allows the slow evaporation of water into space. The water there presently is being protected by the cold temperatures as "stored" ice but still evaporating at a slower rate. So at the equator, there is not so much water, but at the polar caps, there is more. Mars is dry since the atmosphere is limited in $O_2$.

**Evolution**

Whenever we talk about life, the concept of evolution is at the back of our minds. I agree with the concept that galaxies go through cycles and, in that process, evolve from the lighter elements like hydrogen to the heavier elements like gold. And also after three generations, we see complex carbon molecules and other natural but lifeless compounds. There is nothing in the theory of evolution to explain how the random activity of elementary and molecular structures come together to form the most simple cell structure or virus. So the theory as presented has no support at the birth or beginning of life. What explanation is there for inanimate matter to cross over to create a "machine" such as a virus or cell that is very complex and has a DNA with billions of combinations? There is none. Since our DNA has over three billion combinations of sub-base pairs, it is my contention that there is not enough evolutionary time to create such combinations. Solar systems, tectonic machines of the planets, and galaxies do not have the time to allow these variations to evolve from scratch. Hence, the evolutionary time required for variety is diminished within the time allowed by a planetary tectonic machine. It is true that within current science, those genetic codes are viewed as programs. Monsanto patents the DNA of various seeds for commercial sale. Hence, the concept of a program requires a "designer." The genetic code has been referred by other scientists also as a program where various genes can be suppressed in a person, and thus, their effects are eliminated from that person. Not to discount the theory of evolution, I would say that in the DNA of the simplest cell, there exists a "subprogram" for evolution yet to be discovered. After a "designer," this would allow enough time for a tectonic machine to allow a genetic program with an evolution subprogram and with billions of combinations already in place to evolve. It may be that virus and other cellular life-forms traveling on comets and other extraterrestrial objects may act as a source of life for planets beginning a tectonic machine.

**Evolution, God, and Alien Life**

God is the first cause from which all life and creation comes. We can think of all life as alien; *alien* here means another order of being and a part of God's many creations and apart from God. So the objective alien in outer space includes us. We are no more the center than we thought the Earth was the center of the universe or that we are the only ones in it.

I don't think aliens in outer space are descended from Adam. I think each creation has its own separate relationship with God or covenant, and then there is that understanding of God universally. I also believe that the truth isn't any alien or human truth, meaning that an alien truth does not supersede our truth; none has a corner on that. In the end, all you have is faith. So for all the questions, the theory of faith is self-reliance and love, that the greatest part of it requires absolutely no support in anything really. You stand there with nothing but your own merit and capacity, and yet you believe. Faith is the belief in things unseen, but the universe and our existence prompts us to ask the question about God. As Saint Paul said, there is no excuse for not knowing God because of works in creation. That is because, in my opinion, your conscience has a piece of God that knows the works, but it can be suppressed. To say it another way, there is no excuse for not thinking of the idea of God by reviewing the nature and creation or universe around us. So the idea of God has crossed every person's mind; although rejected by some, it has crossed. That crossing is what Saint Paul refers to. The truth may be that in the crossing of our intuition, the greater reality may lie. This is what I refer to as intuitive logic. Many great discoveries have been led by this path, for Einstein and others. With regard to salvation, Saint Paul said there are those that have heard the word of God but do not follow it, and those that never heard it but by their correct actions of which they are unaware, it is accounted to them for salvation. So again, conscience and intuition is important. Christ's example and the forgiveness rendered is an act of God, not of men, and a message about faith and love. From that point, our conscience and actions determine how effectively that forgiveness or offer of love will work in our lives. The presence of God in our lives or conscience is our salvation, the harmonic relationship of creation.

Evil exists with or without Adam, original sin, or even the history of humans. So it exists in the universe. Technology has nothing to do with the evolution of good, which is a spiritual quality. The story of fallen angels tells us that. Original sin is

# Digital Tectonic Activity Map of the Earth

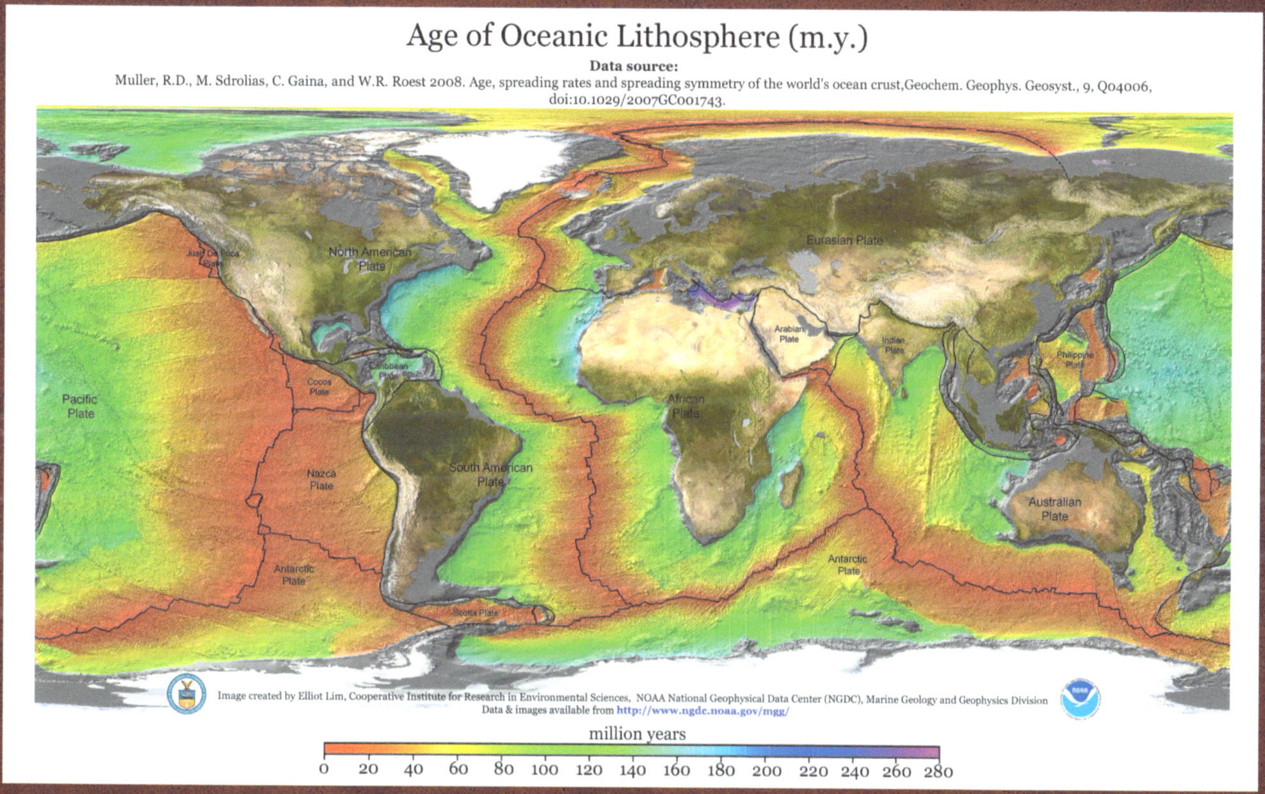

original and specific to the human race and metaphorically to all races or aliens, since abandoning or disowning God wears an aspect of wrath and judgment. So then the commandments, rather than being liberation as naturally intended to be, become a restriction or condemnation. Abandonment can be reviewed as Saul did before the battle; he consulted a witch or medium and so abandoned God. As abandonment of God wears an aspect of wrath and condemnation, God's forgiveness covers that abandonment and subsequent sin and condemnation. That forgiveness is exemplified in Christ, and God's unconditional love is a posture of being receptive to love, redemption, humility, and penance, or more exactly, reconciliation. An invitation to return to the harmony of God's will or an invitation to be reconciled with God is thus presented. There are also exceptions in the lives of many who have represented a good life despite the prevailing example of those around them, such as with Enoch, Mary, Elijah, and maybe even Moses, who were caught up to heaven to be with God.

God is not bound by our interpretations or theories, whether they are scientific or otherwise. Semantics is very limited and limits our understanding of truth and life in general. The bottom line is this: you only get near to God as good gets. So to be good is the harmonic state of the evolved spirit and natural evolvement toward God. Physical evolution and the natural world are a temporary house and experience for the spirit in various states of its spiritual evolution. The high spiritual state of being near to God is an immediate possibility for all for the taking as an epiphany. The truth of God is beyond alien evolution, physics, theories, perceived limitations, hierarchies, and all the books in the world. The greatness that is in life and the goodness that is near to God are no closer than your own heart and can be a creative inspiration, analogous to a creative God. Love—the good that comes from God to all from which we can participate is the exception to all perceived limitations since the realm of the spirit is inspiration—creativity, freedom, and imagination. Evil or that which is measured and limited can be expounded in theories and philosophy via the many limitations of semantics. Evil is perfect for the limited nature of human semantics and so limits our understanding of God. Our language and its use tell us more about our humanity than the truth that is. And hence, for centuries, we struggle with primitive notions and philosophical methodologies such as duality that are more an extension of our humanity than a truth that requires multiple concepts. But God is that which is free will, positive, unmeasured, unconditional love, and preserves good. God's reasoning is as multiple as the universe is diverse. All good things come from God, and the choice of harmony to that energy are those things from which we can participate, such as love, creativity, multiplicity, and other truths by which we understand God about wisdom. Wisdom knows how to live, and that knowledge pervades the universe, and as such, by our conscience and intuition, there is no excuse to deny. Being different does not mean better or even more blessed. It's like when we thought the Earth was the center of the galaxy, and we weren't. Then we thought we were the only life in the universe, and we aren't, so on and so forth. God is diversity and creates all life and that which exists. The ultimate creator (God), the first cause, is not an alien. But the creations under him, all of us, are all aliens and may have some role in creation.

## Emergence and Evolution

Idealized semantics with no strong rational foundation for an "emergent" concept for life devalues the theory of evolution; the idea of evolution ignores the beginning of life and presents a theory that is not comprehensive and lacking scientific explanation. So in the hope of a grand idea, the most important part, the emergence of life from inanimate matter, is ignored. And the truth about the "emergence" of life is the foundation and will determine whether natural selection is natural or artificial. My hypothesis is that all life is engineered, unlike the natural progress of intimate matter and planetary machines. The proposed concept of a primordial soup is a high generalization as an unscientific model. The genetic code is a program as so it is referred to by Monsanto. Therefore, the genome cannot happen by natural random action—the natural action of inanimate objects. And if I can prove that, then there are forces unnatural at work in the universe, and philosophically, that will lead you to the notion of a God.

**Unnatural forces in the universe, you say? That's impossible, you say! Anything that exists is natural, you say!**

It's not, and I'll tell you why. The genetic code only determines a general direction of development; it doesn't even determine everything, and development may change depending on the environmental conditions, but this is not the beginning of life! Since a program is in place, anything is possible. A program can incorporate an allowance for change, adjustment, or as Darwin puts it, natural selection and evolution. But the selection is artificial and programmed; it is transcendental. The galaxy and atoms are created by random action, but there is nothing in evolutionary theory to explain how that becomes the simplest life-form.

**They are created by change, you say, but not by random action!**

Life is not natural like the formation of the planets or the atoms and elements, so if you can see that, then your physics and understanding of the universe becomes both scientific and spiritual.

**He needs life to be unnatural, you say, so that he doesn't need to let go of his God.**

I say you need to prove that the universe exists and for itself, and I suggest jokingly that the universe exists for itself just like any animal in the kingdom of animals ridiculous! Dependency is natural, and since the truths of the universe are an analogy of God and relationship to God, we (the universe) are dependent on God, and life is dependent on God. How is life transcendental and unnatural? The logic states that life does not happen by random events. Life makes everything possible, evolution possible. Because there is no process for random matter to form a genetic code, and there is not enough evolutionary time to compose a sequence with three billion variations in the DNA base pairs of the human genome. There are no simple processes of inanimate objects that result in emergent behavior to be called the birth of life or life. Emergence explains how simple rules can lead to more-complex interactions; this is not perceived in inanimate objects that lead to life. Once we have the program, then it's easier to view; so emergent behavior is a contradiction in this area. It's awfully useful for explaining these things after life has started. Evolutionary time on the Earth will not allow a three billion variation in a genetic code to evolve. This is not addressed in text books.

**You would like to see the math behind that statement?**

It is perfectly possible, and to say Darwin is the best explanation we have is like saying a myth is the best explanation going for the human race for centuries. The haploid human genome occupies a total of just over three billion DNA base pairs, so how does the Earth man evolve with a magnitude in years of less numbers? And in his theory, Darwin omits the beginning of life in the detail required for concrete argument, which then suggests evolution is a program. And a program means a programmer. It is true that mathematics explain our physical world. Yet I am not convinced that it will prove anything with regard to evolution because it is unnatural; like the big bang, math will not explain that occurrence from precreation time. The step from precreation to creation is the realm of the divine and transcendental, and so that too for pre-life to life. The analogy here is what God has allowed, his transcendence in creation. No one has proven in details which conditions existed on the Earth before life arose, just like none will prove what conditions were before the universe came into existence. Dependence is an analogy that describes our dependence on God. Of and for itself can the universe be is a contradiction to the rule of dependence. This then infers the dependency of creation, which leads us to an activity a priori to the mechanisms within the universe, such as emergence. Here we see the difference in a priori and a posteriori arguments. Emergent behavior is present in mathematics and is represented by mathematics, the natural processes of the universe, and the rational mind; generally, emergence is a construct of the universe. However, transcendental activities include life, the creation of the universe, God, the creation of life, love, the spirit, intuition, and those processes that access this universe from alternate realms. So the anomaly of existence in the universe is those processes that transcend, such as life, which also includes free will. Surely math will be unable to explain both the beginning of life and precreation as well as other transcendent processes, except perhaps by intuition, which has access to transcendental existence.

**Unanswered questions are the following:**

- There is RNA or DNA templates, but where did the RNA or DNA bases come from?
- How do RNA and DNA associations evolve to the ribose moieties and phosphates to form nucleotides?
- How did the genetic base pairing rule come to be such as A-T, A-U, and G-C?
- To catalyze the reactions, how did enzymes form?
- Where did the L-amino acids come from, and why limited to L and not D?
- If life started from DNA material on a meteorite from outer space, where and how did that evolve?
- Imidazole binds to the nucleotides to make them more reactive; where does it come from?

Math explains the structure of the universe, but it does not explain life, which is transcendental, just as it cannot explain our relation to God or Creator, the divine transcendental to this existence. But by our transcendental nature and analogies derived through intuition and logic, we can perceive truth. The understanding of the larger spiritual realm is by analogy. But I think scientists are skipping many steps here, not starting at the real beginning of things. Let's hear how math can predict the beginning of the universe or the Earth and atmosphere and the evolution of natural things as this paper proposes for the atmosphere and apparent contradiction to evolution. Then you will realize there is not enough time for the evolution of RNA strands. Chemical compounds form according to the laws of physics, and genetic codes simply cannot form outside of living organisms without violating laws of physics. The mechanisms within this world do not exhibit rules of emergence that explain the beginnings of life. The debate about evolution is really a debate about transcendent and emergent realities. The tectonic machine is an emergent reality, similar to the periodic table and other natural physical processes. Emergent realities are very ordered and do not follow processes that allow the type of latitude of biological coding systems. Emergence is a rule of the natural processes of the universe and includes rational constructs that are used to describe it, such as the sciences. Life is a rule of transcendence and includes precreation activities derived by intuition, the spirit, the divine, and the requirement for a dependency on creation. Non-emergent processes expressing themselves in a universe of emergent processes are transcendental. The mechanisms and definitions of reality and truth go beyond the present and contained experiences; these we call transcendental and are derived by intuitive processes and intuitive mathematical models. These truths are also referred to by analogy from this emergent universe to the transcendental universe. The universe was involved in two big bangs. The first was the creation of the physical universe. After its creation, all natural processes are emergent, but not the creation of the universe itself, which has a transcendent source. This universe, being a particular qualification of infinity, becomes emergent within by that particular qualification. The second transcendent act was the creation of life, the second big bang. And from that creation, we can see emergent characteristics of life's interaction and transformation within this world.

## Nucleosynthesis and Planetary Ocean and Atmosphere Evolution

Continuing from the discussion about the concept of evolution, we need to introduce a model that explains planetary ocean evolution by the life of a tectonic machine and planetary core. The formation of the Earth's ocean or water supply comes before the formation of the oxygen and nitrogen atmosphere. **Phase I** of the tectonic machine delivers the ocean, and **phase II** of the tectonic machine delivers the oxygen and nitrogen atmosphere. All elements in the universe are produced primarily by fusion reactions from lighter to heavier elements. This process by creating heavier elements by nuclear fusion reactions and high temperatures is called nucleosysnthesis. Early in the Earth's evolution, when the planet and core were at much higher temperatures, the vast amount of oxygen needed to produce water came from nucleosysnthesis. During this early period of Earth's history, the vast amounts of water that formed the oceans were produced, along with nitrogen and carbon dioxide. Oxygen and nitrogen were produced by nuclear endothermic reaction (carbon-nitrogen-oxygen cycle) as follows:

$$14N + {}_{1}H \rightarrow 15O + \gamma \quad + 7.35\ MeV$$

$$15O \rightarrow 15N + e^{+} + \nu_{e} + 2.75\ MeV$$

$$15N + {}_{1}H \rightarrow 12C + 4He \quad + 4.96\ MeV$$

The oxygen produced combined with hydrogen to produce water ($H_2O$). After the ocean was created, a less amount of heat energy was required to produce an oxygen-and-nitrogen atmosphere by fusion reaction interacting with the new ocean or water supply. The oxygen produced earlier by fusion and stored in the ocean water was now available for future reactions to produce oxygen and nitrogen by a smaller step process that was mentioned before as follows:

$$12C + {}_{1}H \rightarrow 13N + \gamma \quad + 1.95\ MeV$$

$$13N \rightarrow 13C + e^{+} + \nu_{e} + 2.22\ MeV$$

$$13C + {}_{1}H \rightarrow 14N + \gamma \quad + 7.54\ MeV$$

This model shows how a planetary machine via its tectonic evolutionary process both creates oxygen, stores it in the ocean water and from these two, and reuses the oxygen as a stored supply to re-inflate the atmosphere that supports life. Water and oxygen have in the future evolutionary cycle, a symbiotic relationship.

## Salt

Large amounts of sulfur are found in molten areas such as lava flows. Liquid sulfur is like water. Sulfur occurs widely in nature as $H_2S$ (hydrogen sulfide) and $SO_2$ (sulfur dioxide). There are various chemical reactions that can be classified as acid + base > salt + water. These produce salts and water. Many chemical interactions involving $H_2S$ and an acid produce water and salt. In addition to the chemical interactions involving complex molecules and large amounts of heat, there is the more fundamental nucleosynthesis of fusion reactions producing elements that are available for more complex molecular formation to take place. These reactions include the carbon-nitrogen-oxygen cycle mentioned before and other reactions to include the following:

$$12C + \rightarrow 23Na + \gamma \quad + 1e$$

$$16O + \rightarrow 31S + 1n$$

About 90 percent of the universe is hydrogen and helium, and a vast number of elements are created from these by nucleosysnthesis, such as carbon, oxygen, nitrogen, silicon, chlorine, and others. In the alpha process, helium nuclei 4He are captured either by one or more (multiples of four) with other nuclei and form chlorine. So by this example, you could get 32S + 4He > 36Cl. Other adjustments are possible. So nucleosynthesis allows for a vast array of possibilities for an evolving planet that has a tectonic machine or fusion reactor.

## Global Warming

The ice caps are frozen water, which we can also refer to as oxygen stored in ice for future use. However, when this supply is melted, it allows for increased vapor pressure and loss of the atmosphere into outer space. The original tectonic machine and its sustained creation of our atmosphere do not include the artificial gassing of the planet, which is adding enormous amounts of additional gas to the atmosphere. Hence, the natural processes of the tectonic machine and its relationship to the atmosphere have been altered. As the ice caps cease to exist, there will be a larger quantity of water to displace around the globe with each pass of the moon's gravitation. This shift in mass is bound to the ocean water in relationship to the moon is bound to affect a shift in the axis of spin of the planet Earth. A shift of a few degrees is enough to alter climates across the globe drastically and even the destruction of plant life to a large degree, affecting the world's food supply. In the most serious scenario, 1/3 of the food production could be affected. In addition, the use of nuclear arms on our oceans is also a blatant destruction of the planetary supply of stored oxygen in the water.

## Magnetic Field

The tectonic machine produces the atmosphere in the **phase II** of development mentioned earlier. The magnetic field and gravity assist in creating a balanced atmosphere. A magnetic field prevents erosion of the atmosphere due to solar wind.

The erosion of the atmosphere would enable larger shifts in planetary temperature and unstable climatic conditions. Without a magnetic field, the atmosphere would lose one of its components for stability and would slowly drift into space. The erosion of the oxygen of the planet would create an imbalance with the previous symbiotic relationship with the ocean and eventually lead to the evaporation and erosion of the ocean with vast changes in the vapor pressure. The atmosphere is made possible by the tectonic core, assisted by gravity, the magnetic field, and a favorable distance from the Sun. The magnetic field assisted by gravity and a favorable distance from the Sun stabilizes temperature, the layer transitions in the atmosphere, pressure gradient, and synergy developed with the tectonic machine that continually pumps the atmosphere, translating ocean water into atmospheric gasses.

**Conclusion**

1. My intent is to project a model of the Earth and suggest certain bio/mechanical rules for ecosystem development that could possibly play out in other worlds simply by the possibilities and limitations of natural physics. The possibilities and limitations of natural physics, in my opinion, are rules for evaluation of other worlds and to project possible life models and expectations there. I do not believe they vary that much in our solar system or universally. I believe the balancing effects of natural physics and bio/mechanical processes are quite limiting with respect to the formation and allowances of living systems.

2. Natural physics and natural “machines” or models come before and are precursors to life.

3. Water on Mars indicates the presence of a tectonic machine present on Mars once. Tectonic activity and water in abundance as in an ocean produces oxygen. In the past, water on Mars interacted with the hot temperatures at the core's reaction to produce oxygen and other gases, which I claim is happening on the Earth today. So respect the water, the real oxygen supply. Since there is trace $O_2$ in the atmosphere and water on the surface on Mars, we can assume tectonic activity happened on Mars once. Since water and oxygen are symbiotic, detecting oxygen in the atmosphere is enough to predict water on the surface. Now is there life on Mars? Well, yes. Because if there was a tectonic oxygen atmosphere–producing machine on Mars, there is a great probability that there must have been alien life there once since it was capable of supporting life, whether indigenous or alien. Is there life there now? Almost certainly I would say so because some life may have survived still from the past, even if it is microbial. I like the following phrase: if there is water, then there are plants, and if there are plants, then there are bugs, and if there are bugs, then there are people. So perhaps there are buried cities on Mars or maybe even a library.

4. The evolution of the planet Ocean and planet Atmosphere is a result of different phases of nucleosynthesis and chemical reaction with already-existing environment. The elements (hydrogen and oxygen) by which water came to be came by fusion reaction, and that is how all the water came to be. Our understanding of the carbon-nitrogen-oxygen cycle in a fusion reaction does provide the elements for water. Remember, however, that such great heat also will allow other endothermic reactions, and one of those produces salt. Most of the elements in the universe come about from fusion reactions, even water, and there is plenty of ice floating around in our solar system that has been produced long ago in this manner. Just in the manner large amounts of water was produced by nuclear endothermic reaction, oxygen is being produced today. The idea that plants maintain the balance is much too fragile. And the net effect between plant daily and night breathing is zero since they are not using photosynthesis half the time. Water came in vast amounts by fusion reaction of the carbon-nitrogen-oxygen cycle. And then again oxygen for the atmosphere from the same cycle. There is no other way to produce such great amounts. Nature does it better. Salt is also a by-product of the fusion reaction.

5. I agree with the possibility that the creator can use evolution, but I go on to believe the creator creates and can use intelligent designers. What interests me the most is that theories explain everything, even though they are not tested. I appreciate a logical and complete description or discourse. Unanswered questions about the beginning of life are important to evolution because of the following:

- If evolution cannot be responsible for the beginning of life, then it exists as a compliment with other theories, such as intelligent design and creationism. So let's not discount any ideas. Further, I charge Darwin with being incomplete while using the term *natural* and not defining an explanation for evolution in the "natural" way for the beginning of life. Since if that is not natural, then "natural selection is not natural" but rather artificial, and anything is possible.
- The idea that the genetic code is a "program" makes the use of the term *natural selection* inappropriate. Rather say *programmed evolution* in the revision of evolution terminology, which is not "natural" by way of random events. Plants are artificial programs in my opinion and do not represent natural machines, although their capacity to function is limited by natural forces.

6. The Planetary Atmosphere Tectonic Machine Theory is presented in the manner of natural geological processes as the cause for the atmosphere that excludes life as a nuclear process. Current theory on the evolution of the atmosphere relies on life as a support. This one does not. Also, the concept of a tectonic machine also "predicts" the water-oxygen symbiosis and tectonic activity (ocean) on other planets. Hence, you can predict water, past or current ocean/water tectonic activity, and possibly life from the get-go by observing oxygen in the atmosphere alone.

7. God is presented and may be a stumbling block for some scientists, but since it becomes such a strong metaphysical point and philosophical point—and to me also the tectonic machine also relates to the theory of evolution—I bring it up. In my opinion, science leads us to God, and ultimately, that is what the discussion is about.

The logos or first cause was the source of cognition and motion, and that which was before was everything or endlessness and that which identifies with what came "before form" as we know it is the logos. The reality to which we attribute consciousness form is that which emerged from the logos before which at a time the presence of logos, and space and time are incomprehensible to us. But reflecting on the nature of form itself as we know might allude to that which suggests emergence of form from the logos as a universal awareness. Hence, what is endless and without form, life, and the patterns where by which form manifests is also the very nature of the logos; where the nature and manifestation of form are one and the same. The idea is real since idea and cause are one and the same, and reality and its very nature are the same, love, and so everything is about love and logos. This enables equivalence, hierarchy, and symmetry as form, even before form as we know it emerged, since that which was manifest, was one and the same eternally with the logos. Before time and form as we know there was a universal awareness, where everything is, and there wasn't anything that wasn't this infinite presence. And all the possibilities that could be such as creation and including us were possibilities within that infinite one, the first cause or logos. You or I for a moment within realms of possibilities is as an awakening of consciousness where from awareness the form of consciousness established those possibilities. Within consciousness comes divergence from awareness in motion and thought. This always was, but awakened by action by divergence. That possibility of your awareness was awakened drawn from realms of sleep when manifestation was correct for times of probabilities. And from this divergence of conscious form, the "form" of conscious creation emerged into a level where an internal order manifested by hierarchy and symmetry and projection into a reality of transcendence and free will as an effect of love shared was the logos. This relationship of transcendence and form consciousness which was part of the logos as we know was extended into creation where love was retuned where all reality is shared and also an infinite and propagating resonant action. One day that probability which was you from the very beginning as it always was, became manifest and also etched with the logos that includes all things.

# BIBLIOGRAPHY


Amoroso, Richard. "Derivation of the Fundamental Equation of Consciousness." ***Noetic Journal.*** Vol. 3, no. 1 (January 2000): 1–9.

*Anderton, Roger. "The Devil's Advocate of Einstein: History of Relativity." World Science Database. Video conference. (accessed December 11, 2010).* http://www.worldsci.org/php/index.php?tab0=Events&tab1=Display&id=364.

*Andre, Assis. "Weber's Electrodynamics." World Science Database. Video conference. (accessed November 6, 2010).* http://www.worldsci.org/php/index.php?tab0=Events&tab1=Display&id=366.

Astrometry. "Walter-Russell—Light Does not Travel." University of Science and Philosophy, 2010. http://www.hiddenmysteries.org/mysteries/whatsthis/light.html http://www.youtube.com/watch?v=5RuxUxlcct4.

Bahder, Tomas, and Christian Fazi. "Force on an Asymmetric Capacitor." Army Research Laboratory. (June 2003): 1–34.

Bailey, Patrick. Book review of the ***Free Energy Secrets of Cold Electricity.*** Golden State College SRICF. (August 28, 2005): 1–12.

Bouwmeester, Dik, Jian-Wei Pan, Klaus Mattle, Manfred Eibl, Harald Weinfurter, and Anton Zeilinger. "Experimental Quantum Teleportation." ***NATURE.*** Vol. 390. ***Institut für Experimentalphysik, Universita¨t Innsbruck, Technikerstr. Innsbruck, Austria.*** (December 11, 1997): 1–5.

Canning, Francis, Cory Melcher, and Edwin Winet. "Asymmetrical Capacitors for Propulsion." NASA. (October 2004): 1–23.

Christov, Christo. "*On the Frame Indifferent Formulation of the Governing Equations of the Electromagnetic Field." World Science Database. Video conference. (accessed December 4, 2010).* http://www.worldsci.org/php/index.php?tab0=Events&tab1=Display&id=367.

Coldea, R., Tennant, D., Wheeler, E., Wawrzynska, E., Prabhakaran, D., Telling, M., Habicht, K., Smeibidl, P., and Kiefer, K. "Quantum Criticality in an Ising Chain: Experimental Evidence for Emergent E8 Symmetry." ***Science Magazine.*** Vol. 327, no. 5962. (January 2010): 177–180.

Davis, Eric. "Teleportation Physics Study." Air Force Research Laboratory. (August 2004): 1–88.

Dolphin, Lambert. "Implications of a Non-Constant Velocity of Light." (accessed September 4, 2004). http://ldolphin.org/cdkconseq.html.

Dume, Belle. "Could Sound Move at the Speed of Light." Physicsworld.com. (accessed November 1, 2005). http://physicsworld.com/cws/article/news/23467.

Epstein, Harry. "Could Mass Decrease with Velocity?" Phys. Essays 22, 511 (2009): 1–4.

Evans, Michael R. "The Geometry of Light." Proceedings of the NPA. Vol. 7. (2010): 149–153. http://www.youtube.com/watch?v=C2kq4L28syw; http://www.michaelrevans.com/.

Faith, Curtis. "A Model for Force at a Distance: Gravity." (accessed July 27, 2009). http://www.fusor.net/board/view.php?bn=fusor_theory&key=1248718666.

Gow, David. "Identifying the Gravitational Constant." (March 10, 2005): 1–9.

Hansen, Helmut. "Is Metaphysics the New Frontier for 20th Century Physics?" World Science Database. Video conference. (accessed August 14, 2010). http://www.worldsci.org/php/index.php?tab0=Events&tab1=Display&id=343.

**Haramein, Nassim. "The Schwarzschild Proton." The Resonance Project Foundation. AIP CP. (December 2010): 95–100.**

**Jahn, Robert and Brenda Dunne. "On the Quantum Mechanics of Consciousness with Application to Anomalous Phenomena."** ***Foundation of Physics.*** **Vol. 16, no. *, (August 1986): 1–52.**

**Jacklewski, Tom. "Unified Field Theory of Rotational Energy Physics." (2005): 1–88.**

**LaFreniere, Gabriel. "Spherical Standing Waves." (accessed on December 3, 2009).** http://www.glafreniere.com/sa_spherical.htm http://www.glafreniere.com/matter.htm.

**Lucas, Charles. "The Origin of Symmetry and Beauty in the Universe." World Science Database. Video conference. (accessed October 16, 2010).** http://www.worldsci.org/php/index.php?tab0=Events&tab1=Display&id=359.

**Musser, George. "Faster than Electric Currents Could Explain Pulsars." (accessed June 18, 2010).** http://www.scientificamerican.com/blog/post.cfm?id=faster-than-light-electric-currents-2010-06-18.

**Oberg, Ed and Johnson Jay. "Phi—Pi Product." (accessed April 7, 2002): 1–11.** http://www.goldennumber.net/pi-phi-fibonacci.htm.

**Page, Gary, John Wallin, and David Dixon. "How Well Do We Know the Orbits of the Outer Planets."** ***The Astrophysical Journal.*** **(November 2008): 1–36.**

**Rovelli, Carlo. "Loop Quantum Gravity."** ***Physics World.*** **(November 2003): 1–5.**

**Rushchanskii, K., Kamba, S., Goian, V., Vaněk, P., Savinov, M., Prokleška, J., Nuzhnyy, D., Knížek, K., et al. "A Multiferroic Material to Search for the Permanent Electric Dipole Moment of the Electron."** ***Nature Materials.*** **(2010).**

**Russell, Walter.** ***The Secret of Light.*** **University of Science and Philosophy. Sydney, June 1, 1974.**

**Sankar Hajra.** "Maxwell, Maxwell and Maxwell: Collapse of Relativity, Special and General." World Science Database. Video conference. (accessed November 13, 2010). http://www.worldsci.org/php/index.php?tab0=Events&tab1=Display&id=369.

**Schreiber, Bert. "Heisenberg Uncertainty Principle Is False."** ***General Science Journal.*** **(January 2007): 1–13.** http://www.wbabin.net/science/schreiber15.pdf.

**Tomes, Ray. "Wave Structure Matter." New Zealand. (1994).** http://ray.tomes.biz/ http://www.youtube.com/watch?v=8BTcmuGdLCU.

**Turyshev, Slava. "The Pioneer Anomaly: Effect, Data, and New Investigation." American Physical Society Meeting. (accessed April 13, 2008): 1–60.** http://planetary.s3.amazonaws.com/misc/Pioneer-Anomaly-TPS-20080417.pdf.

**Van Flandern, Tom. "Possible New Properties of Gravity." Astrophysics and Space Science. (1996): 249–261.** http://metaresearch.org/cosmology/gravity/possiblenewpropertiesofgravity.asp.

**Zapato, Lyle. "Lord Kelvin."( 2010): 1.** http://zapatopi.net/kelvin/.

# EPILOGUE

A Poem Dedicated to the First Cause

A. D 5000

# Light as Structure

Lord Christ, God of Heaven and Earth
You are the Tree of Life
This brings forth the fruit
After which my soul hungers

Among storms you are steadfast
As the light you are
And transform all adversity
Into victory and new light

You call out to me and say
"Love me and tend my branches"
And I who realize your grace and glory
Tend the soil in the hope of redemption and salvation

I will sit by my Lord
The Tree of Life
And his greatness will protect me
As my faith works the soil

In the spring
I notice the fruits
They are good
And I am at peace

# INDEX

## Q



## R



## S

**T**



**U**



**V**

www.ingramcontent.com/pod-product-compliance
Ingram Content Group UK Ltd.
Pitfield, Milton Keynes, MK11 3LW, UK
UKHW061028310726
14090UKWH00026B/259
* 9 7 8 1 4 5 6 8 7 1 1 0 9 *